Geoinformatics in Support of Urban Politics and the Development of Civil Engineering

Sérgio António Neves Lousada
University of Madeira, Portugal

IGI Global
PUBLISHER of TIMELY KNOWLEDGE

A volume in the Advances in Civil and Industrial Engineering (ACIE) Book Series

Published in the United States of America by
IGI Global
Engineering Science Reference (an imprint of IGI Global)
701 E. Chocolate Avenue
Hershey PA, USA 17033
Tel: 717-533-8845
Fax: 717-533-8661
E-mail: cust@igi-global.com
Web site: http://www.igi-global.com

Library of Congress Cataloging-in-Publication Data

Names: Lousada, Sérgio, 1981- editor.
Title: Geoinformatics in support of urban politics and the development of civil engineering / edited by Sérgio Lousada.
Description: Hershey, PA : Engineering Science Reference, 2023. | Includes bibliographical references and index. | Summary: "It's the big intention, portray the evolution of the world of Civil Engineering, what happened in recent decades, but what will develop in the future, what will be the path. Thus, this work will seek to characterize the different academic areas of Civil Engineering, demonstrating its evolution. It will be a means to allow the dynamization of publications in the area of Civil Engineering"-- Provided by publisher.
Identifiers: LCCN 2022045615 (print) | LCCN 2022045616 (ebook) | ISBN 9781668464496 (h/c) | ISBN 9781668464502 (s/c) | ISBN 9781668464519 (eISBN)
Subjects: LCSH: Municipal engineering--Planning--Data processing. | Public works--Planning--Data processing. | Geoinformatics. | Civil engineering.
Classification: LCC TD163 .G46 2023 (print) | LCC TD163 (ebook) | DDC 628.0285--dc23/eng/20221128
LC record available at https://lccn.loc.gov/2022045615
LC ebook record available at https://lccn.loc.gov/2022045616

This book is published in the IGI Global book series Advances in Civil and Industrial Engineering (ACIE) (ISSN: 2326-6139; eISSN: 2326-6155)

British Cataloguing in Publication Data
A Cataloguing in Publication record for this book is available from the British Library.

For electronic access to this publication, please contact: eresources@igi-global.com.

Advances in Civil and Industrial Engineering (ACIE) Book Series

Ioan Constantin Dima
University Valahia of Târgovişte, Romania

ISSN:2326-6139
EISSN:2326-6155

Mission

Private and public sector infrastructures begin to age, or require change in the face of developing technologies, the fields of civil and industrial engineering have become increasingly important as a method to mitigate and manage these changes. As governments and the public at large begin to grapple with climate change and growing populations, civil engineering has become more interdisciplinary and the need for publications that discuss the rapid changes and advancements in the field have become more in-demand. Additionally, private corporations and companies are facing similar changes and challenges, with the pressure for new and innovative methods being placed on those involved in industrial engineering.

The **Advances in Civil and Industrial Engineering (ACIE) Book Series** aims to present research and methodology that will provide solutions and discussions to meet such needs. The latest methodologies, applications, tools, and analysis will be published through the books included in **ACIE** in order to keep the available research in civil and industrial engineering as current and timely as possible.

Coverage

- Engineering Economics
- Coastal Engineering
- Productivity
- Materials Management
- Production Planning and Control
- Construction Engineering
- Earthquake engineering
- Ergonomics
- Structural Engineering
- Hydraulic Engineering

IGI Global is currently accepting manuscripts for publication within this series. To submit a proposal for a volume in this series, please contact our Acquisition Editors at Acquisitions@igi-global.com or visit: http://www.igi-global.com/publish/.

The Advances in Civil and Industrial Engineering (ACIE) Book Series (ISSN 2326-6139) is published by IGI Global, 701 E. Chocolate Avenue, Hershey, PA 17033-1240, USA, www.igi-global.com. This series is composed of titles available for purchase individually; each title is edited to be contextually exclusive from any other title within the series. For pricing and ordering information please visit http://www.igi-global.com/book-series/advances-civil-industrial-engineering/73673. Postmaster: Send all address changes to above address.

Titles in this Series

For a list of additional titles in this series, please visit: www.igi-global.com/book-series

Implications of Digital Systems on Mobility as a Service
John M. Easton (University of Birmingham, UK)
Engineering Science Reference • ©2023 • 335pp • H/C (ISBN: 9781799841388) • US $250.00

Artificial Intelligence and Machine Learning Techniques for Civil Engineering
Vagelis Plevris (Qatar University, Qatar) Afaq Ahmad (University of Engineering and Technology, Taxila, Pakistan) and Nikos D. Lagaros (National Technical University of Athens, Greece)
Engineering Science Reference • ©2023 • 315pp • H/C (ISBN: 9781668456439) • US $250.00

Impact of Digital Twins in Smart Cities Development
Ingrid Vasiliu-Feltes (University of Miami, USA)
Engineering Science Reference • ©2023 • 314pp • H/C (ISBN: 9781668438336) • US $260.00

AI Techniques for Renewable Source Integration and Battery Charging Methods in Electric Vehicle Applications
S. Angalaeswari (Vellore Institute of Technology, India) T. Deepa (Vellore Institute of Technology, India) and L. Ashok Kumar (PSG College of Technology, India)
Engineering Science Reference • ©2023 • 288pp • H/C (ISBN: 9781668488164) • US $260.00

Urban Life and the Ambient in Smart Cities, Learning Cities, and Future Cities
H. Patricia McKenna (AmbientEase, Canada)
Engineering Science Reference • ©2023 • 261pp • H/C (ISBN: 9781668440964) • US $215.00

Remapping Urban Heat Island Atlases in Regenerative Cities
Hisham Abusaada (Housing and Building National Research Center, Egypt) Abeer Elshater (Ain Shams University, Egypt) and Marwa Khalifa (Ain Shams University, Egypt)
Engineering Science Reference • ©2022 • 267pp • H/C (ISBN: 9781668424629) • US $250.00

Health and Well-Being Considerations in the Design of Indoor Environments
Roberto A. González-Lezcano (Universidad CEU San Pablo, Spain)
Engineering Science Reference • ©2021 • 286pp • H/C (ISBN: 9781799872795) • US $240.00

IGI Global
PUBLISHER of TIMELY KNOWLEDGE

701 East Chocolate Avenue, Hershey, PA 17033, USA
Tel: 717-533-8845 x100 • Fax: 717-533-8661
E-Mail: cust@igi-global.com • www.igi-global.com

Editorial Advisory Board

Table of Contents

Detailed Table of Contents

The city of Machico grew in an accelerated and unbalanced way. This unplanned lack of control generated urban flooding problems with some damage. Currently, efforts are being made to avoid and mitigate these floods or alluviums, as flash floods are known on the island of Madeira, by investing in studies. The aim of these studies is to understand the problem and its application in the management of prevention and control of this type of risk. In this way and based on the Flood Risk Management Plan for the Autonomous Region of Madeira, the mitigation of alluvial problems for the Ribeira de Machico is idealized.

In a country, Spain, in which drought and desertification mark the environmental reality, water policies acquire a fundamental role when dealing with the management and planning of water resources. From the perspective of what has come to be called "postmodernity," which predicts the end of the territory and the "no place," the refutation of the region concept is established, at a time when no subspace can be separated from reality, complex and disparate, what happens to it, where the processes of globalization and fragmentation mark the future of its activities, relationships, developments, etc., and, therefore, are marked by regionalization and individualization, taking into consideration that, at the present time, the regional ones are shown as the basis and condition of the set of global relations—from the local to the global. In this way, the knowledge and study of the water reality of a country like Spain has as a reference the region and its problems from the territorial perspective, thus laying down what should be the bases of the water policies.

This chapter explains the concept of dams. It mainly focuses on the dams in India. The chapter discusses the purpose as well as history of dams. The author collected data from various sources in order to perform graphical analysis to support the study. A comprehensive review of various scientific and research papers is done by the author to find the gaps in the research in the recent years. Some new dam projects are also briefly mentioned in the chapter. Plus, the author worked in the direction to find whether the current number of dams in the world, including India, is enough or not. Then, would building more dams replace the use of non-renewable sources of energy completely by itself or not? These are the matters that are mentioned in the chapter.

This chapter aims to develop landslide susceptibility maps for the Sikkim state in India by combining

Foreword

There is no doubt and no one questions the very important role that engineering as a whole plays in the development of countries and their societies. Engineers adapt and apply science and technology to improve the infrastructures that guarantee the increase of wealth and quality of life of communities. Society enjoys the great achievements of science and technology more and more rapidly, to the point that we would not understand our way of life today without all the advances that surround us, from the most common activities to the most complex professional ones.

In the context of land use planning, which is a transversal academic discipline with a strong political impact of intervention at many levels and from many scientific areas, engineering is a fundamental pillar of support to give solvency to many of the approaches that arise from sustainability.

In this sense, Civil Engineering has been experiencing in recent times a great capacity to adopt the fundamentals of spatial planning and to adapt to the new technological incorporations that have been put at its service to carry out increasingly precise and demanding projects from the social and environmental point of view.

That is why this book, coordinated by Professor Sérgio Neves Lousada, is an interesting contribution of novelties and experiences that different authors from different specialties show on the most varied matters with direct or indirect influence on the territory, which must be understood as an anthropized space, given that it has, depending on its geological, geomorphological, hydrographical, edaphological and biodiversity characteristics, different uses and uses of the geographical space.

Therefore, this book contains many advances in the integration of risks, whether natural or derived from human activities, and how to respond to different scenarios that may arise in order to minimize damage to populations and the environment, improving both infrastructures and the management of emergencies arising from disasters or negligence.

Throughout this book, the authors investigate how to prevent or intervene using geotechnologies (Geographic Information Systems, LiDAR, GPS, Remote Sensing and computer developments) to fit in each circumstance the correct decision making in different fields:

- Planning and management of natural resources;
- Forest area management;
- Management of communication infrastructures;
- Urban planning;
- Analysis of areas vulnerable to floods and landslides;
- Assessment of surface and groundwater resources;
- Environmental adaptation of water infrastructures;

- Restoration of rivers and riverbanks;
- Environmental Impact Assessment, etc.

At the present time, engineering, with its rapid progress, also involves certain contradictions with respect to its traditional evolution of the last two centuries, which sometimes question the technical-scientific civilization because scientific-technological innovations are positioned above moral, social or political considerations.

However, as reflected in this transdisciplinary work, engineering has become a fundamental tool, with its technological contributions, to meet human needs in environmental and territorial infrastructure matters, aiming at a permanent improvement of quality-of-life indicators in all areas. Therefore, this book becomes a must-consult work for all those specialists who need to respond to the numerous problems they face on a daily basis.

My sincere congratulations to the participating authors and my congratulations to the coordinator of this publication, which I hope will be successful.

Julián Mora Aliseda
University of Extremadura, Spain

Foreword

Urban areas have been at the centre of discussion for many years on different issues from climate change, transport and smart mobility, renewable energy and positive energy neighbourhoods, CO_2 sequestration and water management, as well as social inclusiveness, but their constantly changing nature poses challenges to their adaptability and growth.

Urban planning and development must be driven by sustainability principles to reduce environmental and economic impacts of infrastructure growth or expansion. Geoinformatics have gained great importance, since the combination of geographic information systems (GIS) and remote sensing can generate development models for more accurate assessment, monitoring and decision-making.

These tools provide insight for a more holistic approach to urban planning and better design of infrastructure projects, providing different levels of information for different stakeholders, from city planners to civil engineers. Satellite imaging also helps detect environmental and structural changes in various sites, helping urban planners create safe and sustainable projections. Geoinformatics has played a more relevant role at the level of forecasting and predicting the consequence of natural and anthropogenic risks assessment at different geographical scales, as well as in emergency planning after a catastrophic event. The visual information these tools and approaches provide can also help technicians and experts view visible changes in infrastructure and mitigate structural risks with geo-spatial data.

Evolution of these technologies should be highly aligned with in the maintenance and development of sustainable and resilient cities through safe, environmentally friendly, efficient, and durable construction and infrastructure. This challenge is even greater when considering climate change and environmental impact issues. Numerous applications, not only at the urban planning level, continue to adopt these technologies and increase their use towards safety, sustainability, and data-driven decision-making.

In the long run, were looking at collaborative mapping solutions, that bring together remote sensing and geographic information systems in synergy to provide reliable and systematic processes for many applications in the civil engineering domain.

This publication constitutes a valuable insight on the contribution of geoinformatics in support of urban politics and the development of civil engineering.

Romeu da Silva Vicente
University of Aveiro, Portugal

Preface

In recent years, the world of Civil Engineering has been changing considerably (Liu et al., 2017; Zhou et al., 2020). Within the many obstacles, barriers, and opportunities, significant challenges should be considered for the future of the Civil Engineering, with the contribution of Geoinformatics, several scientific areas associated with Civil Engineering experience rapid technological evolution (Baig et al., 2022; Lousada, Cabezas, et al., 2022; Ray et al., 2023). Thus, evolution is related to several areas, such as structures, civil constructions, geotechnics, foundations, hydraulics, communication routes, regional and urban planning, sustainability, climate change and environmental impact, among others (Chau et al., 2015; Govindan et al., 2016; Han & Thakur, 2015; Jensen, 2020). Considering these challenges, we can work to portray the future of Civil Engineering.

Civil engineering has been greatly influenced by the evolution of Urban Politics and Geoinformatics, as these two fields have had a significant impact on the design, planning, and management of urban infrastructure (Bernhäuserová et al., 2022; Dilworth, 2020; Pineo et al., 2020).

Urban Politics has played a vital role in shaping the development of cities and urban areas (Robinson, 2016). Political decisions made at the local, regional, and national levels have a significant impact on the design and construction of infrastructure projects (Lieberman, 2002; Lousada, Cabezas, et al., 2022). Civil engineers must consider the political environment in which they operate, as well as the political priorities of elected officials and policymakers, when designing and implementing infrastructure projects (Fang et al., 2023; Jensen, 2020; Ray et al., 2023).

Geoinformatics, which involves the collection, analysis, and interpretation of geospatial data, has also had a significant impact on Civil Engineering. Geoinformatics technologies such as geographic information systems (GIS) and remote sensing have revolutionized the way that civil engineers plan and design infrastructure projects. These tools allow engineers to gather and analyze a vast amount of spatial data, which can be used to optimize infrastructure designs and improve the efficiency of construction and maintenance activities (Lousada et al., 2021; Lousada, Gonçalves, et al., 2022; Lousada & Castanho, 2022).

Geoinformatics, also known as Geographic Information Systems (GIS), is a powerful tool for supporting Urban Politics and Civil Engineering development. GIS is a computer-based system that integrates various types of geographic data, such as maps, satellite imagery, and demographic information, to analyze, visualize, and interpret spatial relationships (Baig et al., 2022; Ray et al., 2023; Singh et al., 2011).

In the context of Urban Politics, GIS can be used to support decision-making processes related to land-use planning, transportation, and infrastructure development. For example, GIS can be used to analyze the spatial distribution of population density, income levels, and transportation networks to identify areas that are underserved by public transportation. This information can then be used to guide the development of new transportation routes or to prioritize the allocation of resources for infrastructure development (Dilworth, 2020; Lieberman, 2002; Robinson, 2016).

In the field of Civil Engineering, GIS can be used to support the design and construction of buildings, roads, bridges, and other infrastructure. GIS can be used to analyze the suitability of different sites for construction projects, taking into account factors such as topography, soil type, and environmental conditions. GIS can also be used to monitor construction progress and to identify potential risks or hazards (Chau et al., 2015; Kotis et al., 2023; Lousada et al., 2021; Lousada, Gonçalves, et al., 2022).

Overall, Geoinformatics is a valuable tool for supporting Urban Politics and Civil Engineering development. By providing a comprehensive and integrated view of geographic data, GIS can help decision-makers to make more informed and effective decisions, leading to more sustainable and resilient urban environments (Lieberman, 2002; Ray et al., 2023).

Geoinformatics is the use of geospatial data, tools, and technologies to understand, analyze, and manage geographically referenced information. Geoinformatics has numerous applications in Urban Politics and Civil Engineering, including:

1. **Urban Planning:** Geoinformatics can be used to analyze the spatial patterns and characteristics of urban areas, which can help urban planners make more informed decisions about zoning, land use, and infrastructure development;
2. **Environmental Management:** Geoinformatics can be used to monitor and manage environmental resources, such as water quality, air quality, and natural habitats, which can be critical for maintaining urban sustainability;
3. **Transportation Planning:** Geoinformatics can be used to analyze traffic patterns and optimize transportation routes, which can help reduce congestion, improve safety, and increase efficiency;
4. **Disaster Management:** Geoinformatics can be used to monitor and manage natural disasters, such as floods, earthquakes, and wildfires, which can be critical for minimizing damage and ensuring public safety;
5. **Infrastructure Management:** Geoinformatics can be used to manage critical infrastructure, such as water supply systems, sewage systems, and transportation networks, which can be critical for ensuring public health and safety.

In Civil Engineering, geoinformatics can be used to:

1. **Design and Analyze Infrastructure Projects:** Geoinformatics can be used to design and analyze infrastructure projects, such as roads, bridges, and buildings, by providing accurate and detailed geospatial data;
2. **Monitor Construction Activities:** Geoinformatics can be used to monitor construction activities, such as earthwork, excavation, and foundation construction, by providing real-time data on project progress;
3. **Assess Environmental Impacts:** Geoinformatics can be used to assess the environmental impacts of construction projects, such as soil erosion, habitat destruction, and water pollution;
4. **Manage Construction Materials:** Geoinformatics can be used to manage construction materials, such as aggregates, cement, and steel, by providing data on the location, availability, and quality of these materials.

Overall, Geoinformatics is an important tool for supporting Urban Politics and civil engineering by providing accurate, timely, and actionable information that can inform decision-making and improve project outcomes. Also, can help improve the efficiency, sustainability, and resilience of cities and their infrastructure. As the field of Geoinformatics continues to evolve and new technologies become available, the potential applications and benefits of this field are likely to continue expanding, in the coming years (Iwahashi et al., 2012; Lemenkova & Debeir, 2023; Ohta et al., 2022).

It will be the main intention of this volume, portray the evolution of the world of Civil Engineering, what happened in recent decades, but what will develop in the future, what will be the path. Thus, this volume will seek to characterize the different academic areas of Civil Engineering, demonstrating its evolution. It will be a means to allow the dynamization of publications in the area of Civil Engineering. This volume is essential for academicians, researchers, policymakers, environmentalists, scientists, technicians, decision makers, practitioners, and students. It will be a way to portray what everyone involved is developing in this specific area, but also a way to record this evolution, and allow for future consultation.

With this volume it is intended to portray the theme, according to the framework described above, for this and with the contribution of the various authors, the same was done through its chapters, which will be succinctly exposed, according to the order of their compilation in said volume. A brief summary of each chapter is given below.

Chapter 1: Urban Planning and the Third Landscape in the Field of Local Development – Case Study

In the processes of urban planning, oriented to local development, the concretion of the so-called "geographical spaces" appears to us as an essential issue if the objective is territorial knowledge of landscapes, in general, and, of the so-called "Third Landscape" is intended, in particular. And it is that, knowing the reality of the physical environment as well as the implications and interrelationships with respect to the human and economic environment, allows laying the foundations of the complex -wide and varied- reading and interpretation of landscapes (conceived as a medium or environment, as well as for its aesthetic or artistic values). With all this, throughout the present research the idea according to which the landscape is marked by the valuation of the one who perceives and interprets it is addressed, as well as for its constant transformation, defining a close connection between the viewer and the medium itself, thus specifying the existence of a multitude of landscapes, according to the images perceived by people, societies and even those indefinite ones to which a universal value is given. Thus, when it comes to addressing the so-called "Third Landscape", it is presented to us as a way of approaching the spatial reality of "residual" spaces, just as it happens in a Minor Entity, such as the Cortijo de San Isidro, located in a city of the Community of Madrid, Aranjuez.

Chapter 2-Modeling of Flood Risk: Vulnerability and Adaptation of Machico City Center

The chapter focuses on analyzing the flood risk in Machico City Center, including its hydrological functioning, historical incidence of floods, and potential impacts on the natural and human environment. The authors also discuss human adaptation through flood planning and the use of tools such as hydrological modeling and mapping of flood-prone areas to assess and manage flood risk. The chapter highlights the importance of spatial data in analyzing and planning technical and policy actions related to flood risk management. The keywords associated with the chapter include flood, Machico, Madeira, Machico River, and Machico Watershed.

Chapter 3: Urban Flooding Risk in Machico and Planning Its Territory as a Form of Prevention

This chapter highlights the increasing urbanization and its role in exacerbating the flood risk in Machico, a parish located in Madeira, Portugal. The authors emphasize the need for effective risk management strategies, including identifying risk-prone areas, implementing prevention and damage mitigation measures, and developing early warning systems to protect the population from extreme weather conditions. The chapter stresses the importance of planning the territory to prevent urbanization in risk-prone areas, such as steep slopes or floodplains, and the development of appropriate drainage and flow systems. The authors conclude that the successful management of flood risk in urban areas requires a combination of technical and policy actions, including the involvement of various stakeholders in the decision-making process. The study provides valuable insights into the challenges of urban flood risk management and the importance of planning for prevention.

Chapter 4: Port of Machico, Madeira – Assessment of the Consequences of Overtopping Using AHP Methodology

The chapter proposes the use of the Analytic Hierarchy Process (AHP) to evaluate the consequences of overtopping events and identify measures that can reduce their impact. The AHP is a multicriteria decision-making tool that evaluates alternatives based on a set of criteria and their respective weights. The criteria used in this case are related to the consequences of overtopping, such as economic losses, environmental impacts, and disruption of activities. The results obtained from the AHP can then be used to identify the most effective measures for reducing the impact of overtopping, such as the construction of protective structures. This study provides valuable information for decision-makers and stakeholders involved in the management of the Port of Machico and can contribute to the development of more effective risk management strategies.

Chapter 5: Flood Risk and Vulnerability – An Approach to Urban Floods in Ribeira de Machico in the Autonomous Region of Madeira

The chapter addresses the issue of urban flooding in the city of Machico, located in the Autonomous Region of Madeira. The authors point out that the city's rapid and uncontrolled growth has led to urban flooding problems that have caused damage. To mitigate these floods, studies are being conducted to understand the problem and develop measures for prevention and control. The Flood Risk Management Plan for the Autonomous Region of Madeira serves as a basis for the proposed mitigation plan for the Ribeira de Machico. The chapter highlights the importance of proper urban planning and risk prevention measures to reduce the impact of urban flooding.

Chapter 6: Water Policies – Management and Planning of Water Resources in Spain

In the relationship between economic, social and environmental aspects, the Economic Policy is configured as the basis for global integration of the set of essential elements on which to base the processes of planning and management of water resources in a country of contrasts like Spain. And it is that, the preservation, conservation, regeneration and improvement of water quality, fundamental bases of the so-called "environment" -in which water resources are found-, must be constituted as the main objectives of environmental analysis, as well as policies primary finalists for Economic Policy. In this way, throughout the present study, the concretion of Water Policies is addressed, as the basis of the decision-making process and action in the use of water resources, endowing them with the capacity to specify hierarchies between the set of activities, through the awareness of the set of economic agents of the risks, damages and impacts of environmental degradation processes and, with the scientific argumentation of the expected impact of the use, enjoyment and conservation of water in Spain.

Chapter 7: The Need and Development of Dams in India as Well as in the World for a Better Future

This book chapter explains the concept of dams and its type in general. It mainly focuses on the dams in India. The chapter discusses the purpose as well as history of the dams. The author had collected data from various sources in order to perform graphical analysis, to support the study. A comprehensive review of various scientific and research papers are done by the author to find the gaps in the research in the recent years. Some new dam projects are also briefly mentioned in the chapter. Plus, the author worked in the direction to find whether the current number of dams in the world, including India, is enough or not. Then, building up more dams can replace the use of non-renewable sources of energy completely by itself or not, are the matters which are mentioned in the book chapter.

Chapter 8: Analytical Hierarchy Process-Based Spatial Data Analysis for Landslide Susceptibility Mapping

The purpose of the study is to evaluate the value of a geographic information system (GIS) modelling approach for forecasting effective landslides in various regions of Sikkim state, India. Using Landsat 8 data to create numerous key themed map layers. The authors examine the area's geology, slope, density of drainage, features, and land use using digital elevation model data and Landsat 8 images. For the final mapping of landslip susceptibility, an analytic hierarchy process (AHP) approach was used to integrate the produced landslide occurrence features into a GIS environment. The ROC (Receiver Operating Characteristic) curve is used to assess the effectiveness of landslide maps created using AHP .The area under the curve (Area Under Curve) has been estimated 70% (good accuracy) using the landslide susceptibility map and data from the landslide inventory. The area's most prone to landslides include the Ramgarh thrusts, Ranjit windows, and adjacent MCT-1 and MCT-2 regions.

Chapter 9: Photographic Images Properties and Their Correlation With the Quality Requirements for Aerial Mapping

With the rapid growth in the use of small low-flying Unmanned Aerial Vehicles (UAV) with integrated imaging equipment for mapping of avarious objects and areas, the authors suggest to consider the proper preparation of technical specifications in order to obtain high-quality photographic images. The properties of the resulting aerial photographs must be analyzed in terms of their correlation with the accuracy requirements for aerial mapping products. The application of UAV-Photogrammetry technology, in the realization of the aerial photography mission in many cases depends on the appropriate selection of Ground Sample Distance (GSD), in order to avoid excessive accumulation of photogrammetric data.

Chapter 10: The Preliminary Researches of Pull-Out of GFRP Rods From Mineral Coarse-Grained Soils

The author presented the results of original, model laboratory tests aimed at checking the suitability of GFRP (Glass Fiber Reinforced Polymer) composite rods for reinforcement of selected mineral soils, in this case coarse-grained. The tests were performed in a specially modified direct shear apparatus and determining the relationship between the amount of pulling force and the given normal stresses. The results were presented in tables and illustrated in appropriate graphs. The GFRP rods are light, durable and resistant to corrosion, thanks to which they can be used to effectively strengthen the ground even in an aggressive water environment, minimizing subsidence, which is especially important, for example, in the case of foundations of high buildings, modeling embankments for various purposes or preventing landslides. The research was carried out as the project "The Ground Reinforcement System With GFRP Composite Bars With Fiber Optic Monitoring" at the Rzeszów University of Technology.

I hope the selected chapters will help the possible readers in enhancing their understanding and practicing of *Geoinformatics in Support of Urban Politics and the Development of Civil Engineering*.

Sérgio António Neves Lousada
University of Madeira, Portugal

REFERENCES

Baig, M. F., Mustafa, M. R. U., Baig, I., Takaijudin, H. B., & Zeshan, M. T. (2022). Assessment of Land Use Land Cover Changes and Future Predictions Using CA-ANN Simulation for Selangor, Malaysia. *Water (Basel)*, *3*(3), 402. Advance online publication. doi:10.3390/w14030402

Bernhäuserová, V., Havelková, L., Hátlová, K., & Hanus, M. (2022). The Limits of GIS Implementation in Education: A Systematic Review. *ISPRS International Journal of Geo-Information*, *12*(12), 592. Advance online publication. doi:10.3390/ijgi11120592

Chau, C. K., Leung, T. M., & Ng, W. Y. (2015). A review on Life Cycle Assessment, Life Cycle Energy Assessment and Life Cycle Carbon Emissions Assessment on buildings. *Applied Energy*, *143*, 395–413. doi:10.1016/j.apenergy.2015.01.023

Dilworth, R. (2020). Comparative Case Study Methods in Urban Political Development. *Social Sciences (Basel, Switzerland)*, *10*(10), 183. Advance online publication. doi:10.3390ocsci9100183

Fang, C., Qiu, C., & Zheng, Y. (2023). Shape Memory Alloys for Civil Engineering. *Materials, 16*(2). *Materials (Basel)*, *2*(2), 787. Advance online publication. doi:10.3390/ma16020787 PMID:36676523

Govindan, K., Madan Shankar, K., & Kannan, D. (2016). Sustainable material selection for construction industry – A hybrid multi criteria decision making approach. *Renewable & Sustainable Energy Reviews*, *55*, 1274–1288. doi:10.1016/j.rser.2015.07.100

Han, J., & Thakur, J. K. (2015). Sustainable roadway construction using recycled aggregates with geosynthetics. *Sustainable Cities and Society*, *14*, 342–350. doi:10.1016/j.scs.2013.11.011

Iwahashi, J., Kamiya, I., & Yamagishi, H. (2012). High-resolution DEMs in the study of rainfall- and earthquake-induced landslides: Use of a variable window size method in digital terrain analysis. *Geomorphology*, *153–154*, 29–38. doi:10.1016/j.geomorph.2012.02.002

Jensen, J. S. (2020). Innovative and sustainable operation and maintenance of bridges. *Structure and Infrastructure Engineering*, *16*(1), 72–83. doi:10.1080/15732479.2019.1604772

Kotis, K., Angelis, S., Moraitou, E., Kopsachilis, V., Papadopoulou, E.-E., Soulakellis, N., & Vaitis, M. (2023). A KG-Based Integrated UAV Approach for Engineering Semantic Trajectories in the Cultural Heritage Documentation Domain. *Remote Sensing (Basel)*, *3*(3), 821. Advance online publication. doi:10.3390/rs15030821

Lemenkova, P., & Debeir, O. (2023). Quantitative Morphometric 3D Terrain Analysis of Japan Using Scripts of GMT and R. *Land (Basel)*, *1*(1), 261. Advance online publication. doi:10.3390/land12010261

Lieberman, R. C. (2002). Ideas, Institutions, and Political Order: Explaining Political Change. *The American Political Science Review*, *96*(4), 697–712. doi:10.1017/S0003055402000394

Liu, B., Wu, J., Zhang, J., Wang, L., Yang, J., Liang, D., Dai, Q., Bi, X., Feng, Y., Zhang, Y., & Zhang, Q. (2017). Characterization and source apportionment of PM2.5 based on error estimation from EPA PMF 5.0 model at a medium city in China. *Environmental Pollution*, *222*, 10–22. doi:10.1016/j.envpol.2017.01.005 PMID:28088626

Lousada, S., Cabezas, J., Castanho, R. A., & Gómez, J. M. (2021). Hydraulic Planning in Insular Urban Territories: The Case of Madeira Island-Ribeira Brava. *Water (Basel)*, *13*(21), 2951. Advance online publication. doi:10.3390/w13212951

Lousada, S., Cabezas, J., Castanho, R. A., & Gómez, J. M. N. (2022). Land-Use Changes in Insular Urban Territories: A Retrospective Analysis from 1990 to 2018. The Case of Madeira Island—Ribeira Brava. *Sustainability (Basel)*, *24*(24), 16839. Advance online publication. doi:10.3390u142416839

Lousada, S., & Castanho, R. A. (2022). The Role of Ports in Tourism: Porto Santo Harbour. *Water (Basel)*, *19*(19), 3176. Advance online publication. doi:10.3390/w14193176

Lousada, S., Gonçalves, L., & Atmaca, A. (2022). Hydraulic Planning in Insular Urban Territories: The Case of Madeira Island—São Vicente. *Water (Basel)*, *14*(1), 1–19. doi:10.3390/w14010112

Ohta, R., Matsushi, Y., & Matsuzaki, H. (2022). Use of terrestrial cosmogenic 10Be to quantify anthropogenic sediment yield from mountainous watersheds: Application in reconstructing environmental change in the Tanakami Mountains, central Japan. *Geomorphology*, *405*, 108201. doi:10.1016/j.geomorph.2022.108201

Pineo, H., Zimmermann, N., & Davies, M. (2020). Integrating health into the complex urban planning policy and decision-making context: A systems thinking analysis. *Palgrave Communications*, *1*(1), 21. Advance online publication. doi:10.105741599-020-0398-3

Ray, R., Das, A., Hasan, M. S. U., Aldrees, A., Islam, S., Khan, M. A., & Lama, G. F. C. (2023). Quantitative Analysis of Land Use and Land Cover Dynamics using Geoinformatics Techniques: A Case Study on Kolkata Metropolitan Development Authority (KMDA) in West Bengal, India. *Remote Sensing (Basel)*, *4*(4), 959. Advance online publication. doi:10.3390/rs15040959

Robinson, J. (2016). Comparative Urbanism: New Geographies and Cultures of Theorizing the Urban. *International Journal of Urban and Regional Research*, *40*(1), 187–199. doi:10.1111/1468-2427.12273

Singh, P., Thakur, J. K., Kumar, S., & Singh, U. C. (2011). Assessment of Land Use/Land Cover Using Geospatial Techniques in a Semi-arid Region of Madhya Pradesh, India. In Geospatial Techniques for Managing Environmental Resources (pp. 152–163). Springer Netherlands. doi:10.1007/978-94-007-1858-6_10

Zhou, S., Zhou, M., Wang, Y., Gao, Y., Liu, Y., Shi, C., Lu, Y., & Zhou, T. (2020). Bibliometric and Social Network Analysis of Civil Engineering Sustainability Research from 2015 to 2019. *Sustainability (Basel)*, *17*(17), 6842. Advance online publication. doi:10.3390u12176842

Acknowledgment

This Book represents the materialization of yet another dream: the achievement of yet another goal in my academic and personal life. A goal that has required study, dedication, and sacrifice. I remember that all this was only possible thanks to the contribution of many people and entities to whom I formally thank:

To my parents, who, with the values and principles passed on, always made me want something more and whose personal sacrifices allowed me to get where I am and be who I am. I will always be grateful, without ever being able to repay with the same magnitude, for everything they did for me;

To my family, who never let me give up and always supported me in every way possible, especially my sister and my nephew;

To my wife and daughters, for being the factor of my emotional stability throughout my academic and personal path, and for the time dedicated to helping me, whenever they felt it was necessary;

To all my professional colleagues, reviewers (and friends), for all their availability, guidance and support in the preparation of this book;

To those responsible for different entities, especially the entire editorial staff of IGI Global, always available for any type of support, whether informational or documental;

To all friends and colleagues, for their friendship and support, for their availability and critical view of the work developed in this book.

A good one,

Sérgio António Neves Lousada
University of Madeira, Portugal

Chapter 1
Urban Planning and the Third Landscape in the Field of Local Development:
Case Study

Ignacio Sotelo Pérez
University Institute of Environmental Sciences, Complutense University of Madrid, Spain

Maria Sotelo Pérez
King Juan Carlos University, Spain

ABSTRACT

Urban planning in the community of Madrid (Spain) acquires specific characteristics in the particular case of local development, gaining real importance in the case of minor local entities. A good example is found in the so-called Cortijo de San Isidro, integrated in the area of the municipality of Aranjuez. In the present investigation, the authors approach the reality of the third landscape from the perspective of local development, referring to this administrative situation and which marks a historically-based development model, with geographical perspective and implications both landscape and environmental. The minor local entities have personality and full legal capacity to exercise the powers that the legislation and areas of action recognize in each case. Cortijo de San Isidro is currently made up of different buildings, outbuildings, and a large amount of land dedicated to cultivation in order to serve as a model of colonization (many of them today residual spaces marked by urban speculation and the existence of not a few marginal spaces, a true model of local development).

DOI: 10.4018/978-1-6684-6449-6.ch001

INTRODUCTION: FIRST APPROACH TO THE LANDSCAPE AND ITS GEOGRAPHICAL CONTEXTS

The landscape interpreted from a functional perspective, as a link that unites the different terminologies related to this concept, as well as the human being with the geographical space, and from the value assigned to it as a representative of the historical evolution of human action, and as a conciliator with the environment and vital environments, it has been shaped around the correspondence of both the components resulting from human activity, as well as those other elements derived from nature. Even so, the landscape has managed to be treated as an indicator, molded by its own constituent elements, these are; the climates, the elements settled in rural environments, the reliefs, the vegetation of natural spaces, but, also, by the vegetal and floral masses of the urban spaces (among which are also included those sites in which, naturally, various biological systems emerge due to disuse and abandonment of the action of human). Starting from each and every one of these main elements of the landscape, it is possible to analyze the different scopes and meanings of the landscape concept (and even more, the real implication of the activities of the human being on it) (Sotelo Navalpotro, 1992). In fact, the diachrony of this concept has been presenting different ways of being expressed over the years, and acquiring, in turn, no less consequences with respect to its configuration and relationships. Regarding the first of these issues, we find the diverse conceptualization of the term landscape given by different authors -such as, for example, Pierre Fremont, González Bernáldez, George Bertrand or Claude Bertrand-, in the wonderful course of time, highlights not only the contradictory but also the polyvalence that the landscape concept supposes, since on many occasions it has been treated from a perspective, sometimes ecological, other geographical, other aesthetic, although the differentiation, as it could not be otherwise, it responds to the combination of scientific technicalities, on the one hand, and common language, on the other (to which should be added, of course, those descriptions from literary techniques whose ultimate purposes are incardinated to express the beauty and beauty of the reality that surrounds us) (Sotelo Navalpotro, 2015). In this sense, starting from the definition of Professor Mathieu Kessler, who understands that the landscape represents beauty, and who describes that "this is why the true landscape is beautiful. It is the disinterested aesthetic vision of a geographical space, of an ecumene, of an autarchic residence, protected, peaceful, order without external purpose"(Másmela Díaz, 2010); or, Professor Farinelli who, choosing from among all the possible concepts of landscape, describes as "precisely, the aesthetic character of bourgeois culture imposes [in the words of the professor], that artistic knowledge be transformed into science of nature, of mediation of vision: [and for these purely artistic reasons, the professor citing the Prussian Alexander von Humboldt, as one of the most significant fathers of Geography, saying about him as] for this reason, Humboldt chooses the concept of landscape and uses it as the most suitable vehicle to ensure the transit of the protagonists of the literary public dimension towards the domain of scientific culture" (Farinelli, 1992). Although perhaps the one that best fits within the purposes of this study is the one that Neil Smith introduce by showing how "the problem of course lies in the assumptions since, like that of "nature", the concept of space tends to be considered as given and its meaning without problematizing, while in fact it is a vague one with a multiplicity of meanings, sometimes even contradictory" (Smith, 1984).

Regarding the second of the assessments outlined, that is, the consequences in terms of the conFiguretion and the new relationships originating from the evolution of the landscape, we must start from the remarkable fact that what we understand as landscape is It is closely linked with the human being, with the social group, and with the individuals that make up our modern societies, since the landscapes that surround us are usually influenced to a lesser or lesser extent by the transformations caused by the action of the human being. In this sense, landscapes were initially made up of a series of components of the physical environment, with which the appearance of the human being brings with it a set of new components to the landscape, which meant the advent of sensibly palpable consequences in reality, since human action transforms that initial landscape into a "cluster of physical, biotic and anthropic elements" (in such a way that, when we enter the landscape, what González Hernández described as a situation that is difficult to explain would happen to us, given the complexity of the concept and its possible meanings and origins) (Sotelo Navalpotro, 2015).

In one way or another, even if we simplify the meaning of "landscape" to a single personal and individual approach (an issue that is not intended to be done within these lines), as for example, the one that the artist Claude Monet sample when alluding that for "the landscape does not exist in a strict sense, it changes at every moment; it is the surrounding atmosphere that gives it its true value", it cannot be ignored, an inescapable reality, and that is that when carrying out a certain study, such as the one being carried out here, and in which, among other issues, aspects related to the environment are analyzed, the landscape must be considered within the relevance and incidence that truly corresponds to it, not only because of the central role it plays when putting into practice adequate environmental planning, but also because of its transcendence on matters as important as land use planning, urban planning, and geographic space, ultimately on human activities and chores.

When dealing with landscape issues, and despite all the previously mentioned conceptual itinerary, it must be appreciated that we are dealing with a category that is certainly valuable for areas related to the social sciences, and specifically with the one that most interests us here, Geography. In turn, within the academic field that includes geography, the landscape has been used both to examine regions during the last century, and recently for its use in the framework of sustainability. And it is that the landscape (observed, felt, painted, lived), in accordance with the definition of landscape according to which it refers to the "Part of a territory that can be observed from a certain place", it is conditioned, accommodated to the different needs and demands of the population. Inserted in this same academic field corresponding to geography, the term landscape has often been distorted by its application to the reality that comprises the region, and thus has been reflected by the conceptual juxtaposition that can be seen from the different approaches. Please add some examples, coming from the so-called cultural school of Berkley, in which the American geographer Carl Ortwin Sauer, affirmed in 1925 in his work entitled The Morphology of the Landscape, as "the landscape is a conceptual unit of geography used to characterize the specific geographical association. It is equivalent to terms such as area or region" (Ortwin Saber, 1925), and in which said concept insinuated as a descriptor of the terrestrial morphology (and, whose constitution comes jointly from cultural and physical elements).

This conceptual juxtaposition between the landscape and the region has also been perceived, in some definitions originating around geomorphology in which the landscape is understood as the "main physical-geographical unit of the division (regionalization) of a territory with the same type of relief, geological structure, climate, general character of the surface and groundwater, with sequential conjugations of soils, vegetation and fauna. Each geographic landscape consists of simple physical-geographical units with interrelated boundaries. On the other hand, complex physical-geographical landscapes that are similar in structure can be included in the physical-geographical units of the landscape of higher orders (provinces, regions, zones, etc.)" (Lugo Hubp, 1989). This dynamic concludes with what happened at the turn of the century, when at the beginning of the 21st century, the landscape is understood as the consequence of the environment or rather of the natural environment, transformed by the cultural aspects of human beings (Ramírez Velásquez & López Levi, 2015).

The versatility of the previously shown landscape concept, at first glance, helps us to fully enter the geography related to sustainability and the environment, helping us to discern not only how it works, but also how it is formed. And it is that, as the French geographer Georges Bertrand shows us, in his work entitled "Introduction of the State doctoral thesis", he would express as "nothing is more familiar to the geographer than the changing mosaic of the Earth's landscapes. Nothing is more foreign to the geographical method than the global analysis of these same landscapes. . . . To analyze a landscape is to pose a problem of method. First of all you have to develop the working instrument" (Bertrand, 1974)[1]. In this sense, it is denoted that despite the innumerable intellectual efforts that, throughout the aforementioned time periods, different authors, like George Bertrand, Chantal Blanc-Pamard, Claude Bertrand or Eugene P. Odum, have been making to achieve an accepted sense of landscape, it seems that the inability among us to achieve globally assimilate the natural environment and its corresponding elements (plants, land, etc.), which constitute it. The human being has contributed through his actions on the territory, in the generation of a set of significant impacts on the different landscapes, which has been increasing, in turn, the need to plant new ways to harmonize that "coexistence" of society with its environment, as well as combining the knowledge acquired with the different uses of the territory, the soil, and the natural environment; what ultimately has come to be legally recognized as "enjoying an adequate environment for the development of people" (Spanish Constitution of 1978)[2], as an essential objective for the development of the social community. Arrived, then, at this point we can see the genesis of new meanings that come to cover the less clear points of the interpretation of the intrinsic reality of Environmental Sustainability in relation to the Territory; we are referring to the Third Landscape.

OBJECTIVES AND STUDY METHODOLOGY

Based on the fundamental premise of the importance of the analysis and assessment of the repercussions emanating from the study and interpretation of the landscape, as the primary axis of the implementation of territorial development models (in this specific case, with the aim of improving the environmental, social and economic quality of a space undergoing transformation, in general, and of an entity in the municipality of Aranjuez, in the Community of Madrid), the justification of the present study originates from the initial hypothesis of the evolution of the transformation processes of traditional development models towards others that address the needs, present and future, of territories in which economic changes mark development processes.

With all the above, in the present investigation, the general objective is to know the reality of urban planning processes as the basis of local territorial development -through the case study-, through a new conception, the so-called "Third Landscape" (which is made up of the set of residual spaces of the human being), from an environmental perspective. At the same time, the specific objectives of the research are specified in the study of the reality of the "Third Landscape" in the territorial reality of Spain, and its implications on the environment; know and assess the implications of the analysis of the "Third Landscape" in the processes of protection and conservation of nature; and, finally, these studies are addressed in the so-called "Cortijo de San Isidro", a Minor Entity located in the municipality of Aranjuez, south of the Community of Madrid (given the relevance of this entity with personality and full legal capacity to exercise the powers that the Legislation and the Areas of Action have in the territorial planning process).

For this reason, based on the use and employment of the hypothetical deductive scientific method, this paper starts with a preliminary study of the territorial context in which the realities framed in the so-called "Third Landscape" are framed, to subsequently assess and interpret the real impact of said residual spaces in transformation in the "Cortijo de San Isidro" Minor Entity, through the realization of "field work"; lastly, the results obtained are applied to the concretion of Land Management and Regional Development Models, in a general way, and their direct application in the case under study, in a particular way.

"THE THIRD LANDSCAPE" AND ITS TERRITORIAL REALITY IN SPAIN, FROM THE FIELD OF THE ENVIRONMENT

After the above, we can now approach the praxis of the reflections and postulates collected within a "new" concept, conceived by the French landscaper Gilles Clément, and which he called under the rubric of the signifier "Third Landscape". This process has the objective of developing the implications on the ground of its conceptual meaning; and for this, the theoretical meaning of what the author called the "Third Landscape" will be analyzed and valued on the one hand, and on the other hand, the functional usefulness that this concept can have on the spatial reality that surrounds urban and rural areas will be studied, rural areas of some landscapes and territories of present-day Spain, from the perspective of Sustainability and the Environment.

The objective set in this research work regarding this concept of the "Third Landscape" is framed within two fundamental scientific perspectives: those provided by Regional Geography, and legal knowledge from Environmental Law. To do this, as has been shown in the previous sections, it has been essential to expose the general vision of the relationships and links between urban planning and the rural environment (including those spaces related to industry and sightseeing), complemented by the repercussion that the exploitation of the natural resources present in the Spanish territory usually entails, from the environments and perimeters of ecological, economic and social sustainability.

From this perspective, the concept of the "Third Landscape" has been considered from its practical implication, but not only from its application to urban expansive phenomena, or rural recessive ones, which are evidenced by studies from the landscape discipline, but also, said The concept has been used from the fundamental perspective provided by the applicable legislation within those spaces that make up the complex physical and human reality corresponding to each of the analyzed territories, that is to say, it has been wanted to investigate, if the content of the theoretical concept of the so-called "Third Landscape", has been adapted in reality, to the principles and opinions of what in this investigation has come to be called and conceptualized as the "Unity of the Law" (essential when applying these issues to policies, plans, programs and projects, in relation to Environmental Impact Assessment, Strategic Environmental Assessment and Comprehensive Planning, through the implementation of Territory Planning, the urban planning and socioeconomic planning).

NATURE AND "THIRD LANDSCAPE": PROTECTION OF LIFE AND DIVERSITY

At first glance, nature is configured as the support on which the different human settlements are erected, which subsequently and from a reductionist point of view gives rise to the different urban and rural spaces, which in turn are subjected to a series of regulated guidelines that organize economic activities, political initiatives, and, in general, the behavior of the social body as a whole. In fact, without falling into deterministic approaches, we can affirm that nature is constituted as a space, as a territory in which the various species find their place to form, develop and interact; however, when these circumstances are not fulfilled, that is, when these species are unable to find these spaces, these places and territories in which to display their vital behaviors, they tend to look for others in which to carry out said purposes, to these places and territories are those that we can commonly discern within the thus cataloged as "Third Landscape"[3].

In this sense, when the territories and spaces encompass and integrate a whole set of aspects that concern the relations of the human being with the environment that surrounds him, and being this environment in which he unfolds his biological tasks, converting the terrestrial surface, and its soils in its operative axis, it is then, when we can affirm that these environments correspond to the urban or rural spatial category (according to which cases), in which an endless number of physical and economic interrelationships take place, likely to have a significant impact on the environment. When these places and territories, for one specific reason or another, are abandoned or lose interest for the human being, and become and give rise to spaces comprised of abandoned lots and awaiting new occupations, transposed boundaries between diverse territorial delimitations, to edges and margins of land communication routes, or to constructions and buildings in disuse, it is then when, in the same way as in the previous paragraph, we can state that we are facing the so-called "Third Landscape".

In today's world, in the contemporary era in which we find ourselves, the presence of spaces without a specific identity, without apparent qualification, without a name that allows determining their possible usefulness, performance and interest, have become a common reality generated by the different settlements, whether within territories considered urban or rural, in which the human being carries out his daily activity, as an individual, and as part of the social group; and it is that the "Third Landscape" is closely linked, not only with the territory of diversity, but also with demographic aspects, which immediately affect those spaces, which spontaneously (or not), are reserved for it qualified as "Third Space".

This manifest objectivity is more than recognizable when analyzing urban areas. The city accepted as part of nature, and today more than ever aligned with international trends and interactions of cities from different territorial areas, making it sustainable, artistic, beautiful, and above all functional and competitive; becomes incapable of continuously causing places characterized, not only as having a dubious identity, or lacking any useful meaning, but also because of their eminently residual nature (Pérez Igualada, 2018). These residual spaces comprise, are in turn part of each and every one of the recognizable spaces, these are the tourist spaces, the industrial spaces, the rural spaces, and the urban spaces, all of them generators of spaces considered as residual product of the different tourist, industrial, agricultural, livestock activities, typical of human activities developed in these places. These and some other questions are attempted to be answered through the so-called "Third Landscape".

In this context, in which the "Third Landscape" has been defined, as one that "is made up of all the residual spaces of the human being (and, in turn, shows how), these margins bring together a biological diversity that It is not understood as wealth" (Clément, 2018), the city, and the rural territorial areas, have been strongly influenced by the urbanization processes promoted by the interests of contemporary societies, and which, among other issues, have been the cause of the incorporation of rural spaces and populations into the areas included within urban phenomena, absorbing and inscribing them within the dialectic, and the proper articulation that shapes and configures the city. And it is that these urbanization processes are the cause (at first glance), of the expansion of urban areas (also in those places where the rural predominates), being the city where, when its urban fabric expands more, more spaces of a residual nature can be observed, with the generalized singularity(although not in the same way in all cities), that in the central almond said residual spaces are smaller and scarcer in comparison with what happens in the peripheral areas where these residual spaces are distinguishable by their large number and size.

However, despite all these increasingly constant, intense, and temporarily lasting urbanization processes, it has also been possible to denote the strength of many of the rural spaces, which have managed not to succumb to the stakes of the expansive processes of urban areas, conserving certain economic assets characteristic of these rural places, some of the social traditions typical of each of the areas in question, as well as part of their functional dynamism, additionally managing to prevent their spaces from being harmed by the transformations caused by the dispersion that urban phenomena entail. It is in these rural sectors, in which it can be seen how the residues, or leftovers closely related to the boundaries of the fields, the margins of communication routes and highways, the banks and the hedges; as well as all those reliefs, antagonistic in terms of the coexistence of heavy machinery (for agricultural exploitation), within its territorial extensions; are occupied by spaces classified as residual.

In any case, both in rural areas and in those considered urban, a whole set of activities derived from the action of the human being are developed. Moreover, these circumstances are also appreciated when, after the urbanization processes described above, the city has incorporated, integrated and absorbed within its fabric, the different districts of its surroundings previously recognized as rural, since after these processes, they tend to occur and give way to a series of new activities and certain situations, a direct consequence of that urban annexation, causing an interposition of the urban over the rural, which causes a certain syncretism, which is expressed on the whole territory. These activities, whether those that unfold in the urban environment, as in the rural, or those that arise from the symbiosis of both spaces (the urban with the rural and vice versa), are manifested in each and every one of the related environments with culture, land ownership, productive activity or the environment. When these human activities are consistently neglected, abandoned, it is when we can affirm that we are facing the emergence of the so-called residual spaces, these are spaces, the result of the deliberate inaction of human activity. And it is precisely within these spaces (or around them), the places, and the zones in which the biological systems essentially considered as free appear and flourish.

When analyzing these facts described (and that concern in one way or another the reality of both urban spaces, as well as those considered as rururban, and rural spaces), from the disciplines in which issues related to the territory are conceived as one of the main objects of analysis of the same, that is to say, economics, anthropology, urban planning, in a certain sense law, sociology, but, mainly, geography, one can discern a whole host of relationships that can constantly be distinguished between urban and rural areas (increased even more if possible, if we take into account all that large number of "socio-spatial" processes that have been emerging over the last few centuries, hand in hand with the "metropolization" processes that have led to a transformation series in the traditional ways that the aforementioned spaces had to relate to each other). This set of relationships is not limited exclusively to those symbiotic processes, in which the different activities that, before the expansion processes exposed, only occurred in urban areas, could also be appreciable in rural areas; nor to those relationships that in the same way were only visible in those places considered rural, and that due to processes of social or cultural assimilation, were also no longer seen in urban áreas; but a strong link can also be seen, aside from these two aforementioned relationships, in the essential fact that both in rural and urban spaces (either jointly considered -urban and rural-, well in an individualized way, on the one hand the rural, on the other the urban), can be distinguished, the same processes of evolution, which mark the beginning and the passage to qualify it as the secondary landscape.

These evolutionary processes, start at first from those spaces in which no specific type of exploitation has yet been developed (in fact, they may later in time become exploited, or on the contrary remain in the same circumstances of "apparent uselessness" in which it can be observed with the naked eye). In this first moment, which the cited author describes as *"primary groups"*, the various species that inhabit it (and carry out their daily activities and existence), would have excellent levels of quality of life in correlation with the conditions (climax), from the middle (these primary assemblages host, par excellence, most of the diversity present on the entire planet). Subsequently, and focusing on those relationships set forth in the previous paragraph, we can attend to those spaces caused by disuse, or the abandonment of one or several human activities. These spaces are those that the landscaper Gilles Clément classifies as *"residual spaces"*, which, as it could not be otherwise, would evolve in his consideration to those spaces designated as secondary (characterized in turn by a series of dynamics of powerful transcendence, and for its heterogeneity and disorder). In a last and third place, within this process of evolution, one can differentiate those named as *"reserves"*, which comprise a very varied collection of protected human activities due to their qualitative and quantitative consideration, that is, due to their scarcity and their fragility, thus for its wide and rich diversity in danger of being extinct. These reserves would also be made up of those places respected and esteemed for their sacred nature in some cases, prohibited in others, which the cited author would consider as "territory of the gods" (and among which one could, for example, specify those concerning "fadys" spaces originating in Madagascar, the mountain peaks of the Indian sub-continent, or the complex valleys of "leyaks", located in Bali, among others). These evolutionary moments described, would constitute the three origins of diversity, referred to as "the number of living species that can be distinguished between animals, plants and elemental beings such as viruses, bacteria, etc."; and where the human being would be differentiated and particularized as the only species whose diversity would be externalized through a series of cultural and ethnic differences[4].

Focusing on the reality that the Third Landscape represents, it has to be referred to as this, it understands in a simplified way speaking a fundamentally mental order. When the existence of a series of places is exposed, and of spaces whose identity and definition become diffuse, confused, lacking a specific meaning, the task of elucidating among all the possible applicable denominations (remember the large number of possible proposals granted by many scholars of the subject), the most appropriate interpretation, and that can easily describe the aspects related to those spaces, places and landscapes, fundamental to understand the current states of the territories in which we settle (Pérez Igualada, 2018)[5]. This precision has been made by Gilles Clément by determining, on the one hand, the concept of the "Third Landscape": "it is constituted by the set of residual spaces of the human being. These margins gather a biological diversity that is not understood as wealth" (Clément, 2018); and, on the other hand, when specifying its extension for both rural spaces: "In rural sectors, residual spaces occupy the uneven reliefs, incompatible with the exploitation machinery, as well as all the remains directly linked to the layout: boundaries of fields, hedges, riverbanks, roadsides, etc." (Clément, 2018); as for urban spaces: "In urban sectors, they correspond to land waiting to be assigned, or waiting for the execution of projects that depend on budget provisions or political decisions. The delays, which are often long, allow a tree cover to emerge in the urban wastelands (forests of residual spaces)" (Clément, 2018). In this way, if for the social group, this referred to as "Third Landscape" can be appreciated as: "a space of nature (capture of the Third landscape by the institution); a leisure space; an unproductive space (abandonment of the "Third Landscape" by the institution), a sacred space" (Clément, 2018). It is then, where the relationship of the "Third Landscape" acquires meaning (related as a residual space, that is: "the result of the abandonment of a previously exploited land. Its origin is multiple: agricultural, industrial, urban, tourist, etc. [being in turn, this residual space understood as], synonymous with wasteland, or wasteland" (Clément, 2018); with the mental order to which the landscaper associates it, that is, the "Third Landscape" as a temporary administrative cut, as a permanent ethical domain of the planetary citizen, which is acquiring a political dimension, and is being consolidated in the consciousness of the planetary collectivity.

These spaces are intimately linked to another that in our days continues to be the result of controversy, that of "property", since they reflect, directly or indirectly, what Professor Naredo points out with respect to it when he affirms that we are faced with a notion that was consolidated in the history of humanity as something unified and homogeneous at the same time as the notion of wealth, on which it is based, and the other basic categories of established economic thought-production, work, consumption . . . which came to be related through the economic system that consolidated economics as an independent discipline in the self-sufficient universe of monetary values. So much so that for the economic system to function, it needs "property rights that grant each economic agent the exclusive power of disposal over the economic objects that make up its assets". The domain of applications of the system is thus defined in the homogeneous field of monetary values, establishing its accounting representation from the prices that result from using money as a unit of account. All this worked well in the pre-industrial stages, however, at present, the conventional theory of property seems to be fossilized, apart from the intense mutations that have taken place in the forms of property and social organization, whose direct consequences on the territory we find in the "Third Landscape", among others (we refer to speculative property, specifically). To understand these changes and transformations that have such an impact on Environmental Sustainability, we will stop at the study of the legal consideration of Territorial Planning and Urban Planning.

GOING INTO THE STUDY OF THE "THIRD LANDSCAPE": THE "CORTIJO OF SAN ISIDRO" (ARANJUEZ, MADRID)

Minor Local Entity is the name given in some Autonomous Communities of Spain (Aragón, Castilla y León, Extremadura, Galicia and the Valencian Community) to what the Spanish Legislation and Areas of Action define as an entity with a territorial scope inferior to the municipality, and that in Other Spanish regions are known by the name of Council in Álava and Navarra, rural parish (Asturias), autonomous local entity (Andalusia), decentralized municipal entity (Catalonia) or neighborhood council (Burgos and Cantabria). Likewise, it is the name previously received by local entities whose territorial scope, as we have pointed out, was less than the municipality. The minor local entities have personality and full legal capacity to exercise the powers that the Legislation and Areas of Action recognize in each case. The form of local administration in Spain known as a minor local entity was created by the Municipal Statute of 1924, collecting in its 2nd article that:

Under the denomination of minor local Entities are included the annexes, parishes, places, villages, hamlets and districts that, within a Municipality, and constituting a separate nucleus of buildings, form a group of people and goods, with peculiar and collective rights or interests, distinguishable from the generals of the Municipality. Royal decree-law approving the Municipal Statute.

These entities had a legal personality different from that of the municipality in which they were located. The legal representation of the minor local entities corresponded to their neighborhood council, which had the same capacity in relation to goods, contracts, works and public services and civil, criminal, administrative and contentious-administrative actions as a city council. If they had less than 1.000 inhabitants, they were governed by an open council. In the event that the minor local entity was a parish, the board was called parish instead of neighborhood. If they had a population greater than 1.000 inhabitants, and especially those that formed neighborhoods or urban annexes added to large populations, the neighborhood board was made up of as many members as councilors would correspond to it if it were a city council. This is the specific case of "Cortijo de San Isidro", currently made up of different buildings, dependencies, and a large amount of land dedicated to cultivation created in the time of Carlos III, in order to serve as a model of colonization to through farms, mainly fruit and vegetable.

Figure 1. Rural and urban parcel of the Cortijo de San Isidro
Source: Own elaboration.

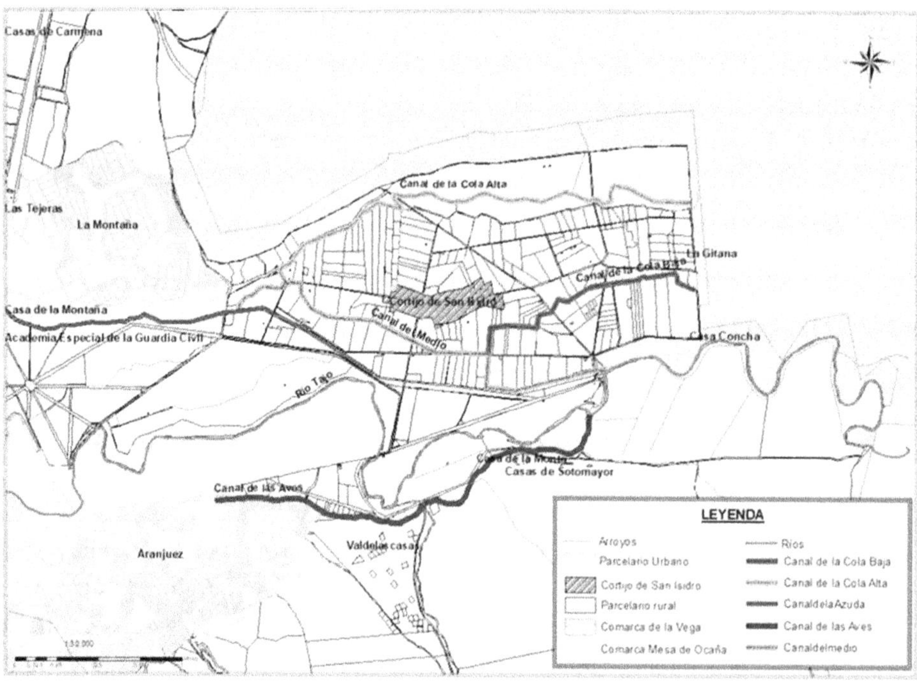

Figure 2. Cortijo de San Isidro: Minor local entity
Source: Own elaboration.

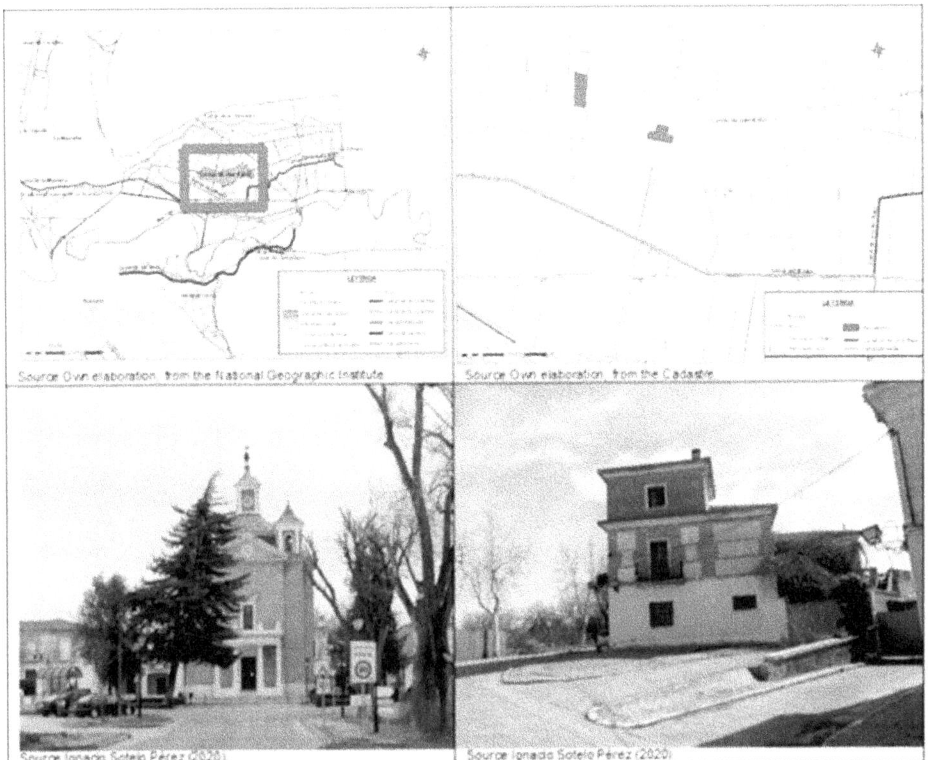

LEGISLATION AND AREAS OF ACTION

- General Urban Planning Plan of Aranjuez of 1996
- Law 9/2001, of July 17, on the Land of the Community of Madrid
- Royal Legislative Decree 7/2015, of October 30, approving the consolidated text of the Law on Land and Urban Rehabilitation
- Special Interior Reform Plan (PERI) of the "Cortijo de San Isidro"

KEY PERFORMANCE INDICATOR (KPI)

Objectives	**Principles**
Economic. Social. Environmental. Growth and Development. Institutional and Landscape Regeneration	Accessibility and vertebration. Growth, development and cooperation. Competitiveness and cohesion
Indicators	**Policies and instruments**
First Generation (FPEIR) Third Generation (triple dimension)	Finalist and Specific Policies (instrumental and sectoral), instr. regulatory and economic

THIRD LANDSCAPE

Cortijo de San Isidro is a Minor Local Entity that responds to the needs that a few centuries ago had to organize geographical spaces for better exploitation, in this case related to agricultural holdings. Following the lines marked by the Enlightenment, the territory was ordered and divided, and it was exploited by parcels. Currently we find a residual space of residential functions, in the manner of *Lost spaces*, maintaining, however, an incipient but prosperous agriculture, thanks to the Community Agrarian Policy. All this is presided over by the Royal Chapel or Hermitage Royal of Cortijo de San Isidro, neoclassical in style, and the ad hoc buildings built for the settlers, by order of Carlos III (today a good part of them are abandoned, becoming part of the residual spaces of the Third Landscape).

Figure 3. Cortijo de San Isidro: Rururban space
Source: Own elaboration.

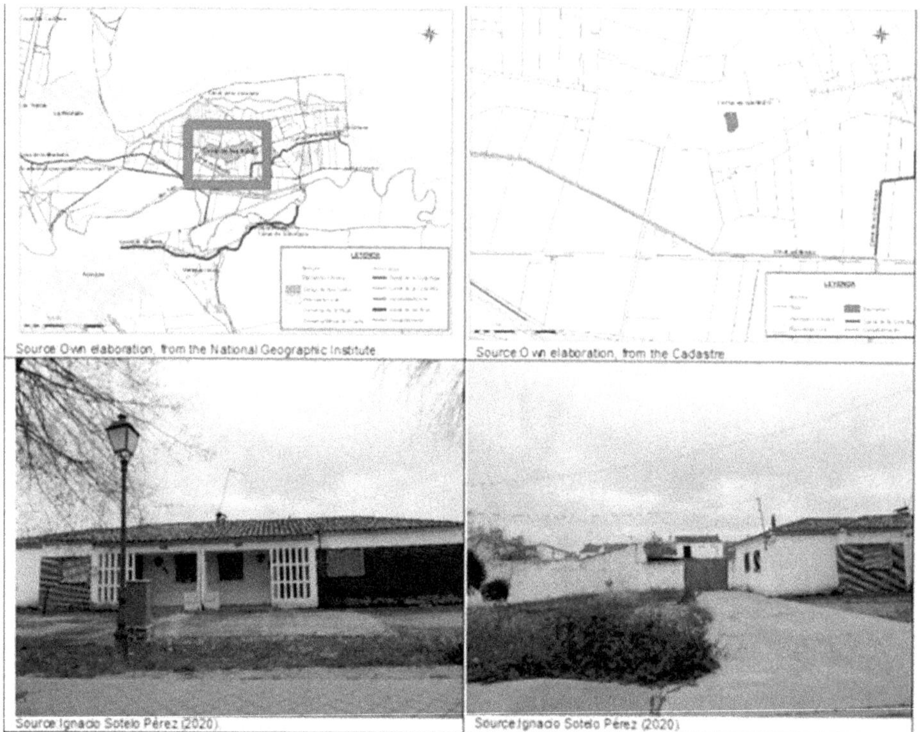

LEGISLATION AND AREAS OF ACTION	
✦ General Urban Planning Plan of Aranjuez of 1996. ✦ Law 9/2001, of July 17, on the Land of the Community of Madrid. ✦ Royal Legislative Decree 7/2015, of October 30, approving the consolidated text of the Law on Land and Urban Rehabilitation. ✦ Special Interior Reform Plan (PERI) of the "Cortijo de San Isidro".	
KEY PERFORMANCE INDICATOR (KPI)	
✦ **Objectives** Economic. Social. Environmental. Growth and Development. Institutional and Landscape Regeneration	✦ **Principles** Accessibility and vertebration. Growth, development and cooperation. Competitiveness and cohesion.
✦ **Indicators** First Generation (FPEIR) Third Generation (triple dimension)	✦ **Policies and instruments** Finalist and Specific Policies (instrumental and sectoral) instr. regulatory and economic
THIRD LANDSCAPE	
The phenomenon of the illegal occupation of housing has caused a large part of the housing -sometimes substandard housing-, which generates rururban spaces, to appear boarded up. Faced with speculative property, there are many who believe that the non-recognition of private property, included in our constitutional text, is the solution. In this way, the lost space ends up being the consequence of a paradox that we could describe as social, by political decision (come on, nonsense).	

Figure 4. Cortijo de San Isidro: Impact of culture
Source: Own elaboration.

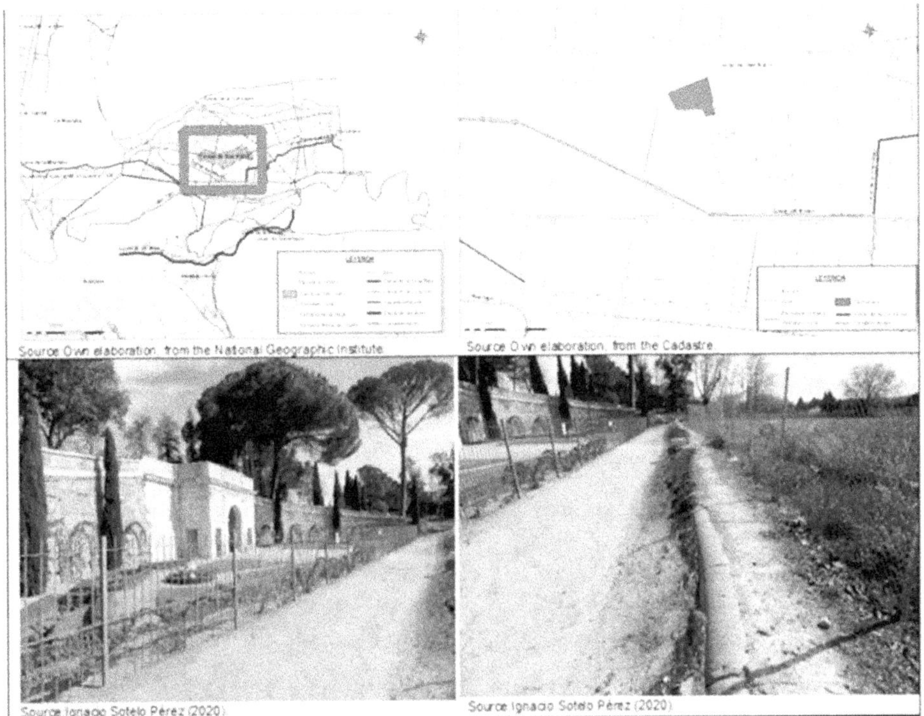

LEGISLATION AND AREAS OF ACTION

- General Urban Planning Plan of Aranjuez of 1996
- Law 9/2001, of July 17, on the Land of the Community of Madrid
- Royal Legislative Decree 7/2015, of October 30, approving the consolidated text of the Law on Land and Urban Rehabilitation
- Special Interior Reform Plan (PERI) of the "Cortijo de San Isidro"

KEY PERFORMANCE INDICATOR (KPI)

Objectives	**Principles**
Economic. Social. Environmental. Growth and Development. Institutional and Landscape Regeneration	Accessibility and vertebration. Growth, development and cooperation. Competitiveness and cohesion
Indicators	**Policies and instruments**
First Generation (FPEIR) Third Generation (triple dimension)	Finalist and Specific Policies (instrumental and sectoral). instr. regulatory and economic

THIRD LANDSCAPE

The wineries have been since time immemorial, the exit to the wine farms; in the case of Cortijo de San Isidro, they were ordered to be built by King Carlos III, so that the organized operations of these spaces could be maintained –providing the Royal House with oil and wine-. At present, although it maintains similar activities of an open nature, its maintenance is linked to wine tourism, with well-preserved infrastructures, although they are far from the quality of its best times (deterioration, degradation, poor conservation,..., are some of its properties).

Figure 5. Cortijo de San Isidro: Rural spaces
Source: Own elaboration.

Source: Own elaboration, from the National Geographic Institute

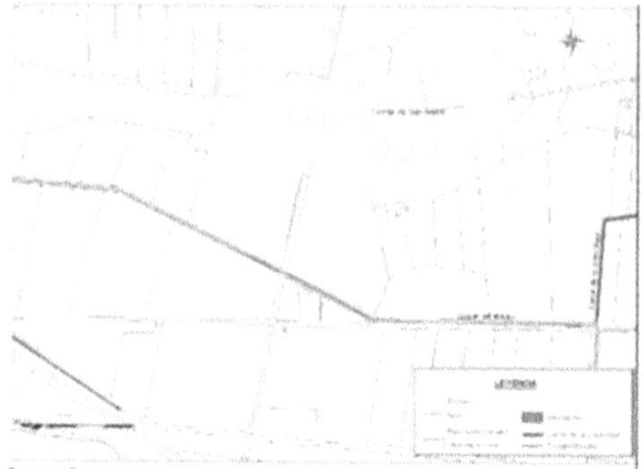

Source: Own elaboration, from the Cadastre

Source: Ignacio Sotelo Pérez (2020)

Source: Ignacio Sotelo Pérez (2020)

LEGISLATION AND AREAS OF ACTION

- General Urban Planning Plan of Aranjuez of 1996
- Law 9/2001, of July 17, on the Land of the Community of Madrid
- Royal Legislative Decree 7/2015, of October 30, approving the consolidated text of the Law on Land and Urban Rehabilitation
- Special Interior Reform Plan (PERI) of the "Cortijo de San Isidro"

KEY PERFORMANCE INDICATOR (KPI)

Objectives	**Principles**
Economic, Social, Environmental, Growth and Development, Institutional and Landscape Regeneration	Accessibility and vertebration. Growth, development and cooperation. Competitiveness and cohesion
Indicators	**Policies and instruments**
First Generation (FPEIR) Third Generation (triple dimension)	Finalist and Specific Policies (instrumental and sectoral) instr. regulatory and economic

THIRD LANDSCAPE

The infrastructures that make up the Aranjuez canals are of high quality, although the changes that have occurred in recent decades in Spain, with the abandonment of agricultural holdings, have reached one of the most prosperous areas of our country. This is reflected, in part, in their maintenance, when not their abandonment, or clogging. In this way, the residual spaces that have been generated present a visual impact of the first magnitude. In spite of everything, a good part of them are still in operation.

Figure 6. Cortijo de San Isidro: Lost space
Source: Own elaboration.

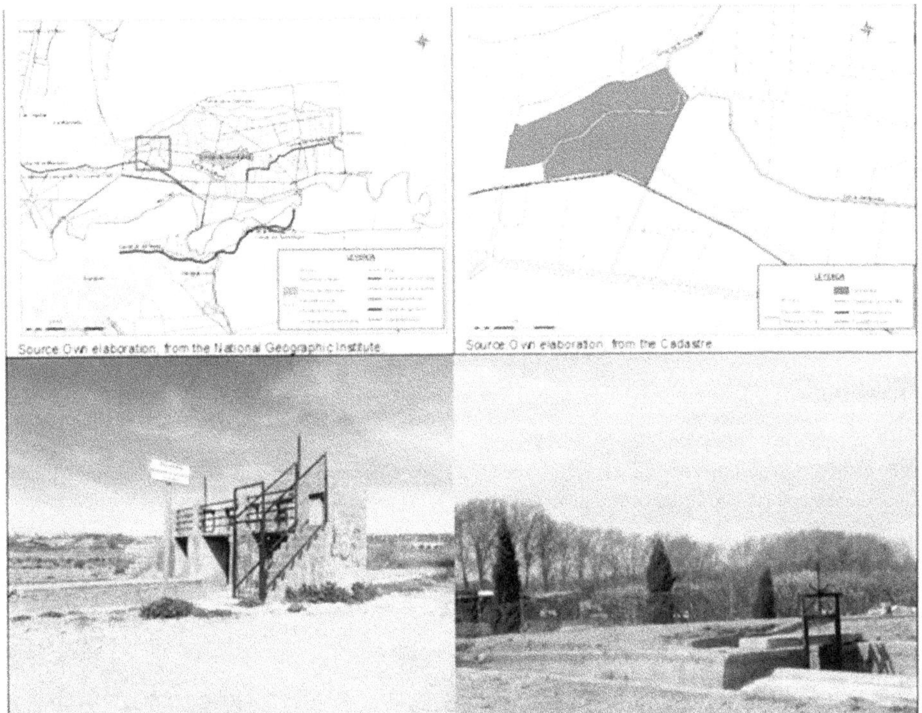

Source Own elaboration, from the National Geographic Institute | Source Own elaboration, from the Cadastre

Source Ignacio Sotelo Pérez (2020) | Source Ignacio Sotelo Pérez (2020)

LEGISLATION AND AREAS OF ACTION	
- General Urban Planning Plan of Aranjuez of 1996 - Law 9/2001, of July 17, on the Land of the Community of Madrid - Royal Legislative Decree 7/2015, of October 30, approving the consolidated text of the Law on Land and Urban Rehabilitation - Special Interior Reform Plan (PERI) of the "Cortijo de San Isidro"	
KEY PERFORMANCE INDICATOR (KPI)	
Objectives Economic, Social, Environmental, Growth and Development, Institutional and Landscape Regeneration	**Principles** Accessibility and vertebration. Growth, development and cooperation. Competitiveness and cohesion
Indicators First Generation (FPEIR) Third Generation (triple dimension)	**Policies and instruments** Finalist and Specific Policies (instrumental and sectoral), instr. regulatory and economic
THIRD LANDSCAPE	
The *lost space* in many cases are the direct consequence of an approach to the rural, rururban and urban issue, from a socio-functional paradigm, according to which the spaces that make up the same are, above all, a support structure for the human activities for life. The combination of the infrastructures of the canals with the copses generate in Aranjuez a landscape in continuous space-time transformation.	

Figure 7. Cortijo de San Isidro: Infrastructures of the third landscape
Source: Own elaboration.

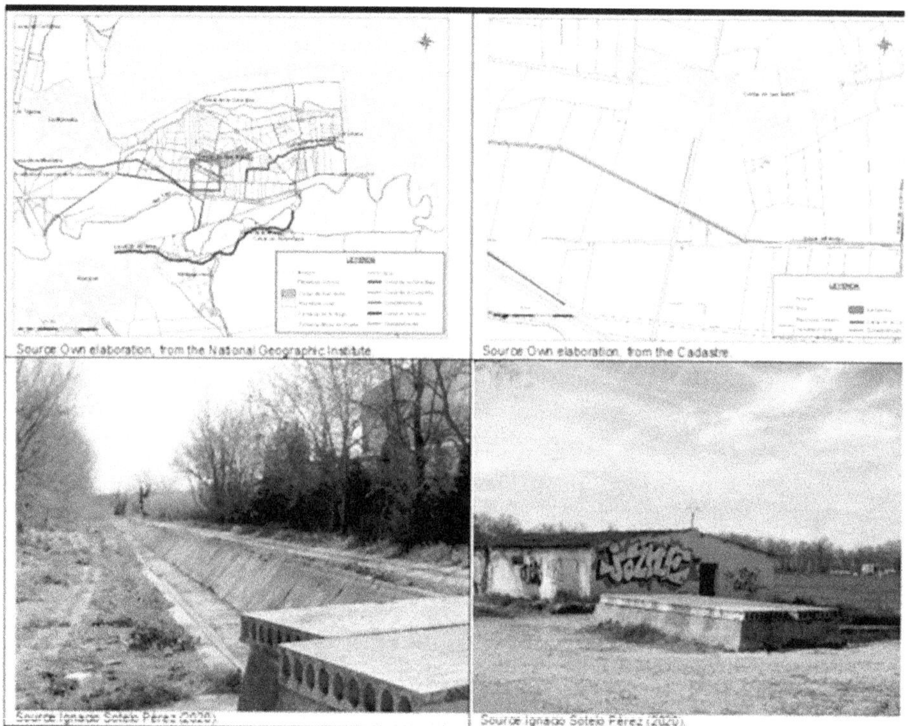

Source: Own elaboration, from the National Geographic Institute | Source: Own elaboration, from the Cadastre
Source: Ignacio Sotelo Pérez (2020) | Source: Ignacio Sotelo Pérez (2020)

LEGISLATION AND AREAS OF ACTION

- General Urban Planning Plan of Aranjuez of 1996.
- Law 9/2001, of July 17, on the Land of the Community of Madrid.
- Royal Legislative Decree 7/2015, of October 30, approving the consolidated text of the Law on Land and Urban Rehabilitation.
- Special Interior Reform Plan (PERI) of the "Cortijo de San Isidro".

KEY PERFORMANCE INDICATOR (KPI)

Objectives	**Principles**
Economic, Social, Environmental, Growth and Development, Institutional and Landscape Regeneration	Accessibility and vertebration. Growth, development and cooperation. Competitiveness and cohesion
Indicators	**Policies and instruments**
First Generation (FPEIR) Third Generation (triple dimension)	Finalist and Specific Policies (instrumental and sectoral) instr. regulatory and economic

THIRD LANDSCAPE

The degradation of the geographical spaces made up of the vegas and the groves in Aranjuez is linked to the abandonment of the traditional farms collected on the plots that housed the homes of the farmers themselves. This is alleviated, however, by the modern infrastructure of the channels that transport the waters, in this specific case, of the Tagus River and its tributaries. Add to this some of the iconic-iconological features of our time, "graffiti", spontaneous expressiveness of a way of interpreting - for others to dirty- reality (at least the magnificent reality of Andy Warhol, among others).

Figure 8. Cortijo de San Isidro
Source: Own elaboration.

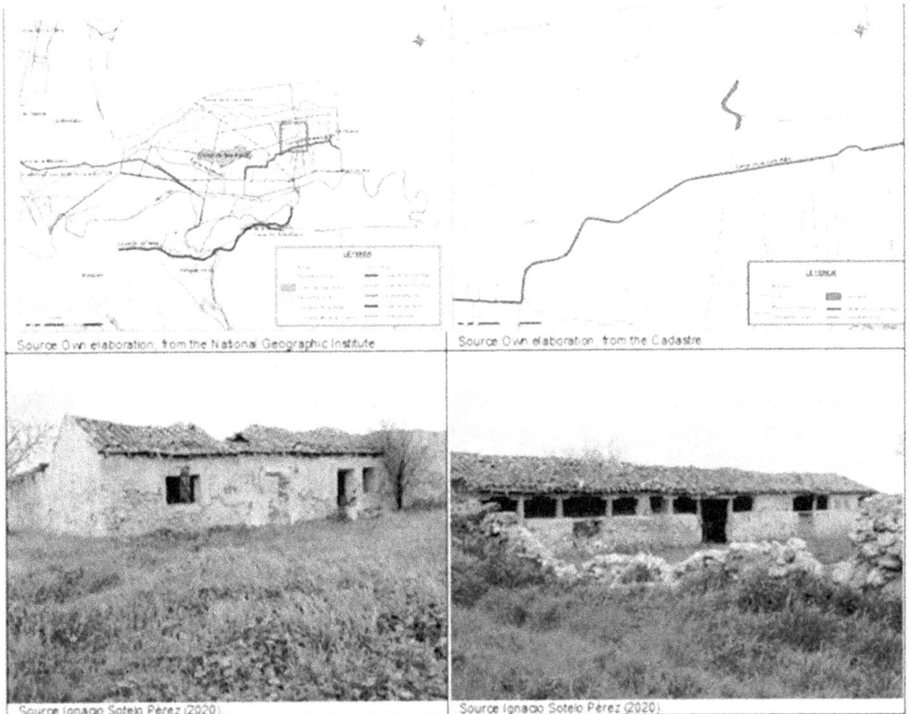

LEGISLATION AND AREAS OF ACTION

- General Urban Planning Plan of Aranjuez of 1996.
- Law 9/2001, of July 17, on the Land of the Community of Madrid.
- Royal Legislative Decree 7/2015, of October 30, approving the consolidated text of the Law on Land and Urban Rehabilitation.
- Special Interior Reform Plan (PERI) of the "Cortijo de San Isidro".

KEY PERFORMANCE INDICATOR (KPI)

Objectives	**Principles**
Economic. Social. Environmental. Growth and Development. Institutional and Landscape Regeneration	Accessibility and vertebration. Growth, development and cooperation. Competitiveness and cohesion
Indicators	**Policies and instruments**
First Generation (FPEIR) Third Generation (triple dimension)	Finalist and Specific Policies (instrumental and sectoral) instr. regulatory and economic

THIRD LANDSCAPE

The Count of Cabarrús and his three thousand settlers are far from us, with part of the remains of those 18th century buildings reaching us. We can imagine what these abandoned spaces would be like, today vacant, barren, in the moments of full occupation with the caballerias, the sharecroppers, the settlers, their families, thinking about their future, not our present. The remains of those constructions, probably maintained until recent times, today make up a space in transformation, if not in reclassification (integral elements of the "Third Landscape").

Figure 9. Cortijo de San Isidro: Functional changes
Source: Own elaboration.

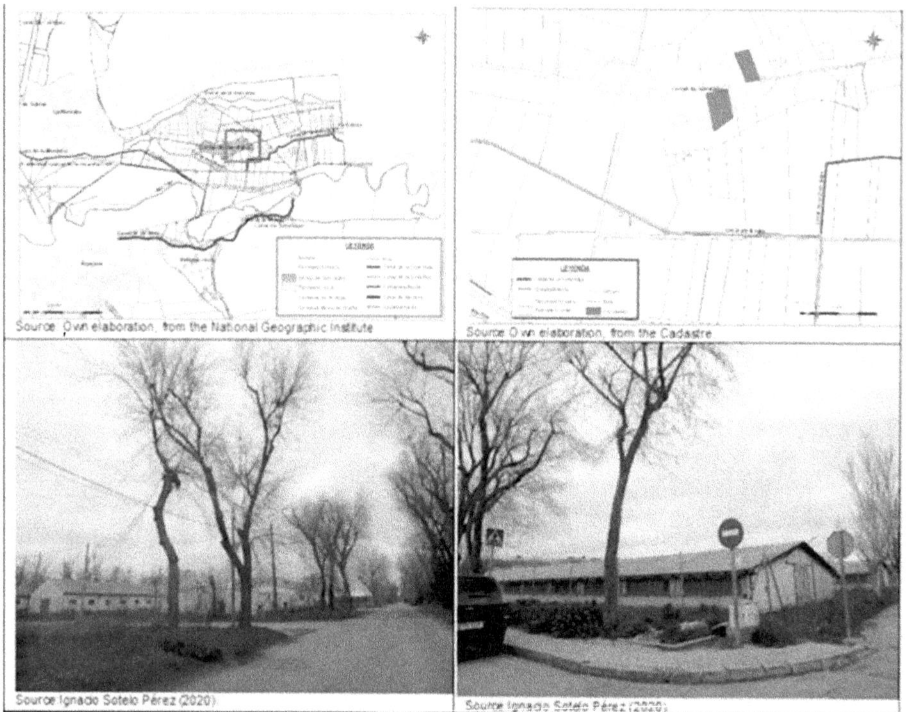

Source: Own elaboration, from the National Geographic Institute

Source: Own elaboration, from the Cadastre

Source: Ignacio Sotelo Pérez (2020)

Source: Ignacio Sotelo Pérez (2020)

LEGISLATION AND AREAS OF ACTION

- General Urban Planning Plan of Aranjuez of 1996
- Law 9/2001, of July 17, on the Land of the Community of Madrid
- Royal Legislative Decree 7/2015, of October 30, approving the consolidated text of the Law on Land and Urban Rehabilitation
- Special Interior Reform Plan (PERI) of the "Cortijo de San Isidro"

KEY PERFORMANCE INDICATOR (KPI)

Objectives	**Principles**
Economic, Social, Environmental, Growth and Development, Institutional and Landscape Regeneration	Accessibility and vertebration, Growth, development and cooperation, Competitiveness and cohesion
Indicators	**Policies and instruments**
First Generation (FPEIR) Third Generation (triple dimension)	Finalist and Specific Policies (instrumental and sectoral) instr. regulatory and economic

THIRD LANDSCAPE

The functional changes of these spaces have not managed to put an end to their abandonment. In this way, the buildings are isolated objects in the landscape, instead of being part of a rural or rururban fabric structured from the public space, through a system of streets,... For this reason, instead of an exterior volume endowed with shape and scale, connected to other spaces, the predominant spatial typology in much of the city of Aranjuez, and its surroundings, is the shapeless anti-space, where even the vegetation appears noticeably degraded.

Figure 10. Cortijo de San Isidro: Agricultural infrastructures
Source: Own elaboration.

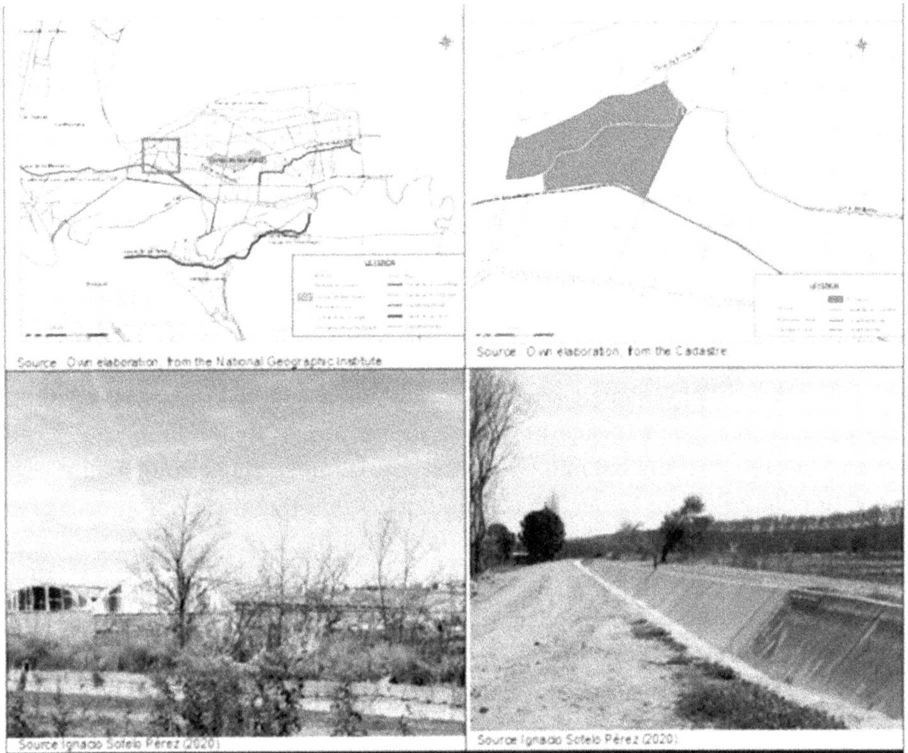

LEGISLATION AND AREAS OF ACTION

- General Urban Planning Plan of Aranjuez of 1996
- Law 9/2001, of July 17, on the Land of the Community of Madrid
- Royal Legislative Decree 7/2015, of October 30, approving the consolidated text of the Law on Land and Urban Rehabilitation
- Special Interior Reform Plan (PERI) of the "Cortijo de San Isidro"

KEY PERFORMANCE INDICATOR (KPI)

Objectives	**Principles**
Economic, Social, Environmental, Growth and Development, Institutional and Landscape Regeneration	Accessibility and vertebration. Growth, development and cooperation. Competitiveness and cohesion
Indicators	**Policies and instruments**
First Generation (FPEIR) Third Generation (triple dimension)	Finalist and Specific Policies (instrumental and sectoral) instr. regulatory and economic

THIRD LANDSCAPE

The introduction of new infrastructures in fruit and vegetable farms have led to a notable modernization through greenhouses, surrounded by hydraulic infrastructures. Its semiological expression in the landscape, however, leaves much to be desired, at the point and time that its maintenance is not adequate, so that Environmental Sustainability with the remains of the materials that make them up, end up being a problem related to waste, in general, and plastics, in particular. All this in a Third Landscape, today consolidated, by its residual spaces.

BY WAY OF OPEN CONCLUSIONS

At this point, after the foregoing, we can conclude that it is interesting to see how the notion of landscape is transformed at present by the idea according to which we must underline the importance of subjectivity in its interpretation for landscape recognition, which would give rise to the cultural changes that have been developed through the mediation of this subjectivity, in the territorial fields, an issue that has taken place mainly in recent years. In fact, in this aforementioned work, regarding this issue, we start from the idea that the landscape of architects and urban planners, of geographers and jurists is one that needs to be created to maintain the sustainability of the planet for future generations. In such a way, that the landscape is an instrument for the preservation of nature and the resources that must be organized again for the future of humanity.

Likewise, in our study it has been shown that when investigating the way in which the term landscape has been influencing the field of scientific knowledge, and regarding what interests us in this research, on geographical knowledge, studies related to the landscape, are presented to us as a positive method to integrate the sciences that are divided into two parts, that is, those concerning nature, and also those related to society. In relation to this issue, it has been understood that the environment in general, and in particular the natural resources associated with it, are in a situation of worrying plundering by the developed model of capitalism, since by recognizing the limitation of these resources and the environment in terms of their level of quality, together with an increasing exploitation, certain consequences have been observed due to this current reality.

REFERENCES

Ávila Sánchez, H. (2005). Introducción Líneas de Investigación y el Debate en los Estudios Urbano-Rurales. In *Lo Urbano-Rural. ¿Nuevas Expresiones Territoriales?* Universidad Nacional Autónoma de México. Centro Regional de Investigaciones Multidisciplinarias de la UNAM.

Bertrand, G. (1974). *Introducción de la tesis doctoral de Estado*. Academic Press.

Bertrand, G. (2010). Itinerario en torno al paisaje: una epistemología de terreno para tiempos de crisis. *Ería: Revista cuatrimestral de geografía, 81*, 5-38.

Clément, G. (2018). Manifiesto del Tercer paisaje. Editorial Gustavo Gili, 21, 82.

Farinelli, F. (1992). I segni del mondo. Immagine cartográfica e discorso geografico in età moderna, Florencia, La Nuova Italia. In El paisaje en la cultura contemporánea. Biblioteca Nueva.

Flint, A., & Jennings, B. (2022). The role of cultural heritage in visitor narratives of peatlands: Analysis of onlineuser-generated reviews from three peatland sites in England. *International Journal of Heritage Studies, 28*(2), 163–179. doi:10.1080/13527258.2021.1941198

Lugo Hubp, J. (1989). Paisaje físico geográfico. In *Diccionario Geomorfológico con equivalentes de los términos de uso más común en alemán, francés, inglés y ruso, Instituto de Geografía*. UNAM.

Másmela Díaz, P. (2010). *El paisaje como elemento de la Ordenación Territorial. Un análisis de Paisaje desde su enfoque visual en el borde centro oriental de Medellín*. Universidad Nacional de Colombia.

Montori Montolar, J. (2019). Cartografía del Tercer Paisaje [Final Degree Project]. Zaragoza University.

Naredo, J.M. (2015). *La Economía en evolución*. Edt. Siglo XXI.

Ortwin Saber, C. (1925). *La Morfología del Paisaje*. Estados Unidos de América.

Pérez Igualada, J. (2018). Los Nombres de los Lugares sin Nombres. *ACE: Architecture, City and Environment = Arquitectura, Ciudad y Entorno, 13*(38), 129-150. http://dx.doi.org/ doi:10.5821/ace.13.38.5419

Ramírez Velásquez, R., & López Levi, L. (2015). *Espacio, paisaje, región, territorio y lugar: la diversidad en el pensamiento contemporáneo. Geografía para el Siglo XXI. Serie: Textos Universitarios*. Universidad Autónoma Metropolitana. Méjico.

Rayner, J. (2022). The city (as) archive: Are your memories in place? *Landscape Research, 47*(7), 829–839. doi:10.1080/01426397.2022.2077923

Smith, N. (1984). *Uneven development: nature, capital and the production of space*. Blackwell.

Sotelo Navalpotro, J. A. (1992). *Paisaje, semiología y análisis geográfico. Anales de Geografía de la Universidad Complutense. Nº 11. Págs. 11 a la 23*. Edición Complutense.

Sotelo Navalpotro, J. A. (2015). Paisajes para el Medioambiente. Estudio de caso. *Observatorio Medioambiental, 18*, 97-128.

Sotelo Pérez, I. (2021). Groundwater in the Field of Water Policies, in Spain. In Sustainable Policies and Practices in Energy, Environment and Health Research. Springer.

Sotelo Pérez, M., Galvani, A., & Sotelo Navalpotro, J. A. (2021). "Water Footprint" and Territorial Sustainability in Spain. In Sustainable Policies and Practices in Energy, Environment and Health Research. Springer.

ENDNOTES

1 The author was making during the sixties of the last century (and specifically between the years 1960 to 1967), and that would finally lead to his reading and defense in 1974. Work in which the above-referenced statements are quoted verbatim, these statements being perfectly applicable today, essentially in terms of what refers to the principles defended in its text. Therefore, a long era begins for the conceptual development of the landscape concept, but always at the dawn of the contexts of geography, since, although it may seem implausible, the aforementioned statement and despite its outdated content, and that the terminology in use has evolved notably in recent years, is perfectly usable in terms of the descriptive value that is collected in it.

2 Spanish Constitution of 1978. Published in: "BOE" no. 311, of 12/29/1978. Referencia: BOE-A-1978-31229. Permalink ELI: https://www.boe.es/eli/es/c/1978/12/27/(1)/con. In Spain, this harmonizing need, society-natural environment, contemplated by the 1978 constituent, in which it regulate within its articles as "1. Everyone has the right to enjoy an adequate environment for the development of the person, as well as the duty to preserve it" (art.45.1). In relation to this article, it is worth mentioning the work of Galindo Elola-Olaso, F. (2017). Synopsis article 45, published in

"Congress of Deputies", in which in summary account and in relation to the concept of landscape, but in its constitutional application, show how the environment, interest and concern for its protection (among which natural resources would be included), is one of the issues of greatest importance and depth, which houses the dogmatic part of our Basic Text. With this recognition, the Magna Carta would become one of the first in which this matter would be collected. For some authors, said constitutional recognition has been increased by the growth of sensitivity for the environment, at the same time that it has been increasingly harmed, and deteriorated mainly due to human action (being some problems such as climate change, or the destruction of numerous natural environments, the greatest concern that must be addressed). In this sense, "the right to enjoy an adequate environment for the development of the person, as well as the duty to conserve it", consolidated as one of the most important issues at the time of drafting the current Constitution of 1978. After this article 45 cited, another series of rights is intuited, such as the one corresponding to the enjoyment of an adequate quality of life, or that of the dignity of the person (an issue that our constitution mentions in article 10 that shows how "1. The dignity of the person, the inviolable rights that are inherent to him, the free development of the personality, respect for the law and the rights of others are the foundation of political order and social peace"), among others. These two rights, in turn, would be related, and recognized by the jurisprudence of the Constitutional Court, and specifically through its Judgment 102/1995, of June 26, in which it would be ruled that "The protection consists of an action of protection, help, defense and promotion, guardianship and custody, both preventive and repressive, as clearly indicated by the constitutional text so many times mentioned in its last paragraph, a protective action in short that, due to its own condition, is condenses into another indeterminate legal concept whose concretion corresponds both to the norms and to the actions for their fulfillment. However, it would not be good to forget that protection is always posed against "something", the dangers suggested above and against "someone" whose activity is potentially or actually harmful to protected assets or interests. Well, in the case of the environment there is the paradox that it has to be defended by the human being from the actions of the human being, author of all the outrages and outrages that degrade it, also for the benefit of other human beings, and successive generations. Protection is thus a belligerent activity that aims to avert danger and, where appropriate, restore the damage suffered and even improve the characteristics of the environment, to ensure enjoyment by all. Hence its ambivalent conFiguretion as a duty and as a right, which implies the demand for citizen participation at the level of each one, with leading roles played by women, youth and indigenous peoples, as stated in the Declaration of River (10, 20, 21 and 22). This leads us by the hand to the dignity of the person as a transcendent constitutional value (art. 10.1 C.E.), because everyone has the inalienable right to live in their environment according to their cultural characteristics" (F.J.7). Around this pronouncement, is where the true relevance of the aforementioned article 45 of our Basic Text resides.

3 Concept defined by Clément, G. (2018), in his Third Landscape Manifesto. In this sense, the author states in this cited work that "By nature, the Third Landscape constitutes a territory for numerous species that do not find a place elsewhere. The remaining part of species that are not found in the Third Landscape is represented by cultivated plants, farmed animals and all those beings whose existence depends on cultures and breedings". The author gives the example of the poppy plant, showing how the harvests are directly dependent on the different practices related to agriculture. The poppy is visible in the damaged, damaged and labeled territorial groups, but nowhere else. At the same time, it can also be observed how the species considered as adventitious and the mesicolas

from the crops, are constituted as part of the herbs that, due to their apparent unproductiveness, are discarded, and that despite their high regenerative degree, they are in danger of extinction, although reality shows a very distant image.

4 Within the twelve points that author develops as the "Manifesto of the Third Landscape", these three origins of what are called "spaces of diversity", are opposed and clarified as follows: In the first place, it shows how the so-called "primary assemblages" house most of the planet's diversity. Secondly, it assimilates and identifies in a similar way the "primary groups" and the so-called "reserves", clarifying how it would be a "climax", which have levels considered stable and whose aspects are transformed by the passage of time. the times. Thirdly, it shows us how flora is concerned, that which develops within the so-called "primary groups" and reserves, simultaneously exclude the rest, prohibiting access to those beings considered exogenous, mainly due to that the medium maintains its own closure. Fourth, it is shown how with respect to the primary media, these form part of a single territory of resistance to the "planetary hodgepodge", which is nothing more than the sum of the residual spaces par excellence. As far as this question is concerned, the species that emerged from the so-called planetary hodgepodge are unknown regarding the nature and number of the total that can coexist in this balance; balance, which in turn is closely dependent on the number of human beings operating in these places. In fifth place, it refers to how the end of this entire procedure entails the total disappearance of these primary media, and consequently the generalization of the media considered as secondary. Sixthly, as far as residual spaces are concerned, the cited author shows us how these residual spaces can at no time benefit from the status available to reserves. In this sense, the cycles are mentioned, which consist of the pioneer species preparing the coming of the subsequent species, whose cycles are extended until they are finally installed. Seventh, when said pioneer species mentioned above disappear, the stable species take advantage of this situation, constituting this circumstance, an element to consider of the so-called residual space. In the eighth place, it is shown how the existence in these residual spaces is characterized by its brevity, since while it is closing the terrain, the different dynamics of conquest are accentuated. In the ninth place, it is inferred, how the residual space is in turn recycling itself, since each of the accidents understood as natural characteristics, contribute to reopening those previously closed lands. In tenth place, the flora of the residual spaces, do not exclude their indigenous natural procession, and states that probably each and every one of the exotic floras compatible with the biome or environment is gathered. In addition, the flora of these residual spaces and that of the primary groups are understood as almost identical on many occasions, even in many cases mixing slowly or even not mixing at all. The author continues with the description of the existing relations between the different characters that comprise the spaces of diversity, in order to be able to reinforce his hypothesis according to which if the territory constitutes an appropriate territory for the diverse species that do not manage to find a place in other different places, then that remaining part of the species, which are not found in said Third Landscape, would correspond to the cultivated plants, the farmed animals, and all those species that depend for their existence on the breeding and cultures.

5 In this sense, it is worth remembering how many authors –like Eugent Clement, Serge Briffaud, Claude Bertrand…from various disciplines have focused over time on responding to an ideal conceptualization for the interpretation of reality, of what should be understood by place. Thus evaluations and readings: "Lost spaces; Non-lieux; Junkspaces; Shadowed Places; Terrain vague; SLOAP; Drosscapes; Spazi rifiuto, spazi scoria, spazi-scarto; Tiers Paysage; Hyper-lieux; Non-place urban realm. These names, which are added to the more generic ones of vacant lots, vacant lots,

urban voids, vacant land or residual spaces, are associated with different readings and assessments of the places they designate, carried out from different paradigms (formal, socio-functional or environmental), and, also to different proposals for intervention strategies -or non-intervention- in the same". Similarly, in other works by different authors like, for example, Montori Montolar, J (2019). In which in his final degree project, entitled Cartographies of the Third Landscape. The Zaragoza case. And whose tutors were: Ávila, C, & Bambó, R; cites how, in addition to Gilles Clément, the shelter space that this author deals with has been similarly studied by other authors such as: Ignasi de Solà Morales; Enric Batlle; Marc Augé; Michel Desvigne; or Rem Tails. Being on the part of each of these, treated the relative subject, in general to the places, or spaces considered as residual, and in particular the landscapes from this perspective.

Chapter 2
Modeling of Flood Risk: Vulnerability and Adaptation of Machico City Center

Raul Manuel Costa Alves
https://orcid.org/0000-0002-3680-6626
University of Extremadura, Spain

Sérgio António Neves Lousada
https://orcid.org/0000-0002-8429-2164
University of Madeira, Portugal

José Cabezas
https://orcid.org/0000-0002-7209-3731
University of Extremadura, Spain

José Manuel Naranjo Gómez
https://orcid.org/0000-0001-7998-9154
University of Extremadura, Spain

ABSTRACT

The objective of this work is to analyze the hydrological functioning of the Machico watershed and the historical incidence of floods in the historical center of Machico. The potential impacts, exposure, and vulnerability on the natural and human environment are also analyzed, as well as human adaptation through flood planning. Tools such as hydrological modeling, flood risk assessment, and mapping of flood-prone areas are used for this purpose. Spatial data is important for analyzing and planning technical and policy actions.

DOI: 10.4018/978-1-6684-6449-6.ch002

INTRODUCTION

Alluvial processes and floods are natural phenomena that are increasingly subject to human influence and interference. Throughout history, human beings have adapted land distribution and usage to reduce exposure and vulnerability to local alluvial patterns. However, there has recently been a significant increase in exposure to risk factors due to inadequate planning in relation to local realities. In this context, a case study was carried out in the Machico River basin in Portugal, which included:

1. Hydrological modeling of the basin, calculation of peak flow, flow modeling, calculation of runoff surface, and flood areas for different return periods;
2. Description and characterization of flood history, identification, inventory, and characterization of exposed elements, as well as research of property owners and residents for vulnerability analysis of the historical center of Machico in flood situations;
3. Evaluation of the degree of land use adaptation to flood occurrence in relation to spatial and environmental planning options.

The size and diversity of conditions, along with changes in land occupation and usage in river basins, lead to floods in riverside areas, where the main regional and local urban centers are usually located. Alluvial processes expose a range of natural and human heritage elements, including the built and cultural heritage of Machico's historical zone (Alves et al., 2022b). These more vulnerable spaces require greater knowledge, organization, and training, reflected in municipal planning and licensing options, as well as in the planning and management of classified and protected environmental spaces, in order to anticipate, manage, and recover from alluvial/flood events. Technological instruments, particularly geographic information systems (GIS), can be central elements in describing, evaluating, and supporting risk management, implementing collaborative risk governance systems at regional and local scales, and managing territorial risks, particularly those related to floods.

Flood risk management is essential to ensure the safety and well-being of populations, protect the environment, and sustain economic development. Goal 13 of the UN's 2030 Agenda highlights the importance of adaptation and resilience measures to address challenges related to climate change and minimize the impact of natural disasters such as floods (RSENR, 2022).

THE RISK OF ALLUVIAL PROCESSES AND FLOOD AREAS

Considering the complexity of territorial systems, it is important to implement an evaluation of the possible harmful consequences of future floods for human health, the environment, cultural heritage, and economic activities. This evaluation should take into account issues such as topography, the position of watercourses and their general hydrological and geomorphological characteristics, the location of populated areas and economic activities, and long-term evolution, including the impacts of climate change on flood occurrences, as foreseen in Directive 2007/60/EC of 23 October, transposed into national legislation through Decree-Law No. 115/2010 of 22 October (Directive on the Assessment and Management of Flood Risks).

Through Directive 2007/60/EC of the European Parliament and of the Council, of 23 October 2007, a framework was established for the assessment and management of flood risks, in order to reduce the consequences associated with floods that are harmful to human health, the environment, cultural heritage, and economic activities. This directive was transposed into domestic law through Decree-Law No. 115/2010 of 22 October, thus establishing a framework for the assessment and management of flood risks.

The scope of Directive No. 2007/60/EC defines flooding as "the temporary covering by water of land which is normally not covered by water. It includes floods resulting from rivers, mountain torrents, and ephemeral Mediterranean watercourses, and floods resulting from the sea in coastal areas, and may exclude floods originating from sewerage systems."

Water resources have great importance as a factor for socio-economic development and leisure activities. However, the perception that there is not sustained abundance of water, combined with the spatial and temporal variability of water flows in the hydrological cycle and the sensitivity of aquatic ecosystems, demands rigorous management. This includes the adoption of specific measures to prevent, protect, recover and improve the state of water resources, as well as large investments in infrastructure.

To manage water resources efficiently, it is essential to understand the natural dynamics of the water system and its relationship with human activities in the hydrological cycle, especially at the scale of the river basin. In this way, it is possible to predict flood risk situations and identify threatened areas, in order to make the decision-making process more efficient in adopting measures for prevention, protection, recovery and improvement of the state of water resources.

Flood risk is defined as the possibility of flooding occurring combined with its potential negative consequences for human health, the environment, cultural heritage, and economic activities. This definition was established by Directive 2007/60/EC of the European Parliament and Council, dated October 23, transposed into national legislation through Decree-Law no. 115/2010, of October 22.

Flood management involves processes at different spatial and temporal scales, as well as multiple actors with different interests and conflicts. This is because floods can have significant impacts on broad areas and in different sectors, such as public health, agriculture, industry, tourism, and the environment. In addition, flood management generally involves the coordination of multiple levels of government, community organizations, and other interest groups that may have different perspectives on how floods should be managed. Collaboration and communication between these actors are essential to ensure that management measures are effective and socially acceptable (Wendy, 2007).

Over time, humans have learned to deal with floods through adaptation and management measures. One form of adaptation is the proper use of land, such as avoiding construction in flood-prone areas or constructing buildings with water-resistant materials (Wheater & Evans, 2009). Additionally, the implementation of management instruments, such as territorial planning, emergency and early warning plans, drainage and retention systems for rainwater, has been important in reducing the impacts of floods. However, the increasing complexity of territorial systems and the intensification of extreme events due to climate change require more improvement in these adaptation and management measures (Khatibi, 2011).

The analysis of spatial and temporal patterns of flood risk is essential for proper management of water resources and for the adoption of prevention and mitigation measures for flood impacts (Apel et al., 2004; Wheater et al., 2005). It is important to consider not only historical data, but also future projections, taking into account climate change and possible changes in hydrological regimes.

The thematic, temporal, and geographical scope is very important in the choice and development of spatial models for flood assessment and inundation areas (Büchele et al., 2006). For example, the chosen model may vary depending on the analysis objective, whether it is to predict inundation areas in a particular watershed or to evaluate the impacts of climate change on sea level in a coastal area. Similarly, the studied time period and geographical scale also influence the choice of model, as different models are more suitable for different time periods and geographical scales.

The increasingly comprehensive and accurate databases (Devillers & Jeansoulin, 2010; Honrado et al., 2011), along with the development of modeling and simulation technologies, allow for the creation of more advanced tools for planning and managing water resources, including flood risk management (Vieira et al., 2005; Charneca, 2006; Alonso et al., 2011a). Geographic Information Systems (GIS) are often used as a platform for these tools, allowing for the analysis and visualization of complex spatial data. Additionally, the availability of open-source technologies, such as QGIS and GRASS, makes these tools more accessible and cost-effective for many users (Fotopoulos et al., 2010).

Flood risk mapping plays a crucial role in identifying risk areas and evaluating associated vulnerability, enabling the adoption of appropriate preventive and management measures. The delimitation of flood-prone areas and the analysis of flow magnitudes and occurrence frequencies are key elements for assessing and predicting flood risk. Additionally, the identification of exposed elements, including infrastructure, buildings, population, and natural resources, is crucial for assessing the vulnerability and adaptability of these elements to flood risks (UNISDR, 2009; Jonkman et al., 2015).

The cartographic delimitation of risk zones synthesizes the hydrological, geomorphological, hydraulic, and socioeconomic knowledge of the floodable surface (Mateu, 1989; Rosselló, 1989) and implicitly recognizes the importance of assessing the level of risk (Lahr & Kooistra, 2010). Flood mapping can be implemented using two approaches: numerical modeling with spatial and statistical models (Liu et al., 2003; Correia et al., 1998) or delimitation by recording occurrences (Wolski et al., 2005). These approaches can be complementary and serve as calibration, cross-validation, and continuous improvement.

According to Camarasa (1995), the analysis of specific cases is important to understand the processes of flood formation. However, to delimit flood-prone areas, it is necessary to analyze a greater number of phenomena. Studies of specific events may not ensure the statistical robustness necessary for land use allocation and construction of protective structures according to current regulations. In this sense, Aronica et al. (2012) state that it is common to associate statistical parameters, such as return periods, with different risk categories in flood-prone areas.

The study of flood risk is an area of great relevance, as floods represent one of the main natural threats to populations worldwide. Several studies have highlighted the importance of analyzing the vulnerability of populations and their socioeconomic, political, and cultural aspects for a better understanding of flood-related risks (Cunha & Dimuccio, 2001; Saraiva & Carvalho, 2009). Urban concentration and disorderly growth of peri-urban areas are factors that increase the vulnerability of populations to floods. Studies by Aronica et al. (2012) have highlighted the association between statistical parameters on the probability of occurrence of a certain flow and different risk categories in flood-prone areas. Other studies, such as that of Camarasa (1995), have highlighted the importance of analyzing specific cases to understand the processes of flood formation. To allocate land use and construct protective structures in accordance with current regulations, it is necessary to consider the statistical robustness of the studies conducted. This is because studies of specific events may be overly specific and not ensure the statistical robustness that the planner requires (Aronica et al., 2012).

The study of flood risk is complex and involves the analysis of various factors. According to Gouldby and Samuels (2005) and Vrijling (1995), risk is understood as a dangerous and uncertain event that can be characterized in probabilistic terms regarding its potential occurrence and physical magnitude. Exposure is defined as the set of personal, material or environmental assets that are susceptible to being affected or damaged by the impact resulting from the dangerous event in question. Vulnerability, on the other hand, is the expected degree of damage or loss of resistance of the exposed impact resulting from the dangerous event. According to Saraiva and Carvalho (2009), risk always has a probabilistic dimension close to a feeling of uncertainty, although not of fatality, felt and perceived differently by the population depending on historical time. The interaction of the dangerous flood with the vulnerability of the assets exposed to the impact is the origin of risk. The flood, as a danger, is characterized by its physical characteristics or magnitude, such as peak flow, volume and duration, as well as by the probability of occurrence. In some approaches, exposure to risk is synonymous with vulnerability, a concept used by several sciences and, therefore, polysemic (Paiva, 2005). In the social sciences, vulnerability is understood as the capacity of a society to anticipate, resist and recover, in the short and long term, from a disaster situation (Blaikie et al., 1994). Vulnerability to floods depends mainly on human activities and behavior, the location of buildings and infrastructure, the existence of warning systems and emergency planning in their relation to legal and institutional frameworks (United Nations, 2009). The analysis of vulnerability, with reference to a specific territory and a given hazard, is developed considering physical and social dimensions, including parameters associated with human elements, the building stock (such as volume, density and types of uses) and the population at risk.

IMPACTS, VULNERABILITY, AND ADAPTATION TO RISKS AND FLOOD AREAS

The occupation and human activities are closely related to the search for productive potential, in accordance with the management of environmental and technological risks, depending on population and demographic contexts. Floods can cause potential damage to human health, the environment, cultural heritage, and economic activities, as established in Directive 2007/60/EC of the European Parliament and of the Council of 23 October. The exposure and vulnerability of the present elements imply adaptive processes by the populations and the local biodiversity. The occurrence of floods contributes to the distribution and evolution of natural and human elements in the territory. The frequency and size of floods may be on the rise due to global phenomena, such as climate change, as indicated by (Lehner et al., 2006), as well as by changes in land cover, infiltration, and runoff resulting from human actions, in a more local or regional perspective, as referred to by (Fitzpatrick et al., 1999).

The influence and response of humans to risk perception can occur at different levels, from the watershed scale to the local level (Fuchs et al., 2020). The suitability of occupation and land use in the watershed is generally complemented by local interventions, including the construction of containment infrastructure, regulation of channels, and drainage (Fuchs et al., 2020). In terms of human action, as a result of individual and collective interventions in (in)formal planning processes, efforts are made to maintain flood-prone areas with as few human elements as possible and low vulnerability (Fuchs et al., 2020). Additionally, it is important to highlight the adaptive capacity of floral and faunal communities and habitats to these cyclical processes of recurrent or exceptional nature (Fuchs et al., 2020; Moss et al., 2019).

Floods are a natural phenomenon that can bring significant negative impacts, requiring the development and application of technical, technological, organizational, and legislative tools to prevent, manage, and mitigate risks (Lousada et al., 2021b). In this context, Directive 2007/60/EC of the European Parliament and of the Council, transposed into domestic law by Decree-Law no. 115/2010, stands out, aiming at the assessment and management of flood risks. This directive requires the development of flood risk management plans, which must be integrated with spatial and environmental planning, as well as with

Table 1. National legislation related to risk management and floods

Legislation	Description
Law no. 16/2003, of June 4th (AR, 2003)	Legal regime for water public domain lands
Decree-Law no. 166/2008, of August 22nd (MAOTDR, 2008a)	Mandatory elaboration of flood zone maps in municipalities with urban clusters affected by floods
Decree-Law no. 364/1998, of November 21st (MEPAT, 1998)	Obrigatoriedade de elaboração da carta de zonas inundáveis nos municípios com aglomerados urbanos atingidos por cheias
Law no. 54/2005, of November 15th (AR, 2005a)	Ownership of water resources
Law no. 58/2005, of December 29th (AR, 2005b)	Water Law, transposing into national law the Directive no. 2000/60/CE of the European Parliament and of the Council, of October 23rd, and establishing the bases and institutional framework for sustainable water management
Law no. 16/2003, of June 4th (AR, 2003)	Legal regime for water public domain lands

Source: Author.

integrated and active territorial management, ensuring the coordination of means and tools to effectively reduce the risk of flooding. In addition, other national legislation exists in this context (Table 1).

After the transposition of the relevant European Directive, the Resolution of the Assembly of the Republic no. 15/2008, of April 21, established the urgent need to carry out a preliminary assessment of flood risk situations in each district, prepare reports on the main flood risk situations, develop a National Plan for the Reduction of Flood Risk, and promote awareness campaigns for citizens to adopt responsible behaviors in preventing flood risks. Decree-Law no. 54/2005 and Law no. 58/2005 (Water Law) establish that the government has the prerogative to classify as a threatened zone by floods the one designated as an area adjacent to the contiguous area to the edge of a watercourse, and which extends to the line reached by the greatest flood, with a return period of 100 years.

The Flood Risk Management Plan of the Autonomous Region of Madeira (PGRI-RAM) presents a survey of the main vulnerabilities in each hydrographic basin, regarding the flood risk, including elements at risk such as affected populations, buildings, agricultural areas, and infrastructure (Alves et al., 2022a). In the hydrographic basin of Ribeira de Machico, seven critical points were identified, as shown in Table 2, with Baixa de Machico being the most affected urban area by floods in this basin (Alves

Table 2. Various study zones

Zones	Area (ha)
Baixa de Machico (Figure 1)	30,90
Paraíso (Figure 2)	31,56
Moinho da Serra (Figure 3)	43,77
Ribeira Seca (Figure 4)	29,01
Ribeira Grande (Figure 5)	12,81
Maroços (Figure 6)	8,47
Madeira da Igreja (Figure 7).	14,95

Source: Alves et al. (2023)

Table 3. Historical summary of the main alluviums on the island of Madeira

Date	Source
Winter of 1467	(Almada, 1865)
November 9, 1678	(Almada, 1865)
November 18, 1724	(Almada, 1865)
October 9, 1803	(Almada, 1865)
October 23, 1815	(Almada, 1865)
October 23, 1842	(Almada, 1865)
February 22, 1853	(Almada, 1865)
November 29, 1901	(Silva & Menezes, 1921; Quintal, 1999)
February 25 e 26, 1920	(Silva & Menezes, 1921; Quintal, 1999)
March 5 e 6, 1921	(Quintal, 1999)
November 3, 1956	(Quintal, 1999)
September 27, 1989	(Quintal, 1999)
October 24 e 29, 1991	(Quintal, 1999)
October 29, 1993	(Quintal, 1999)
October 19 e 20, 1997	(Quintal, 1999)
February 20, 2010	(IST & UMA & LREC, 2010)
November 6, 2012	(Jornal Público, 2012)
November 29, 2013	(DROTA, 2016)
March 27, 2021	(Radio TSF, 2021)

Source: Alves et al. (2022a).

et al., 2023a). It is important to emphasize that the identification of critical points is essential for the implementation of appropriate preventive measures to minimize the negative effects of floods.

The specific characteristics of the city of Machico, together with the frequent occurrence of flooding events in the Autonomous Region of Madeira (Table 3), provide a unique opportunity to understand the exposure, vulnerability, and collective and individual adaptation to the human, social, economic, and environmental consequences resulting from such events (Hartmann et al., 2020; Ramos et al., 2018). Understanding exposure and vulnerability to floods is fundamental to identifying risk areas and planning prevention and adaptation measures (IPCC, 2014). In Machico, risk areas are often identified in the lower parts of the city, such as Baixa de Machico (Alves et al., 2023a). Furthermore, collective and individual adaptation to floods can be better understood through case studies such as that of Machico, which highlight the specific needs and challenges of local communities (Fuchs et al., 2021; Ramos et al., 2018).

METHODOLOGY

The study adopts a methodological approach that integrates the implementation of a Geographic Information System (GIS) project in QGIS, which incorporates reference and thematic geographic information

Table 4. Methodological approach

Methodological Approach	Use of the Databases For
GIS project in QGIS	a) Application of hydrological and hydraulic models
	b) Identification of potential harmful consequences of floods
	c) Evaluation of historical and prospective human adaptation

Source: Authors

for the Machico River Basin (Table 4). The organization of these data follows the guidance and data structures of international standards (INSPIRE Directive). These databases were used to: a) Apply hydrological and hydraulic models for the elaboration of maps of flood-prone zones, such as susceptibility maps for flood occurrences and flood risk maps for the Machico River Basin (Mehran & AghaKouchak, 2014); b) Identify potential harmful consequences of major floods for the biophysical environment, such as natural resources and functions, habitats and protected spaces, and for the human environment, such as human health, buildings, cultural heritage and economic activities, associating them with the respective vulnerability of Baixa de Machico (Cancelliere & Turco, 2014); c) Evaluate the historical and prospective adaptation of humans to these recurrent episodes, considering the current land use and the intended uses in spatial and environmental planning at the municipal or local level (Burch, 2010).

This study adopts a hierarchical approach, both spatially and temporally, which considers the entire basin to the area of Baixa de Machico (Alves et al., 2023a), covering a historical perspective and characterization of the current situation, as well as prospective analyses (Alves et al., 2022a). The variation of scale and the use of case studies allow the general challenges to be contextualized in relation to the objectivity of the local reality. The implementation and achievement for each area seek to evaluate local specificities for the understanding of global complexity (Fuchs et al., 2021; Ramos et al., 2018).

According to the Methodological Guide for the production of municipal risk cartography and the creation of municipal-based geographic information systems of the National Civil Protection Authority (ANPC) (Julião et al., 2009), it is recommended that the assessment of susceptibility to flooding occurrence be performed using distinct methodological approaches for urban and rural areas. The assessment should be based on the analysis of the entire basin, taking into consideration the variables that directly influence the susceptibility of the areas. According to Julião et al. (2009), the use of different methodological approaches to assess susceptibility to flooding occurrence in urban and rural areas is essential. The assessment should be based on the analysis of the entire basin, considering variables that directly affect susceptibility. The first methodological model consists of considering the potential runoff accumulation area, permeability, and slope as fundamental variables. In this case, precipitation is considered constant along the watersheds and is represented by the area accumulated in each terrain unit. Additionally, the effects of evapotranspiration, interception, and retention by the vegetation cover are not considered. The second approach requires adequate hydrometric and pluviometric data, allowing for the application of different hydrological or hydraulic models, provided that they are calibrated and validated (Julião et al., 2009). Table 5 presents the methodological approaches:

Considering the geographic databases collected and organized for the Ribeira de Machico watershed area, which include information on physiography, lithology, geomorphology, hydrography, hydrology (Alves et al., 2022a), land use and occupation, as well as data on current plans, economic activities, and human elements present (Alves et al., 2023a), a methodological scheme/model was established to be ap-

Table 5. Methodological approaches

Approach	Fundamental Variables	Considerations
1	Potential runoff accumulation area, permeability, slope	Constant precipitation along the watershed. Does not consider effects of evapotranspiration, interception and retention by vegetation cover
2	Appropriate hydrometric and pluviometric data	Allows application of different hydrological or hydraulic models, provided they are calibrated and validated

Source: Authors

Table 6. Phases and respective activities in the modeling process

Phase	Description of Activity
I	Statistical analysis of precipitation information, including the description of precipitation intensity based on return periods (T) and pluviometric data. This phase assumes that precipitations of a certain duration occur uniformly throughout the watershed with constant intensity or distributed in a linear gradient with variable intensity (Smith et al., 2015).
II	Use of a hydrological model, including the estimation of flood peak discharge through the Synthetic Unit Hydrograph (HUS) method according to the Hydrologic Engineering Center - Hydrologic Modeling System (HEC-HMS) model, with support from the HEC-GeoHMS extension for data integration in a GIS environment (Jones & Williams, 2012).
III	Use of a hydraulic model, including the modeling of surface flow and the resulting definition of flood-prone areas based on the application of the HEC-RAS model, integrated with the QGIS software and RiverGis extension (Wang et al., 2018).

Source: Authors

plied in modeling (Figure 1). These data allow for their direct integration into spatial analysis processes (according to the presented methodological scheme) and the validation/verification of the obtained results.

Based on data availability, this study adopted the use of hydrological and hydraulic models to produce maps of flood-prone areas (susceptibility maps). According to the methodology proposed for this study, three distinct phases can be identified in the modeling process (Table 6).

The application of the proposed methodology, which comprises the three presented phases, results in the generation of flood polygons/areas and, consequently, in the production of flood susceptibility maps, indicative of the risk of such events occurring in a particular area. Additionally, the methodology allows for the identification and location of exposed/sensitive elements to events of this nature through data cross-referencing. Considering the geographical region of interest, the historical center of Machico, located in the Baixa de Machico, was selected to carry out the functional inventory of buildings and population (Alves et al., 2023a). The choice of this location is justified by the fact that this area has historically been affected by flooding events, as stated in Article 4, point 2, chapter II of Directive 2007/60/EC of 23 October 2007, and has also been identified as a risk zone in institutional documents such as the Flood Risk Management Plan of the Autonomous Region of Madeira (PGRI-RAM). Based on the analysis of the degree of exposure and vulnerability of the population and the different affected elements, such as heritage, activities, and services, spatial databases were developed and analyzed, and detailed cartography related to flooded areas was produced in a GIS environment. The evaluation of the degree of exposure of a community should be carried out through quantitative parameters such as the number of inhabitants and the type and quantity of potentially affected economic activities, allowing for categorization on a qualitative scale. The mapping of assets at risk, as envisaged in the European Floods Directive, requires the provision of information on elements potentially affected by a flood, such as population, assets, and economic activities, as well as environmental issues (European Environment

Agency, 2007). Territorial systems are subject to changes resulting from natural processes and human actions. The perception and experience of environmental risks by the population contribute to individual and collective decision-making (Carter et al., 2009). In this study, the following were carried out: a) an assessment of land use and occupation within the flood-prone area in relation to the possible impact on natural resources and functions, habitats and protected areas, as well as impacts on human health, cultural heritage, and economic activities (Alves et al., 2023a); b) an evaluation of the perception of historical flood patterns and their relation to municipal and environmental planning options, including urban use and management of heritage in territorial planning, national ecological network, national ecological reserve, guidelines for the protection of historical and natural heritage (Alves et al., 2022a).

PRESENTATION AND DISCUSSION OF RESULTS

Figure 1 shows the U-shaped Machico River watershed, located in the municipality of Machico in the eastern region of Madeira Island. It is an open valley landscape that extends from altitudes of 1000 meters above sea level to the coastline, composed of several secondary valleys that receive smaller tributaries (Alves et al., 2022a).

Figure 1. Map of the Machico River watershed
Source: Authors

The Ribeira de Machico watershed has a drainage area of 24.58 km^2. The determination of the drainage area is carried out through planimetric techniques or geographic information systems (GIS), and corresponds to the horizontal projection of the watershed surface (Hipólito & Vaz, 2011). The relationship between the watershed area and the peak flow rate, calculated by the rational method, is direct, meaning that the larger the area, the greater the peak flow rate (Barreto, 2013). The perimeter, in turn, is defined as the length of the boundary line that delimits the watershed in the horizontal plane (Barreto, 2013). The Ribeira de Machico watershed is composed of the intermediate (CVM) and upper

(CVS) volcanic complexes, which have medium to good soil permeability (Caetano, 2014). The region is rich in Andosols (Caetano, 2014), and according to the Soil Map of Madeira Island (Ricardo et al., 1992), it has a diversified vegetation. The Ribeira de Machico is located in the subtropical region of the North Atlantic, where the mild climate results from the moderating effect of the Atlantic Ocean on the temperature amplitude, fitting within the regional climatic framework associated with the Macaronesia region. During the winter months, the arrival of depression systems in the Ribeira de Machico results in the formation of local depressions that generate abundant precipitation (DROTA, 2016). The climatic characteristics of the Ribeira de Machico, in addition to its geographical location, are significantly influenced by the relief. The distribution of precipitation and temperature values is a reflection of this influence. The depression systems that cross the Atlantic Ocean tend to originate abundant precipitation when they reach the Ribeira de Machico, especially on the north face, due to their typical movement in a north-south direction (DROTA, 2016).

The interaction between physiography and topography influences the morphology of the hydrographic basin and its flood patterns, even allowing for the automatic classification of natural units based on their behavior during floods (Cook and Merwade, 2009; Robinson et al., 2010). According to Alves et al. (2022), the Machico river basin has a perimeter of 24.97 km (P), an average width (b) of 2.45 km, a compactness index (KC) of 1.42, a shape index (Kf) of 0.16, and an elongation index (KL) of 4.10. These data indicate that the Machico river basin has a low shape factor, which makes it prone to the occurrence of smaller flood peaks. In addition, the basin is elongated, which implies a lower risk of floods: longer paths and longer travel times result in greater response times. These characteristics demonstrate a marked tendency for flooding to occur. The classification of the shape of the hydrographic basin as elongated rectangular (KC = 1.42) highlights its irregularity, and the flows occur predominantly through the main channel, making the path longer to the main section and resulting in damping of the flows. The hydrographic basin has good drainage.

Rapid urban growth and disorderly consumption demand the realization of studies to support public policies aimed at building a socio-economic and environmentally sustainable environment. One of these studies concerns problems related to water resources, such as surface runoff. When precipitation exceeds the interception and infiltration capacities of the soil, the excess water volume flows superficially to the nearest waterline, following a natural drainage hierarchy that leads to the formation of rivers that generally flow into the sea (ISEL, 2015 & Lencastre, 1972). Population growth results in the impermeabilization of soil and subsurface, as well as the pollution of water resources, since rivers, streams, and coastal lines are used for building construction and urban expansion. This results in the narrowing of these water sources, which are often altered to meet population demands. The impermeabilization of the soil and the overload of artificial water drainage systems, combined with river canalization works using materials such as concrete, increase surface runoff and reduce rainwater infiltration, resulting in floods in urban areas, as well as causing an increase in flood peaks and a decrease in the response time of drainage basins (Tucci, 2005; Ferreira, 1996).

The study conducted by (Alves et al., 2023a) indicates that the Baixa de Machico (Figure 2), delimited based on the flood zone maps of the Ribeira de Machico contained in the Flood Risk Management Plan of the Autonomous Region of Madeira (PGRI-RAM) (DROTA, 2016), has an area of 30.90 hectares. Between 1991 and 2001, there was a 10% decrease in the resident population, while between 2011 and 2021, a new decrease of 10% was observed. However, the number of buildings in Baixa de Machico has increased considerably, from 182 in 1991 to 342 in 2021. In 1991, Baixa de Machico already had the highest number of inhabitants, with a total of 1,097 residents, while in 2001 there were around 983

Figure 2. Map with Machico downtown
Source: Adapted from Alves et al. (2023a)

residents and 286 buildings in the area. Additionally, a 4% increase (0.72 hectares) in the impermeable area of Baixa de Machico was observed, which went from 17.87 hectares in 1990 to 18.59 hectares in more recent data.

The application of the recommended methodology indicates patterns and the risk of flooding mainly in the lower zone of the hydrographic basin (Baixa de Machico). Figure 3 shows 5 levels of risk associated with the occurrence of flooding, with red representing Very High Risk and dark green representing Very Low Risk. In Figure 3, there are only 3 levels of risk, where cream and yellow represent High Risk and Medium Risk, respectively.

The periodic occurrence of floods directly affects the value and function of exposed human activities and elements, with economic losses according to the level of vulnerability and local adaptation (Hansson et al., 2008). Floods can also potentially impact the physical-chemical (Lindenschmidt et al., 2009; Oeurng et al., 2010) and even (micro)biological quality of water by altering flow conditions and promoting the dispersion of effluents and waste from human activities (Posthumus et al., 2008), as well as affecting water treatment and supply systems (Ten Veldhuis et al., 2010), thus having a direct relationship with environmental and public health.

In the area known as Baixa de Machico (Figure 4), high-risk zones intersect with highly vulnerable elements, both due to the concentration of people and the typology of associated activities, such as residential buildings, wastewater treatment plants, recreational areas and sports fields, commercial areas, among others (Alves et al., 2023b). It is also important to highlight the large number of exposed elements related to urban infrastructure, especially buildings and communication routes. This set of po-

Figure 3. Risk of flooding in downtown Machico
Source: Adapted from DRAAC (2022)

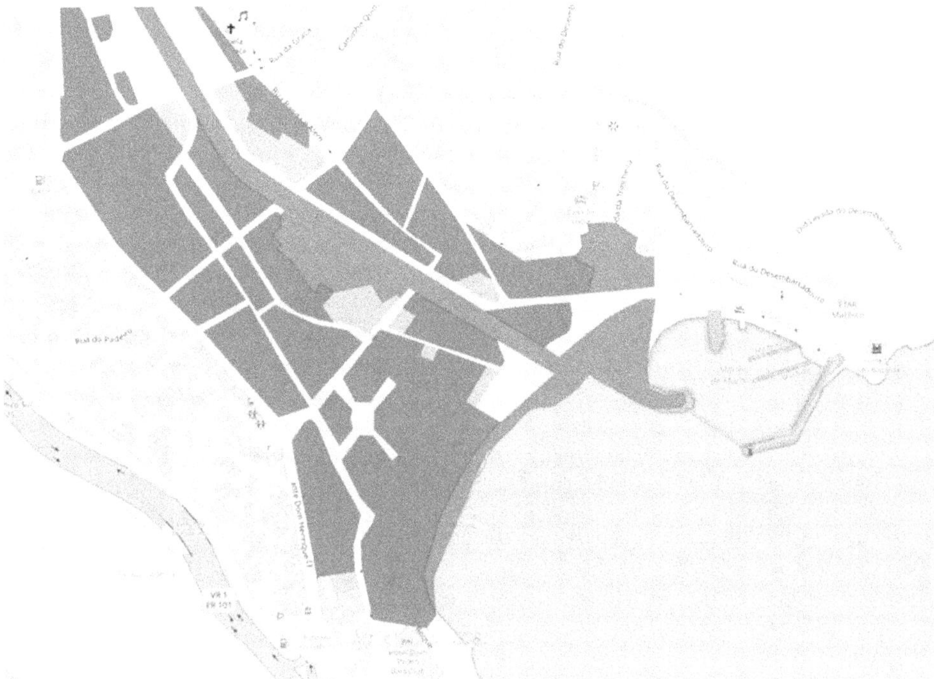

Figure 4. Machico downtown and areas and elements exposed to floods/inundations

Figure 5. Quantity of constructed buildings by time period
Source: INE (2021)

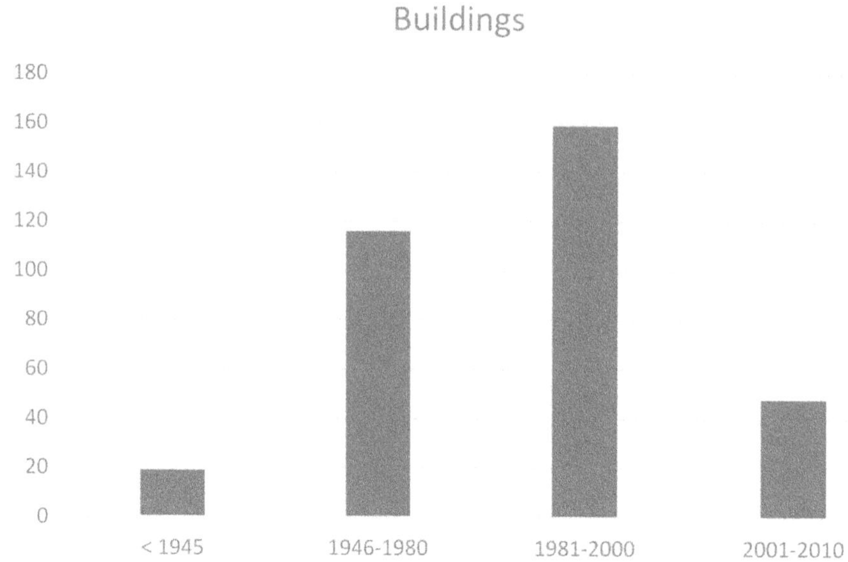

tentially exposed elements raises serious problems from the point of view of management, preparation, and organization of civil protection means (Alves et al., 2023b).

Floods can result in loss of life, displacement of populations, and environmental damage, seriously compromising economic development and harming economic activities. Floods are a natural phenomenon that cannot be avoided. However, certain human activities and climate change contribute to increasing the likelihood of flooding occurrence and its negative impact. It is possible and desirable to reduce the risk of harmful consequences associated with floods, especially for human health and life, the environment, cultural heritage, economic activities, and infrastructure (Directive 2007/60/EC of the European Parliament and of the Council of 23 October). The Baixa de Machico area is characterized by the predominant presence of buildings built between 1981 and 2000 (Figure 5), as well as by the resident population between 25 and 64 years old (Figure 6). In addition, ground floor occupancy is mainly composed of commercial establishments, which makes these places highly vulnerable to the occurrence of a potential flood event. It is important to note that the area also has several public and private institutions, such as the Municipal Chamber, nurseries, children's schools, churches, Municipal Civil Protection Service, municipal market, museums, chapels, conservatories, supermarkets, hotels, and pensions, which, due to their social attraction characteristics, may punctually worsen the degree of vulnerability by concentrating a large number of people and high-value goods.

Based on the concept of vulnerability and the degree of exposure of Baixa de Machico to the risk of flooding, as evaluated by (Alves et al., 2023b), quantitative parameters (Table 7) were used to assess the degree of exposure in relation to Architectural Heritage, Special Sites, Buildings, Altitude, Road Network, Land Use and Population.

According to the results obtained, the degree of exposure in Baixa de Machico is considered high. Floods in the region affect some homes, but mainly the ground-floor commercial establishments, as

Figure 6. Resident population by age
Source: INE (2021)

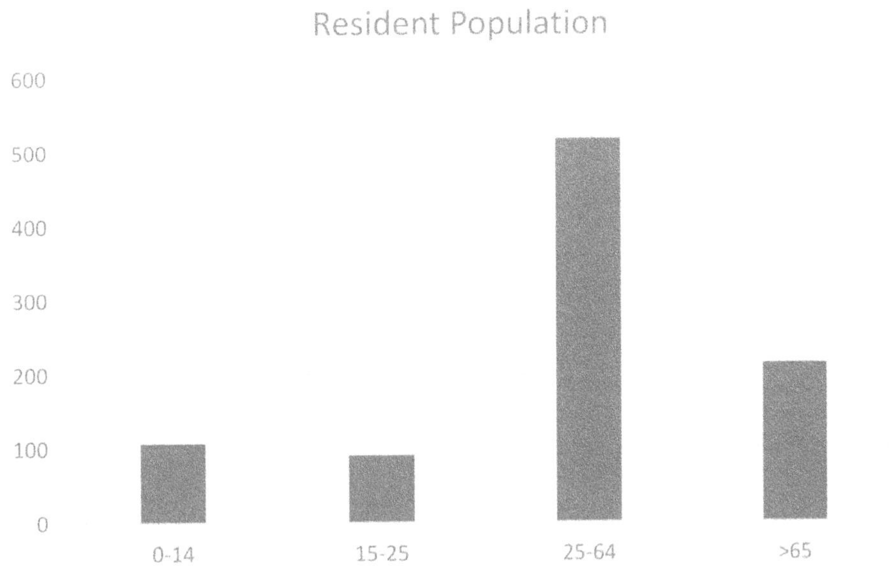

Table 7. Calculation of the eigenvector

Indicator	Weight	Weight (%)
Architectural Heritage	0,221	22,06
Special Sites	0,256	25,63
Buildings	0,121	12,10
Altitude	0,038	3,83
Road Network	0,050	4,98
Land Use	0,050	4,98
Population	0,264	26,42

Source: Alves et al. (2023b)

illustrated in Figure 7 and Figure 8, which are more vulnerable to variations in the flow of the Ribeira de Machico. In addition, there is potential damage to architectural heritage as indicated in Table 7. Insufficient drainage and urban stormwater management infrastructure are identified as the main factor of flooding in Baixa de Machico. It can be stated that Baixa de Machico has become safer since the channeling of Ribeira de Machico, carried out between 2001 and 2006, at an estimated cost of €5,135,000.

Using QGIS software, the area occupied by the building susceptible to flood risk for a 100-year return period was calculated, as shown in Figure 7. The selected return period is similar to that used in previous studies by (Lousada et al., 2021a; Lousada et al., 2022).

Figure 7. Flood in the municipality of Machico in 2010
Source: Facebook - António Nóbrega

Figure 8. Flood in the municipality of Machico in 2012
Source: Facebook - Foto Mundo

Figure 9. Exposure to floods in Baixa de Machico
Source: Adapted from DRAAC (2022)

The bed of the Machico stream corresponds to the flood-prone area. Along both banks of the stream, especially on the left side, there is a high concentration of built heritage, including classified buildings, as well as strong investment in the creation of new green infrastructure and equipment for municipal management (Alves et al., 2022b). In this region, it is possible to find government buildings, as well as facilities related to civil protection, sanitation and treatment of waste and effluents. In addition, there are also buildings for collective use, such as hotels, churches, and public parking areas.

Despite facing strong pressure for constructions close to the flood-prone area and flood zones, which cannot be avoided through municipal planning (PMOT and PDM), the local council is working together with other socio-economic agents, especially those linked to civil protection. They are developing emergency and civil protection plans that include a detailed analysis of susceptible areas, mapping of risks, and organization of resources and response capacity in emergency situations.

CONCLUSION

The complexity of territorial systems demands knowledge and responsiveness from social and economic agents, resulting from organization, sharing of elements, and communication in the context of implementing territorial governance systems. In this scenario, environmental and technological risks present a multidisciplinary, systemic, integrative, and prospective dimension. Risk assessment and management are instruments and processes that allow for understanding and intervening in the territory towards sustainability, especially in the management of the hydrological cycle and water resources, with a focus on knowledge of hydrological risks and their impact on the quantity and quality of water, as well as on human adaptation to the environment.

In this study, a Geographic Information System (GIS) project was developed for the Machico River watershed, which incorporates reference and thematic geographic information organized according to the Annexes of the INSPIRE Directive (Alves et al., 2022a; Alves et al., 2023a). This project supported the characterization of the watershed, as well as providing support and feeding of the hydrological and hydraulic models for flood risk assessment and mapping of flood-prone areas. It also allowed for the identification of exposed elements and the vulnerability of natural and human elements present in downtown Machico, as well as the evaluation of adaptation and governance of flood risks at the local level.

In this context, it is necessary to consider the inclusion of flood risk in local and municipal planning processes, as well as strategic environmental assessment, with a focus on maintaining agricultural and forest management areas, installing specific hydraulic infrastructure, and renaturalizing the environment. Additionally, it is important to promote the participation and adaptation of the population and adequacy of social and institutional organization to the implementation of risk management plans and governance systems for flood risk.

Flood risk management and mitigation plans should focus on prevention, protection, and preparation for both populations and exposed assets, as well as the protection civil agents involved. These plans should be periodically reviewed and updated, taking into account climate changes or the level of transformation of land use and occupation on flood occurrences.

It is important that the information and geographic information system developed support the development and adoption of collaborative-based systems and support for local administration, civil protection agents, security, and health. This system should also support communication and interaction with the local population and socio-economic agents, aiming to convey information on the incidence, probability, and possible impacts resulting from flood occurrences.

REFERENCES

Almada, J.A. (1865). Annals of the Municipality of the Old Villa de Machico. *Flor do Oceano*, 247-286.

Alonso, J., Castro, P., Ribeiro, J., Martins, I., Mamede, J., Machado, A., & Brito, A. (2011). The information and decision support system [SI.ADD] of the Northern Regional Water Authority, I.P.: Objectives and development. *Water Resources Journal, 32*(1), 5-12.

Alves, R., Lousada S., Cabezas, J., & Gómez J. (2022a). *Flood Risk and Vulnerability An Approach to Urban Floods in Ribeira de Machico in the Autonomous Region of Madeira*. Geoinformatics in Support of Urban Politics and the Development of Civil Engineering.

Alves, R., Lousada, S., Cabezas, J., & Gómez, J. (2022b). The Importance of the Strategic Urban Rehabilitation Plan in the Sustainable Development of the Municipality of Machico. *Sustainability, 14*, 16816. doi:10.3390/su142416816

Alves, R., Lousada S., Cabezas, J., & Gómez J. (2023a). *Urban Flooding Risk in Machico and Planning its Territory as a Form of Prevention*. Geoinformatics in Support of Urban Politics and the Development of Civil Engineering.

Alves, R., Lousada S., Cabezas, J., & Gómez, J. (2023b). *Port Of Machico, Madeira: Assessment of the Consequences of Overtopping Using AHP Methodology*. Geoinformatics in Support of Urban Politics and the Development of Civil Engineering.

Apel, H., Thieken, A. H., Merz, B., & Blöschl, G. (2004). Flood risk assessment and associated uncertainty. Natural Hazards and Earth System Sciences, (4), 295–308.

AR. (2003). Law No. 16/2003, of the Assembly of the Republic, of June 4, 2003, revises, updates and unifies the legal regime of lands in the water public domain. Diário da República n.º 129/2003, Série I-A de 2003-06-04, pages 3343-3349.

AR. (2005a). Law no. 54/2005, National Assembly, of November 15, regarding the ownership of water resources. Diário da República n.º 219/2005, Series I-A of 2005-11-15, pages 6520-6525.

AR. (2005b). Law no. 58/2005, National Assembly, of December 29, regarding the framework for the management of surface waters, namely inland, transitional, and coastal waters, and groundwater. Diário da República n.º 249/2005, Series I-A of 2005-12-29.

AR. (2008). Resolution of the National Assembly no. 15/2008, of April 21, regarding flood risks. Diário da República n.º 78/2008, Series I of 2008-04-21.

Aronica, G., Candela, A., Fabio, P., & Santoro, M. (2012). Estimation of flood inundation probabilities using global hazard indexes based on hydrodynamic variables. *Physics and Chemistry of the Earth, 42-44*, 119-129.

Barreto, V. H. (2013). *Study of the Machico River Basin - Vulnerability and Floods* [Master's thesis]. University of Madeira.

Blaikie, P., Cannon, T., Davis, I., & Wisner, B. (1994). *At Risk: Natural Hazards, People's Vulnerability and Disaster*. Routledge Editions.

Büchele, B., Kreibich, H., Kron, A., Thieken, A., Ihringe, R. J., Oberle, P., Merz, B., & Nestmann, F. (2006). Flood-risk mapping: Contributions towards an enhanced assessment of extreme events and associated risks. *Natural Hazards and Earth System Sciences, 6*(6), 485–503. doi:10.5194/nhess-6-485-2006

Burch, S. (2010). Transforming barriers into enablers of action on climate change: Insights from three municipal case studies in British Columbia, Canada. *Global Environmental Change, 20*(2), 287–297. doi:10.1016/j.gloenvcha.2009.11.009

Caetano, C. (2014). *Evaluation of flash flood risk in the rivers of Madeira Island* [Master's thesis]. Instituto Superior Técnico.

Camarasa, A. (1995). Genesis of floods in small semiarid watersheds: Barranc del Carrsáxet and Rambla de Poyo. MOPTConfederadón Hidrográfica del Júcar.

Cancelliere, A., & Turco, M. (2014). Multi-criteria flood vulnerability assessment of buildings at urban scale. *Journal of Flood Risk Management*, *7*(4), 325–336.

Carter, J. G., White, I., & Richards, J. (2009). Sustainability appraisal and flood risk management. *Environmental Impact Assessment Review*, *29*(1), 7–14. doi:10.1016/j.eiar.2008.06.003

Charneca, N. (2006). *Geographic information management in the implementation of the Water Framework Directive*. National Laboratory of Civil Engineering, Department of Hydraulics and Environment.

Cook, M., & MeRwade, L. (2009). Morfometria e hidrologia da bacia hidrográfica. In *Manual de Hidrologia da Engenharia* (pp. 1-20). McGraw-Hill.

Correia, F. N., Rego, F. C., SaRaiva, M. G., & Ramos, I. (1998). Acoplamento de SIG com modelagem hidrológica e hidráulica de enchentes. *Gerenciamento de Recursos Hídricos*, *12*, 229–249.

Devillers, R., & Jeansoulin, R. (2010). Qualidade de dados espaciais: conceitos. In Fundamentos da Qualidade de Dados Espaciais. Sistemas de Informação Geográfica. ISTE - Série GIS.

Direção Regional do Ambiente e Alterações Climáticas - DRAAC. (2022). Cartas de Zonas Inundáveis e Cartas de Riscos e Inundações. PGRI-RAM_2022-2027. Funchal.

Diretiva INSPIRE: European Parliament and Council. (2007). Diretiva 2007/2/CE do Parlamento Europeu e do Conselho, de 14 de março de 2007, que estabelece uma infraestrutura de informação geoespacial na Comunidade Europeia (INSPIRE).

European Environment Agency. (2007). *Handbook on good practices for flood mapping in Europe, EXCIMAP European exchange circle on flood mapping*. European Environment Agency.

Ferreira, F. P. M., & Cunha, S. B. (1996). *Urban Hydrology: Flooding in Rio de Janeiro – Jacarepaguá*. Federal University of Rio de Janeiro/CNPq.

Fitzpatrick, F. A., Knox, J. C., & Whitman, H. E. (1999). USGS Water-Resources Investigations Report: Vol. 99–4083. Effects of Historical Land-Cover Changes on Flooding and Sedimentation, North Fish Creek, Wisconsin. Academic Press.

Fotopoulos, F., Makropoulos, C., & Mimikou, M. A. (2010). Flood forecasting in transboundary catchments using the Open Modeling Interface. *Environmental Modelling & Software*, *25*(12), 1640–1649. doi:10.1016/j.envsoft.2010.06.013

Fuchs, S., Keiler, M., Sokratov, S., & Shnyparkov, A. (2021). Flood risk and urban planning in Russia: Lessons learned from Krymsk. *Natural Hazards and Earth System Sciences*, *21*(2), 567–583. doi:10.5194/nhess-21-567-2021

Fuchs, S., Kellermann, P., Kreibich, H., & Thieken, A. H. (2020). Natural hazard events in Europe: A regional diachronic analysis of changes in impacts. *Regional Environmental Change*, *20*(4), 1–14.

Gouldby, B. P., & Samuels, P. G. (2005). Probabilistic assessment of coastal flood risk. *Journal of Coastal Research*, 214–228.

Hansson, K., Danielson, M., & Ekenberg, L. (2008). A framework for evaluation of flood management strategies. *Journal of Environmental Management, 86*(3), 465–480. doi:10.1016/j.jenvman.2006.12.037 PMID:17292530

Hartmann, T., Wendler, J., & Jülich, S. (2020). Flood Risk Management: Challenges and Strategies for Enhancing Resilience. *Sustainability, 12*(14), 5782. doi:10.3390u12145782

Hipólito, J. R., & Vaz, Á. C. (2011). *Hydrology and Water Resources* (1st ed.). IST Press.

Honrado, J., Alonso, J., Guerra, C., Pôças, I., Gonçalves, J., & Marcos, B. (2011). Deliverable No: D4.1 Report on pre-existing in situ and ancillary datasets for sites (FP7-SPA-2010-1-263435). Academic Press.

Intergovernmental Panel on Climate Change (IPCC). (2014). *Synthesis Report. Contribution of Working Groups I, II and III to the Fifth Assessment Report of the Intergovernmental Panel on Climate Change.* IPCC.

ISEL. (2015). *Hydrology.* Lisbon: ISEL. Retrieved from https://issuu.com/joanadecastrocortesao/docs/hidrologia_resumo_luis_mendes

IST, UMA, & LREC. (2010). *Alluvial risk assessment study on the island of Madeira.* Academic Press.

Jones, A., & Williams, B. (2012). *Hydrologic Engineering Center - Hydrologic Modeling System (HEC-HMS): User's manual, version 4.0.* US Army Corps of Engineers.

Jonkman, S. N., Kelman, I., & van Veen, A. (2015). Challenges in flood risk management–analysis and modelling. *Natural Hazards, 76*(2), 967–977.

Khatibi, R. (2011). Evolutionary systemic modelling of practices on flood risk. *Journal of Hydrology (Amsterdam), 401*(1-2), 36–52. doi:10.1016/j.jhydrol.2011.02.006

Lahr, J., & Kooistra, L. (2010). Environmental risk mapping of pollutants: State of the art and communication aspects. *The Science of the Total Environment, 408*(18), 3899–3907. doi:10.1016/j.scitotenv.2009.10.045 PMID:19939435

Lehner, B., Döll, P., Alcamo, J., Henrichs, T., & Kaspar, F. (2006). Estimating the Impact of Global Change on Flood and Drought Risks in Europe: A Continental, Integrated Analysis. Springer. *Climatic Change, 75*(3), 273–299. doi:10.100710584-006-6338-4

Lencastre, A. (1972). *Manual of General Hydraulics.* Academic Press.

Lindenschmidt, K. E., Pech, I., & Baborowski, M. (2009). Environmental risk of dissolved oxygen depletion of diverted flood waters in river polder systems – A quasi-2D flood modelling approach. *The Science of the Total Environment, 407*(5), 1598–1612. doi:10.1016/j.scitotenv.2008.11.024 PMID:19106003

Liu, Y. B., Gebremeskel, S., De Smedt, F., Hoffmann, L., & Pfister, L. (2003). A diffusive transport approach for flow routing in GIS-based flood modeling. *Journal of Hydrology (Amsterdam), 283*(1-4), 91–106. doi:10.1016/S0022-1694(03)00242-7

Lousada, S., Cabezas, J., Castanho, R. A., & Gómez, J. M. N. (2021a). Hydraulic Planning in Insular Urban Territories: The Case of Madeira Is-land—Ribeira Brava. *Water (Basel), 2021*(13), 2951. doi:10.3390/w13212951

Lousada, S., & Castanho, R. A. (2021b). GIS-Based Assessment of Morphological and Hydrological Parameters of Ribeira Dos Socorridos and Ribeira Do Vigario Basins, Madeira Island, Portugal. *Current World Environment, 2021*(16), 408–426. doi:10.12944/CWE.16.2.08

Lousada, S., Gonçalves, L., & Atmaca, A. (2022). Hydraulic Planning in Insular Urban Territories: The Case of Madeira Island—São Vicente. *Water (Basel), 2022*(14), 112. doi:10.3390/w14010112

MAOT. (2010). Decreto-Lei n. º115/2010 do Ministério do Ambiente e do Ordenamento do Território, de 22 de outubro de 2010, relativo à avaliação e gestão dos riscos de inundações, com o objetivo de reduzir as suas consequências prejudiciais. Diário da República n.º 206/2010, Série I de 2010-10-22, páginas 4757 – 4764.

MAOTDR. (2008). Decreto-Lei n.º 166/2008, do Ministério do Ambiente e do Ordenamento do Território, de 22 de agosto de 2008, relativo ao Regime Jurídico da Reserva Ecológica Nacional. Diário da República n.º 162/2008, Série I de 2008-08-22, páginas 5865 – 5884.

Mateu Belles, J. F. (1989). *Crescidas e Inundaciones*. Guía de la Naturaleza de la Comunidad Valenciana.

Mehran, A., & AghaKouchak, A. (2014). A generalized framework for deriving nonparametric standardized drought indicators. *Advances in Water Resources, 63*, 125–133.

MEPAT. (1998). Decreto-Lei n.º 364/1998, Ministério do Equipamento, do Planeamento e da Administração do Território, de 21 de novembro, relativo à obrigatoriedade de elaboração da carta de zonas inundáveis nos municípios com aglomerados urbanos atingidos por cheias. Diário da República n.º 270/1998, Série I-A de 1998-11-21, páginas 6280 – 6281.

Moss, R. H., Edmonds, J. A., Hibbard, K. A., Manning, M. R., Rose, S. K., van Vuuren, D. P., ... Zhao, Z. C. (2019). The next generation of scenarios for climate change research and assessment. *Nature*, 463–471. PMID:20148028

Oeurng, C., Sauvage, S., & Sánchez-Pérez, J.-M. (2010). Temporal variability of nitrate transport through hydrological response during flood events within a large agricultural catchment in south-west France. *The Science of the Total Environment, 409*(1), 140–149. doi:10.1016/j.scitotenv.2010.09.006 PMID:20951409

Paiva, I. (2005). *Flood risk in Coimbra: physical factors and human action. In Urban floods and the floods of the Mondego (1950/51 - 2003/04)*. Faculty of Letters, University of Coimbra.

PECUE. (2007). Directive 2007/60/EC of the European Parliament and of the Council of 23 October 2007 on the assessment and management of flood risks. Official Journal of the European Union no. L 288 of 06/11/2007 p. 27-34. Brussels.

Posthumus, H., Hewett, C. J. M., Morris, J., & Quinn, P. F. (2008). Agricultural land use and flood risk management: Engaging with stakeholders" in North Yorkshire. *Agricultural Water Management, 95*(7), 787–798. doi:10.1016/j.agwat.2008.02.001

Publico Newspaper. (2012). *Four dozen displaced people and roads closed in Madeira*. Author.

Quintal, R. (1999). *Alluviums of Madeira. 19th and 20th centuries*. Territorium Magazine.

Radio TSF. (2021). *Tree falls, landslides, and floods. Heavy rains promise not to give truce in Madeira.* Author.

Ramos, M. H., Cea, L., Ferreira, C. S. S., & Coelho, C. (2018). Estimation of flood hazard and risk in Portugal. *Natural Hazards*, *90*(3), 1263–1282. doi:10.100711069-017-3103-9

Ramos, T. B., Ribeiro, C., & Sousa, S. (2018). Climate change and flooding: Exploring gendered differences in impacts and vulnerabilities in Portugal. *Regional Environmental Change*, *18*, 2117–2130. doi:10.100710113-018-1366-2

Regional Secretariat for the Environment and Natural Resources (DROTA) – Regional Directorate for Spatial Planning and Environment. (2016). *Flood Risk Management Plan for the Autonomous Region of Madeira*. Author.

Regional Secretariat for the Environment and Natural Resources – Regional Directorate for Spatial Planning and Environment. (2022). *Flood Risk Management Plan for the Autonomous Region of Madeira*. Author.

Ricardo, R. P., da Câmara, E. M. S., & Ferreira, M. A. M. (1992). Soil Map of the Island of Madeira. RAM.

Robinson, D. A., Jones, S. B., & Glendening, J. W. (2010). Geomorphology and hydrology of a mountainous tropical island, Kaua'i, Hawai'i. *Geomorphology*, *116*(1-2), 52–64.

Ronica, R., Nadim, F., Kaynia, A. M., & Kumar, P. (2012). The use of stochastic models for flood hazard assessment: A review. *Natural Hazards*, *63*(2), 609–641.

Rosselló, V. M. (1989). Flood plains. In River floods and floods in the Mediterranean. University of Alicante.

Saraiva, M. G. E., & Carvalho, L. (2009). *Flood risk, Methodology for the formulation of the Riprocity indicator no. 4. Workshop "Rivers and Cities"*. Opportunities for Urban Sustainability.

Silva, F. A., & Menezes, C. A. (1921). *Madeiran Elucidário*. Esperança Typography.

Smith, J. M., Brown, P., & Johnson, T. (2015). Flood Risk Analysis. In Handbook of Hydrology (pp. 18.1-18.21). McGraw Hill Education.

Ten Veldhuis, J. A., Clemens, F. H., Sterk, G., & Berends, B. R. (2010). Microbial risks associated with exposure to pathogens in contaminated urban flood water. *Water Research*, *44*(9), 2910–2918. doi:10.1016/j.watres.2010.02.009 PMID:20227742

Tucci, C. (2005). *Urban Flood Management*. Urban Flood Management Handbook.

UNISDR. (2009). *Terminology on disaster risk reduction.* United Nations Office for Disaster Risk Reduction.

United Nations. (2009). *Global Assessment Report on Disaster Risk Reduction: Risk and poverty in a changing climate*. United Nations Development Programme.

Vieira, J. M. P., Pinho, J. L., & Lima, M. M. (2005). A hydroinformatics environment for decision support in water management in a river basin. *ODEANA. 8th Water Congress. Portuguese Association of Water Resources (APRH).*

Vrijling, J. K. (1995). Flood risk assessment: Concepts and applications. *Journal of Flood Risk Management*, *2*(2), 77–84.

Wang, Z., Qin, C., Li, X., Zhang, Y., & Li, Z. (2018). A new GIS-based framework for integrated flood risk assessment: A case study in Chaohu Lake Basin, China. *Natural Hazards*, *91*(2), 695–714.

Wendy, K. (2007). Evaluating flood risk management options in Scotland: A participant-led multi-criteria approach. *Ecological Economics*, *64*(1), 70–81. doi:10.1016/j.ecolecon.2007.06.011

Wheater, H., & Evans, E. (2009). Land use, water management and future flood risk. *Land Use Policy*, *26*(Supplement 1), S251–S264. doi:10.1016/j.landusepol.2009.08.019

Wheater, H. S., Chandler, R. E., Onof, C. J., Isham, V. S., Bellone, E., Yang, C., Lekkas, D., Lourmas, G., & Segond, M.-L. (2005). Spatial-temporal rainfall modelling for flood risk estimation. *Stochastic Environmental Research and Risk Assessment*, *19*(6), 403–416. doi:10.100700477-005-0011-8

Wolshi, P., Masaka, T., Raditsebe, L., & Murray-Hudson, M. (2005). Aspects of Seasonal Dynamics of Flooding in the Okavango Delta. *Botswana Notes and Records*, *37*.

KEY TERMS AND DEFINITIONS

Flood: A flood is an overflow of water that submerges land which is normally dry. Floods can be caused by a variety of factors, including heavy rainfall, snowmelt, coastal storms, and storm surge from hurricanes or tropical cyclones. Flooding can cause extensive damage to infrastructure, homes, and businesses, and can result in loss of life.

Machico: Machico is a municipality on the eastern coast of the island of Madeira, Portugal. It is located along the Machico River and encompasses the Machico Watershed. Machico is known for its historic sites and natural beauty, including rugged coastline and mountainous terrain.

Machico River: The Machico River is a river that flows through the municipality of Machico on the island of Madeira, Portugal. It is approximately 11 km long and runs from its source in the mountains to the Atlantic Ocean.

Machico Watershed: The Machico Watershed is the geographic area that drains into the Machico River. It includes the land and water resources of the surrounding area, including the streams, rivers, and wetlands that contribute to the flow of water in the Machico River. The watershed plays a critical role in regulating the flow of water, supporting ecological functions, and providing important resources for the local community.

Madeira: Madeira is an archipelago and a region of Portugal located in the Atlantic Ocean, about 400 kilometers north of the Canary Islands and 850 kilometers southwest of Portugal. The archipelago is made up of two inhabited islands, Madeira and Porto Santo, and two uninhabited islands, the Desertas and the Selvagens. Madeira is known for its rugged volcanic landscapes, subtropical climate, lush forests, and levadas, the island's extensive system of irrigation channels that carry water from the mountains to the agricultural fields below. Madeira is also a popular tourist destination, attracting visitors from around the world with its natural beauty, rich cultural heritage, and outdoor activities such as hiking, surfing, and paragliding.

Chapter 3
Urban Flooding Risk in Machico and Planning Its Territory as a Form of Prevention

Raul Manuel Costa Alves
https://orcid.org/0000-0002-3680-6626
University of Extremadura, Spain

Sérgio António Neves Lousada
https://orcid.org/0000-0002-8429-2164
University of Madeira, Portugal

José Cabezas
https://orcid.org/0000-0002-7209-3731
University of Extremadura, Spain

José Manuel Naranjo Gómez
https://orcid.org/0000-0001-7998-9154
University of Extremadura, Spain

ABSTRACT

Machico, a parish located in Madeira, Portugal, is prone to recurring floods during periods of heavy rainfall, which can cause significant damage to property, infrastructure, and even put the lives of people and animals at risk. The increasing urbanization has worsened this problem since urban growth often occurs in risk-prone areas, such as steep slopes or floodplains. Additionally, the development of drainage and flow systems often cannot keep up with the pace of urban expansion, leading to greater vulnerability to flooding. Therefore, it is essential to conduct studies to better understand how urbanization contributes to flooding to develop more effective risk management strategies. This involves identifying risk-prone areas, establishing prevention and damage mitigation measures, and implementing early warning systems to protect the population in the event of extreme weather conditions.

DOI: 10.4018/978-1-6684-6449-6.ch003

INTRODUCTION

Floods are one of the phenomena that become more and more frequent and destructive due to the disorderly growth of cities and the lack of adequate urban planning (Freitas & Andrade, 2007). Thus, it is clear that floods are the most common type of natural disaster in Europe, taking into account the international database on disasters, representing 43% of all disasters that occurred during the period between 1998 and 2002 (EEA, 2005). The potential effects of climate change and changing environmental conditions on flooding events threatens to exacerbate urban flood risk worldwide (Miller & Hutchins, 2017; Skougaard Kaspersen et al., 2017), requiring adaptation and further lot-scale mitigation to protect property from water damage (Leandro et al., 2020). Floods can be defined as extreme hydrological phenomena of variable frequency. These events could be natural or human-induced, consisting of the overflow of a watercourse (Chow, 1956) e (Guedes, 2014). With the increase of this problem at a global level and, mainly in tropical regions, it became necessary to develop measures and devices to mitigate the impacts of floods (Lousada & Castanho, 2021; Lousada et al., 2021).

Floods, depending on the size of the flow in the surrounding area, affect settlements and agriculture by damaging their areas, lower and upper structures, facilities, and living things, and they interrupt human life and socio-economic activities. Issues concerning urban flooding have been identified in regions across the globe (Zhu & Chen, 2017; Mobini et al., 2020; Pagliacci et al., 2020). Sociological effects on humans from floods, psychological disorders, and the like, are also seriously affected (Lousada et al., 2022).

Considering that the elevation of the Machico stream occasionally causes flooding in the Machico parish and that these events affect the lives of citizens, it was decided to study the occurrence of this phenomenon. The characterization of the runoff factors in the Machico parish and the surrounding region (from the Machico stream mouth to the mountains at 1000 m altitude), together with the historical analysis of floods, the study of the urbanization process and the constraints on surface runoff (Ramos, 2009; DGRAH, 1984; Quintal, 1999; Barreto, 2013; Hipólito & Vaz, 2011) and the calculation of flood risk allow for a better understanding of the conditions under which the phenomenon occurs and how it manifests itself. Based on this information, the objective of this study is to project the future of floods (Deltares, 2010; SRARN, 2016) in the city of Machico, creating scenarios that allow for the definition of mitigation measures to reduce the harmful effects that these floods can cause. This can include citizen participation in hazard mitigation policy (Oulahen & Doberstein, 2012) which engages the public in the area of flood risk and vulnerability and the adoption of technologies to reduce this risk. This aids in developing an awareness for homeowners of the infrastructure present in the home and the requirements for homeowner maintenance (Sandink, 2011; Owusu et al., 2015).

The methodology of this study involves the characterization of the urbanization process, through the analysis of the variation in the number of buildings and residents in the Machico parish, as well as the development of maps of impermeable areas at different temporal moments, allowing an evaluation of how this process has occurred. It is observed that the urbanization process in the Machico parish is related to the number of buildings and residents in the region, as well as the evolution of impermeable areas over time.

To analyze the variation in the number of buildings and residents in Machico, statistical data provided by the National Institute of Statistics (INE, 1981; INE, 1991; INE, 2001; INE, 2021) and cartography are used. This analysis is carried out for the municipality as a whole and for the Machico parish, delimiting the study area based on physical characteristics of the terrain and the spatial occurrence of floods since

Table 1. Historical summary of the main floods in Madeira Island

Date	Source
Winter of 1467	(Almada, 1865)
November 9, 1678	(Almada, 1865)
November 18, 1724	(Almada, 1865)
October 9, 1803	(Almada, 1865)
October 23, 1815	(Almada, 1865)
October 23, 1842	(Almada, 1865)
February 22, 1853	(Almada, 1865)
November 29, 1901	(Silva & Menezes, 1921; Quintal, 1999)
February 25 e 26, 1920	(Silva & Menezes, 1921; Quintal, 1999)
March 5 e 6, 1921	(Quintal, 1999)
November 3, 1956	(Quintal, 1999)
September 27, 1989	(Quintal, 1999)
October 24 e 29, 1991	(Quintal, 1999)
October 29, 1993	(Quintal, 1999)
October 19 e 20, 1997	(Quintal, 1999)
February 20, 2010	(IST & UMA & LREC, 2010)
November 6, 2012	(Jornal Público, 2012)
November 29, 2013	(DROTA, 2016)
March 27, 2021	(Radio TSF, 2021)

Source: Authors

the 19th century. In addition, the occurrence of floods in Madeira is mentioned based on news published in local, regional or national newspapers on the following dates (Table 1).

The delimitation of the impermeable area in the study area was based on the configuration of the urban area at different temporal moments. For this purpose, available cartographic means were used, such as vectorized cartography from 1990, raster format orthophotomaps from 2018, raster format military maps from 2003, and the Geometric Registry of Rustic Property of Madeira Island from different years, in raster format. This approach allowed for an understanding of the evolution of the impermeable area over more than thirty years.

The relationship between flood risk and land-use planning aspects has been the subject of several studies (Costa, 2009a; Teles, 2010; Costa, 2009b; Cunha & Taveira-Pinto, 2011), as well as the production of legislation at different territorial levels, from European to local. An example of this situation is the transposition of European Directive no. 2007/60/EC of October 23rd into national law through Decree-Law no. 115/2010 of October 22nd.

Existing legislation is broad and comprehensive, aiming to mitigate risk through the mandatory development of flood history and consequences, flood-prone area mapping, special emergency plans, risk mapping, among other measures, including the creation of flood scenarios. However, the application of the law has not been very favorable, as floods caused by inadequate land-use planning and the non-application of rules that minimize or overcome the harmful consequences of a flood continue to occur. Several authors focus on this subject, and their contributions are valuable for improving the living conditions of the population. There is a critical importance to engage with the public on flood risk and vulnerability in order to improve flood risk communication and response to flooding events (Sampson et al., 2019), however, challenges exist with respect to effectively engaging the public in flood risk reduction options (Grothmann & Reusswig, 2006; Terpstra et al., 2009; Meyer et al., 2014).

So far, the studies carried out show that there are several methods for calculating flood risk, which also differ in terms of the variables that should be considered. However, despite some controversies, all authors agree that the study of flood risk should be taken into account and integrated into all processes related to land-use planning, and should be carried out by multidisciplinary teams.

URBANIZATION PROCESS IN THE CITY OF MACHICO

Over the last 40 years, there have been variations in the resident population and number of buildings in the municipality of Machico. Between 1981 and 2001, there was a decrease of 379 residents, which was partially compensated for by an increase of 81 residents in the following decade. However, between 2011 and 2021, there was a decrease of 2235 residents. Over this period, the resident population in Machico decreased by 11.4% (see Figure 1).

Over the last 40 years, from 1981 to 2021, there has been a constant growth in built-up areas in the municipality of Machico (Figure 2). In the first 30 years, the variation was positive, reaching 63.3%, which corresponded to an increase of 3393 buildings. However, in the last 10 years, there was a reduction of 46 buildings, resulting in a negative variation of 0.5%.

Figure 1. Resident population of the municipality of Machico
Source: INE (1981, 1991, 2001, 2011, 2021)

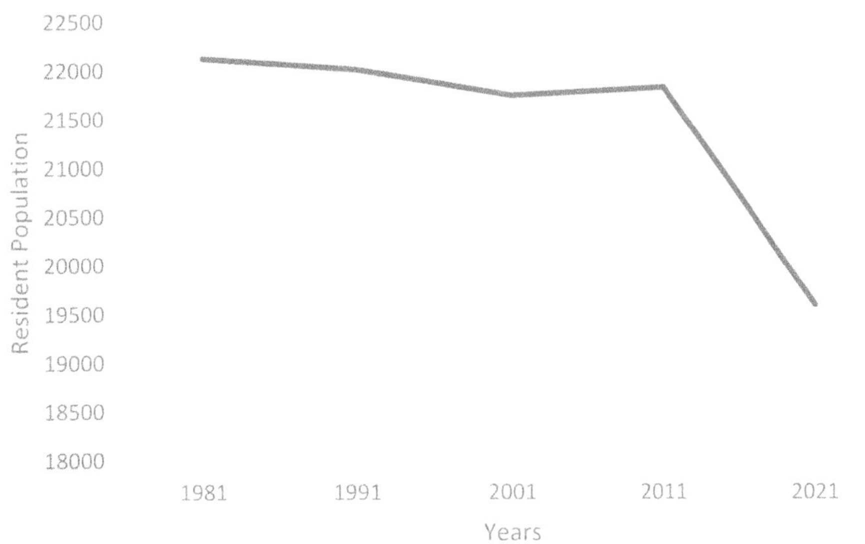

Figure 2. Number of buildings in the municipality of Machico
Source: INE (1981, 1991, 2001, 2011, 2021)

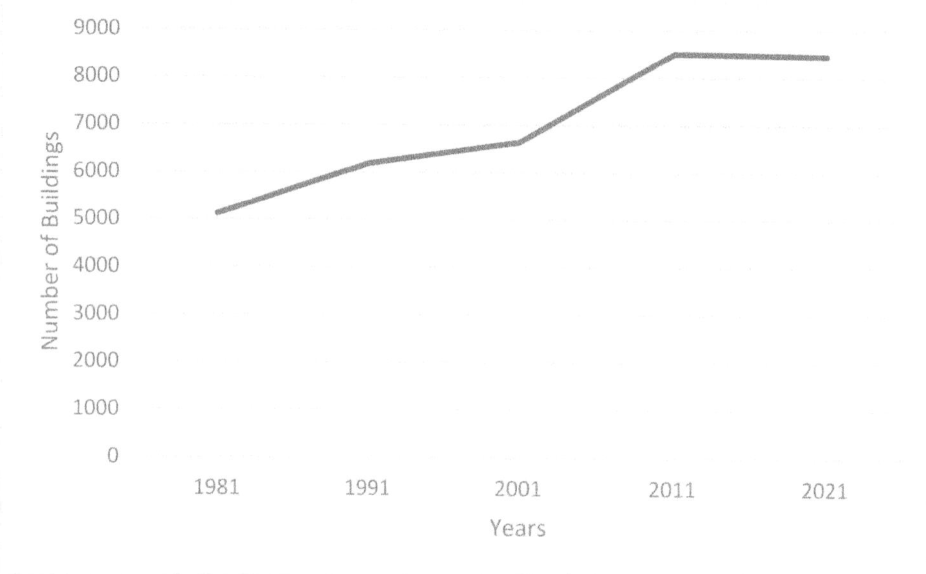

Between 1981 and 1991, the population of the parish of Machico decreased, contrary to what happened in the municipality of Machico, with a variation rate of -2.9%, which represented a reduction of 31 residents. Subsequently, between 2001 and 2021, the population also decreased, losing 2119 residents in 20 years (Figure 3). It is important to note that this trend of population decline in the parish of Machico

Figure 3. Resident population of the Parish of Machico
Source: INE (1981, 1991, 2001, 2011, 2021)

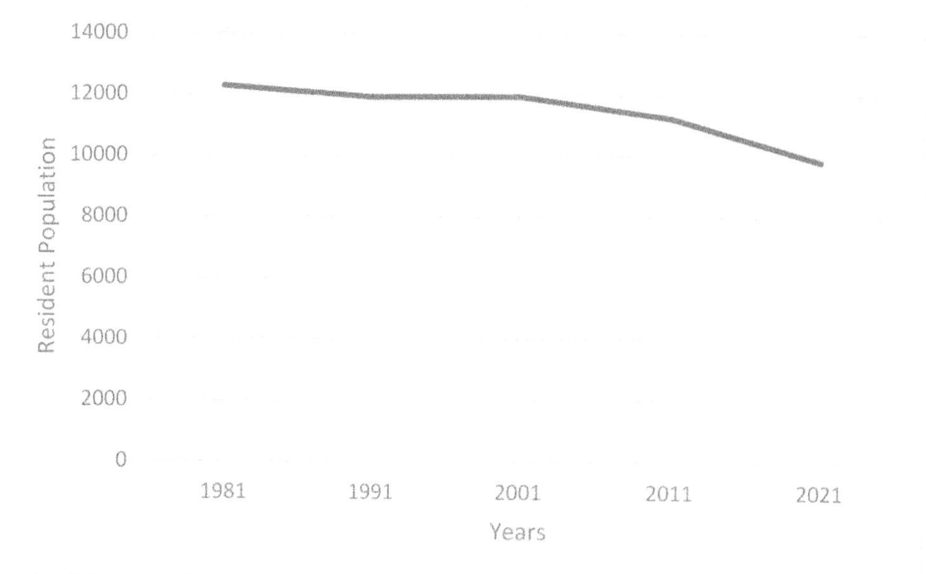

is similar to the trend observed throughout the municipality, which presented a negative variation rate of 11.4% over the last 40 years (1981-2021).

This variation is due to a combination of factors, including the decrease in the natural balance (i.e., the difference between births and deaths) in almost all municipalities in the region, as well as the negative migratory balance in Machico and some other municipalities. In addition, the natural growth rate in the region decreased overall, while the migratory growth rate, although positive, decreased in many municipalities. The proportion of people of working age remained relatively stable, but there was a decrease in the number of marriages and an increase in the number of divorces in some municipalities, which may have impacted the birth rate. All of these factors, along with other socioeconomic factors, contributed to the decrease in population in Machico over time.

Over the 40 years between 1981 and 2021, there was continuous growth in the number of buildings, except for the last decade where there was a 0.5% decrease. However, this decrease did not affect the overall growth rate of 57.1%, which represents an absolute increase of 1524 buildings (Figure 4). This increase is similar to what occurred in the municipality, with a rate of about 65.4%, resulting in 3346 buildings constructed.

Figure 4. Number of buildings in the Parish of Machico
Source: INE (1981, 1991, 2001, 2011, 2021).

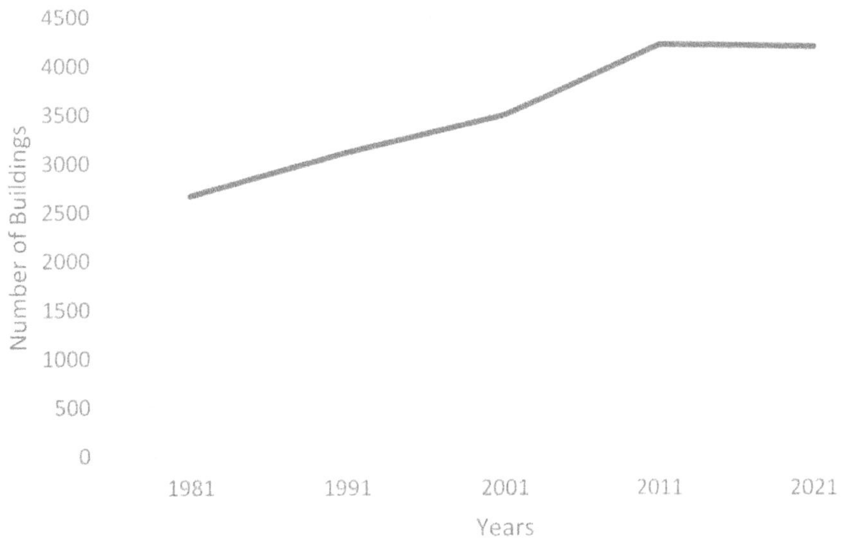

The study area was defined based on the flood zone maps of Ribeira de Machico present in the Flood Risk Management Plan of the Autonomous Region of Madeira PGRI-RAM (RSENR, 2017). This area covers 171.47 hectares and is composed of seven continuous spaces. The area follows the course of the Ribeira de Machico from the parish of Santo António da Serra, to the west, to the sea, to the east (Table 2).

Table 2. Various study areas

Zones	Area (ha)
Baixa de Machico (Figure 1)	30,90
Paraíso (Figure 2)	31,56
Moinho da Serra (Figure 3)	43,77
Ribeira Seca (Figure 4)	29,01
Ribeira Grande (Figure 5)	12,81
Maroços (Figure 6)	8,47
Madeira da Igreja (Figure 7).	14,95

Source: Author

Figure 5. Study area: Baixa de Machico
Source: Adapted from DRAAC (2022)

Between 1991 and 2001, all areas showed a population decrease, ranging from -10% in Baixa de Machico to -34% in Madeira da Igreja. However, between 2001 and 2011, some areas showed a population increase, such as Paraíso (18%) and Moinho da Serra (15%), while others showed a decrease, such as Maroços (-19%). Between 2011 and 2021, all areas showed a population decrease, ranging from -10% in Baixa de Machico to -27% in Maroços (Table 3).

Figure 6. Study area: Paraíso
Source: Adapted from DRAAC (2022)

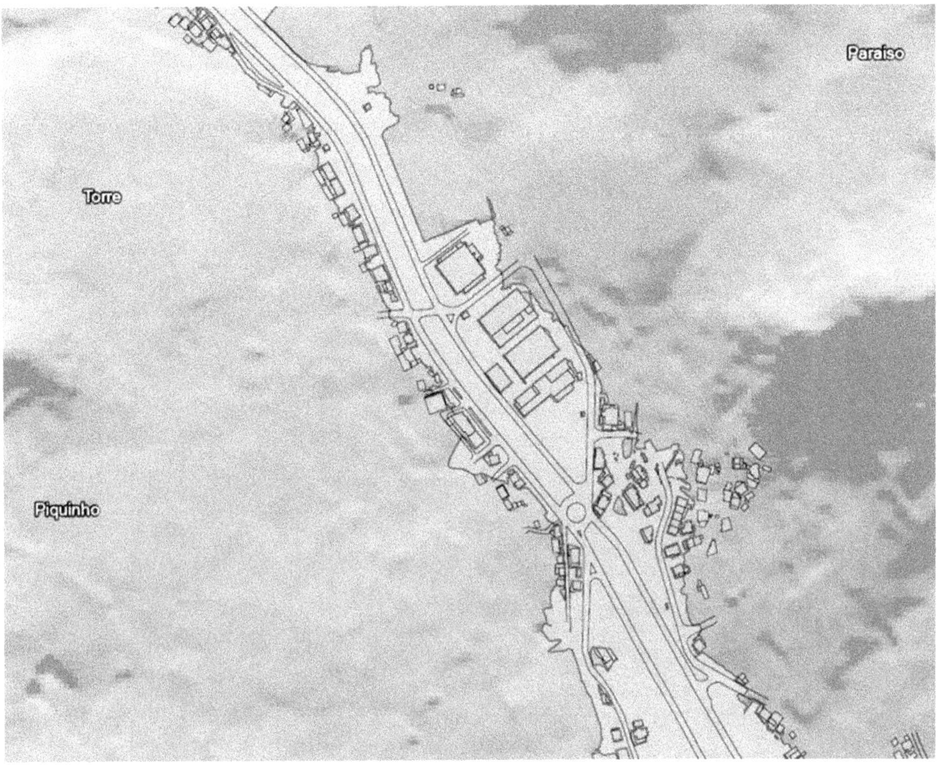

Figure 7. Study area: Moinho da Serra
Source: Adapted from DRAAC (2022)

Figure 8. Study area: Ribeira Seca
Source: Adapted from DRAAC (2022)

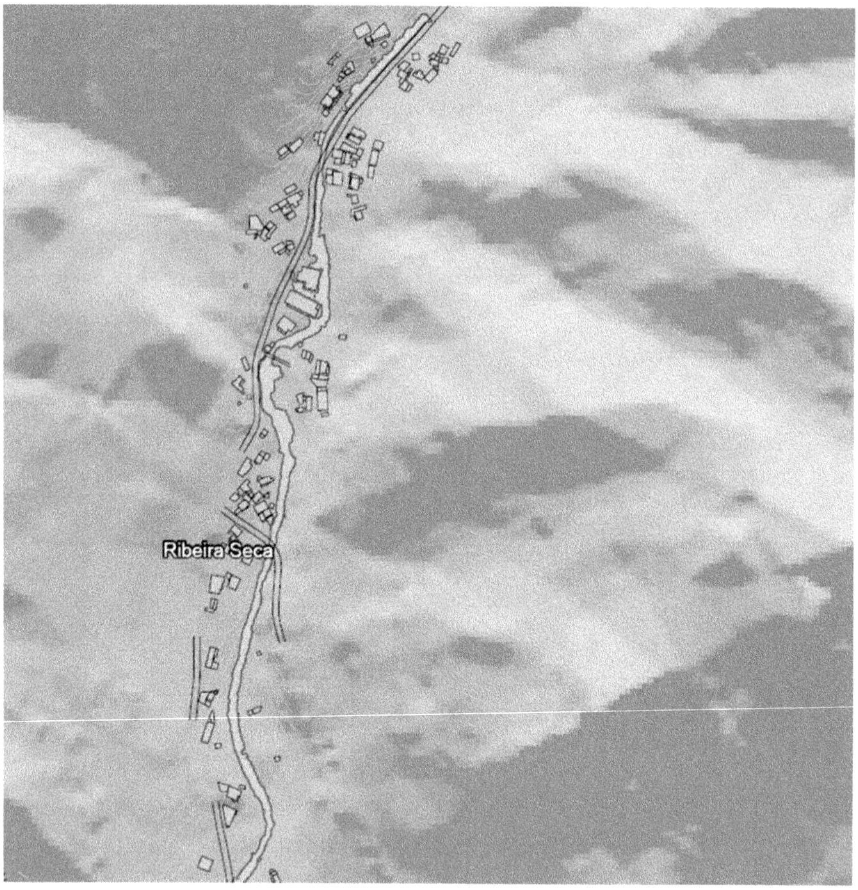

Figure 9. Study area: Ribeira Grande
Source: Adapted from DRAAC (2022)

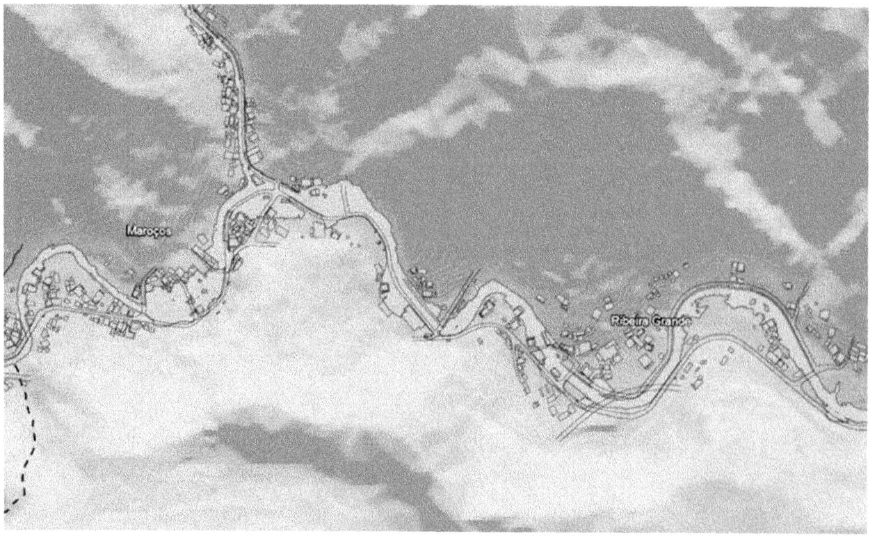

Figure 10. Study area: Maroços
Source: Adapted from DRAAC (2022)

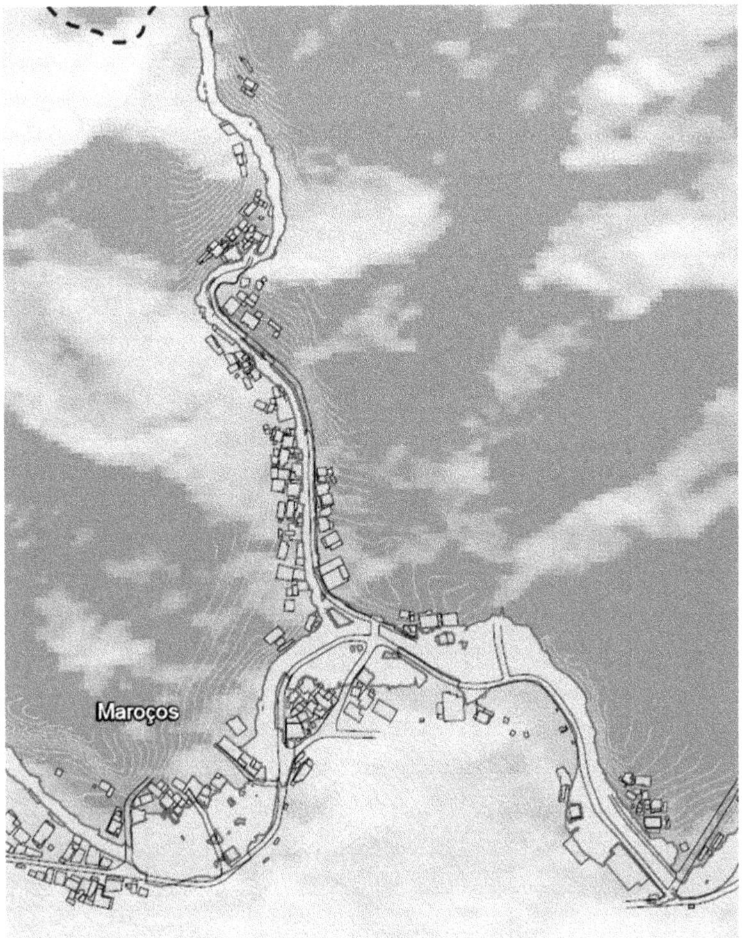

Figure 11. Study area: Madeira da Igreja
Source: Adapted from DRAAC (2022)

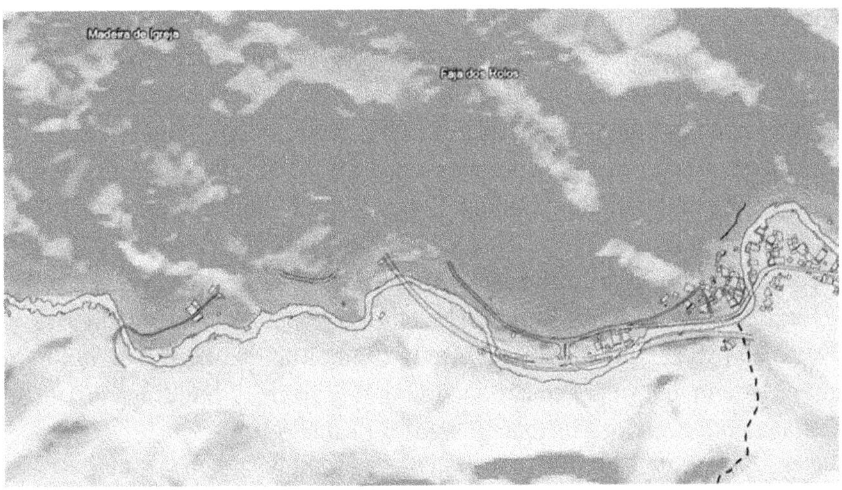

Table 3. Resident population in the study area

Zones	Resident Population			
	1991	2001	2011	2021
Baixa de Machico	1097	983	1033	933
Paraíso	923	848	998	801
Moinho da Serra	631	588	674	569
Ribeira Seca	779	897	836	722
Ribeira Grande	304	249	239	212
Maroços	276	385	312	228
Madeira da Igreja	66	44	40	32

Source: INE (1991, 2001, 2011, 2021)

Table 4. Variation in resident population in the study area

Zones	Resident Population Variation		
	1991-2001	2001-2011	2011-2021
Baixa de Machico	-10%	5%	-10%
Paraíso	-8%	18%	-20%
Moinho da Serra	-7%	15%	-16%
Ribeira Seca	15%	-7%	-14%
Ribeira Grande	-18%	-4%	-11%
Maroços	39%	-19%	-27%
Madeira da Igreja	-34%	-8%	-20%
Median	-8%	-4%	-16%
Machico Municipality	-1,2%	0,4%	-10,2%
Machico Parish	0,3%	-5,8%	-12,7%

Source: INE (1991, 2001, 2011, 2021)

Overall, the population of the municipality of Machico showed a negative variation of 1.2% between 1991 and 2001, an increase of 0.4% between 2001 and 2011, and a reduction of 10.2% between 2011 and 2021. Meanwhile, the parish of Machico showed an increase of 0.3% between 1991 and 2001, a decrease of 5.8% between 2001 and 2011, and a reduction of 12.7% between 2011 and 2021. The median population variations for the areas over the three periods were -8%, -4%, and -16%, respectively. The median population variation coincided with the Municipality of Machico for the period of 1991-2001, with the parish of Machico for the period of 2001-2011, and with both for the period of 2011-2021 (Table 4).

There has been an increase in the number of buildings in all areas between 1991 and 2021, except for Madeira da Igreja. Ribeira Seca and Baixa de Machico showed the largest increases, going from 144 buildings in 1991 to 364 buildings in 2021 and from 182 buildings in 1991 to 342 buildings in 2021. On the other hand, Madeira da Igreja had a reduction, going from 17 buildings in 1991 to 32 buildings in 2021. The other areas showed intermediate increases (Table 5).

Table 5. Number of buildings in the study area

Zones	Number of Buildings			
	1991	2001	2011	2021
Baixa de Machico	182	273	364	342
Paraíso	242	244	247	239
Moinho da Serra	187	231	241	194
Ribeira Seca	144	286	366	364
Ribeira Grande	79	119	172	85
Maroços	61	111	124	92
Madeira da Igreja	17	17	18	16

Source: INE (1991, 2001, 2011, 2021)

Between 1991 and 2011, there was a significant increase in the number of buildings in all areas except for Paraíso and Madeira da Igreja, with Ribeira Seca showing the largest increase of 154%. On the other hand, the Madeira da Igreja zone had a decrease of 11% between 2011 and 2021, despite having increases in the previous periods. The other zones had intermediate increases, ranging from 4% to 88%. Between 1991 and 2021, the median of the percentage variations in the three periods was 50%, 34%, and -11%, respectively. Overall, the Machico municipality had an increase of 7.3% in the number of buildings between 1991 and 2001, an increase of 28.2% between 2001 and 2011, and a slight reduction of 0.5% between 2011 and 2021. The Machico parish had an increase of 12.2% between 1991 and 2001, an increase of 20.7% between 2001 and 2011, and a slight reduction of 0.5% between 2011 and 2021. The median of the variations coincided with the Machico municipality and the Machico parish for the periods of 1991-2001, 2001-2011, and 2011-2021 (Table 6).

Table 6. Variation in the number of buildings in the study area

Zones	Variation in the Number of Buildings		
	1991-2001	2001-2011	2011-2021
Baixa de Machico	50%	34%	-6%
Paraíso	1%	1%	-3%
Moinho da Serra	23%	4%	-20%
Ribeira Seca	98%	28%	-1%
Ribeira Grande	51%	45%	-51%
Maroços	81%	12%	-26%
Madeira da Igreja	1%	6%	-11%
Median	50%	12%	-11%
Machico Municipality	7,3%	28,2%	-0,5%
Machico Parish	12,2%	20,7%	-0,5%

Source: INE (1991, 2001, 2011, 2021)

Through the presented graphs and tables, it can be seen that the resident population in the municipality of Machico decreased slightly from 22,126 in 1981 to 19,593 in 2021, with negative variations of -0.5%, -1.2%, 0.4%, and -10.2% between the periods of 1981-1991, 1991-2001, 2001-2011, and 2011-2021, respectively. In the parish of Machico, the resident population decreased from 12,278 in 1981 to 9,828 in 2021, with negative variations of -2.9%, 0.3%, -5.8%, and -12.7% between the periods of 1981-1991, 1991-2001, 2001-2011, and 2011-2021, respectively. In the case of the Municipality and Parish of Machico, the reduction in the resident population is related to migration abroad, population aging, and a decrease in the birth rate. Other factors that also influence this decline are changes in economic activities and job availability. Unlike the resident population, which showed negative variations in some periods, the number of buildings in the area of the Municipality and Parish of Machico has shown a constant increase over the years. According to the presented data, the number of buildings in the Municipality increased by 20.8% between 1981 and 1991, 7.3% between 1991 and 2001, 28.2% between 2001 and 2011, and practically remained stable with a slight reduction of 0.5% between 2011 and 2021. In the Parish of Machico, the number of buildings increased by 16.7% between 1981 and 1991, 12.2% between 1991 and 2001, 20.7% between 2001 and 2011, and also practically remained stable with a slight reduction of 0.5% between 2011 and 2021. The National Institute of Statistics provided data on the number of buildings and resident population for the years 1991, 2001, 2011, and 2021, at the level of statistical subsections. It was then decided to spatialize this data in the study area, using the same scale for the three years and analyzing the modifications that occurred over 30 years (INE, 2021). However, when overlaying the study area on the statistical subsections, it was realized that the boundaries were not coincident (Figure 12, Figure 13, Figure 14, and Figure 15). For the subsections that exceeded the limits of the study area, the percentage of representation in the study area was calculated, and it was decided to eliminate all statistical subsections whose representativeness was less than 10%, since statistical subsections represented in less than 10% in the study area have no expression in cartographic production (Gouveia & Lourenço, 2017).

After calculating the percentage of representation of each statistical subsection in the study area, that same percentage was applied to the total number of inhabitants and buildings in the statistical subsection. When the result obtained was less than one building, that record was eliminated, since without buildings there are no residents. In this way, the existence of data-free spaces was explained, which do not appear represented in the cartography (Gouveia & Lourenço, 2017).

Table 7 presents data on the resident population and the number of buildings in the Study Zone, Machico Municipality and Machico Parish, for the years 1991, 2001, 2011, and 2021. It is possible to note that over these thirty years, there were significant variations in the numbers presented.

Regarding the resident population in the Study Zone, there was a reduction between 1991 and 2021, from 4076 to 3497 inhabitants. However, it is important to note that in some sub-regions, there was an increase in the population, such as in Machico Parish in 2011.

Regarding the number of buildings, there was a general increase over the analyzed period, with emphasis on Machico Parish, which had an increase of more than 1000 buildings between 1991 and 2021. It is important to emphasize that this increase may be related to the urban expansion of the study area, which may have influenced both the construction of new buildings and the increase of the resident population in some sub-regions.

In 1991, the study area of Machico housed a total population of 4,076 inhabitants, which corresponded to 6.1% of the resident population in the municipality of Machico and 34.3% of the resident population in the parish of Machico. The population density was 322 inhabitants per km^2. As shown in Figure 16,

Figure 12. Limits of Baixa de Machico 1991
Source: Authors

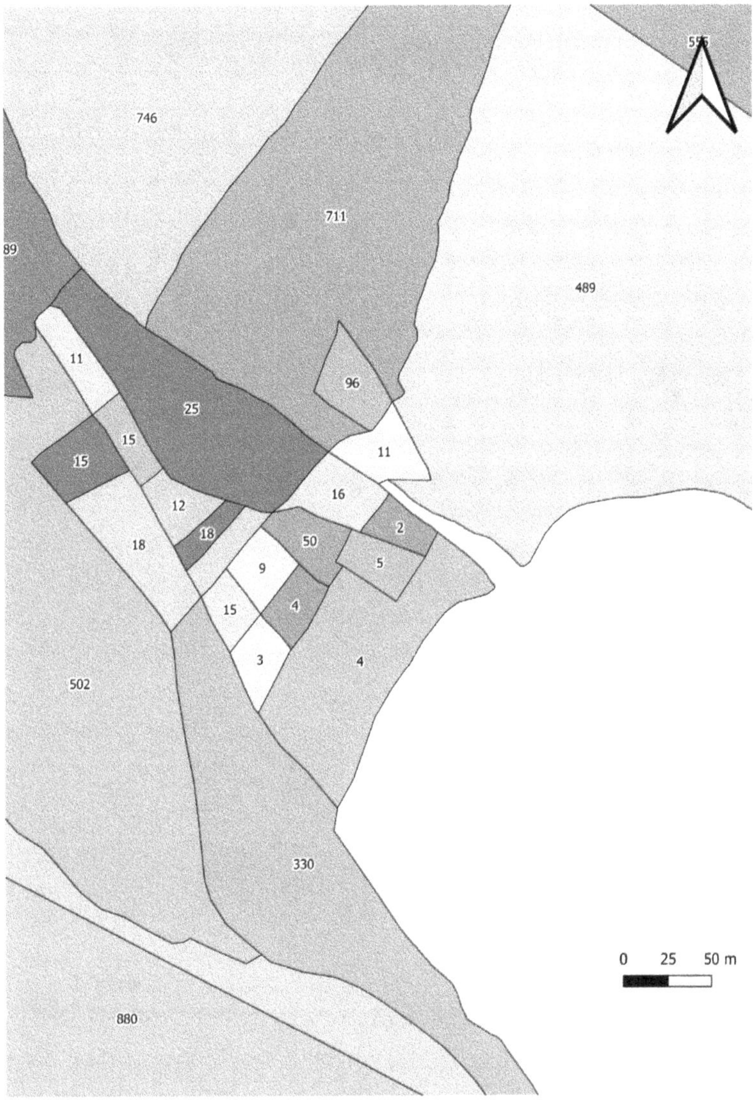

the area with the highest number of inhabitants was Baixa de Machico, with a total of 1,097 residents. Regarding the number of buildings, the area with the highest quantity was Paraíso, with a total of 242 buildings (Figure 17).

In 2001, the total population in the study area of Machico was 3,994 inhabitants, which represented about 5.9% of the resident population in the municipality of Machico and about 33.6% of the resident population in the parish of Machico. The population density was 318 inhabitants per km2. As in 1991, in 2001, out of the 3,994 inhabitants and 1,280 buildings in the study area of Machico, about 983 inhabitants and 286 buildings were concentrated in Baixa de Machico and Ribeira Seca, respectively (Figure 18 and Figure 19).

Figure 13. Limits of Baixa de Machico 2001
Source: Authors

In the years 1991 and 2001, the statistical subsections presented differences in their configuration, allowing for a comparison of the variation in the location of the resident population and buildings during that period. However, the cartography of the statistical subsections for the years 2011 and 2021 are significantly different from the previous ones, making comparative analysis difficult. Therefore, separate analysis was conducted for these two years (2011 and 2021).

In 2011, the area with the highest population concentration continued to be the same as in 1991 and 2001. In that year, the resident population in the study area was 4,132 inhabitants, which corresponded to 19.0% of the total population of Machico municipality and 37.0% of the population of Machico parish. The population density was 320 inhabitants/km2 (Figure 20). As for the concentration of buildings, in 2011 it remained in the same area as in previous years, Ribeira Seca (Figure 21), where 366 buildings were concentrated.

Figure 14. Limits of Baixa de Machico 2011
Source: Authors

Analyzing the year 2021, it can be seen that the largest concentration of population still corresponds to the same area as in 1991, 2001, and 2011. The total population residing in the study area was 3,497 inhabitants, representing 18.0% of the resident population in the Machico municipality and 36.0% of the resident population in the Machico parish. The population density of the area in 2021 was 287 inhabitants/km^2 (Figure 22).

As in previous years, the largest concentration of buildings in 2021 is located in the Ribeira Seca area (Figure 23), with about 364 buildings, remaining identical to 1991, 2001, and 2011.

It was decided to create a single polygon to map the configuration of the urban area in the study area, taking into account that soil impermeability is a result of the urbanization process. This polygon encompasses both buildings (visible on the map in brown color) and existing roads (visible on the map

Figure 15. Limits of Baixa de Machico 2021
Source: Authors

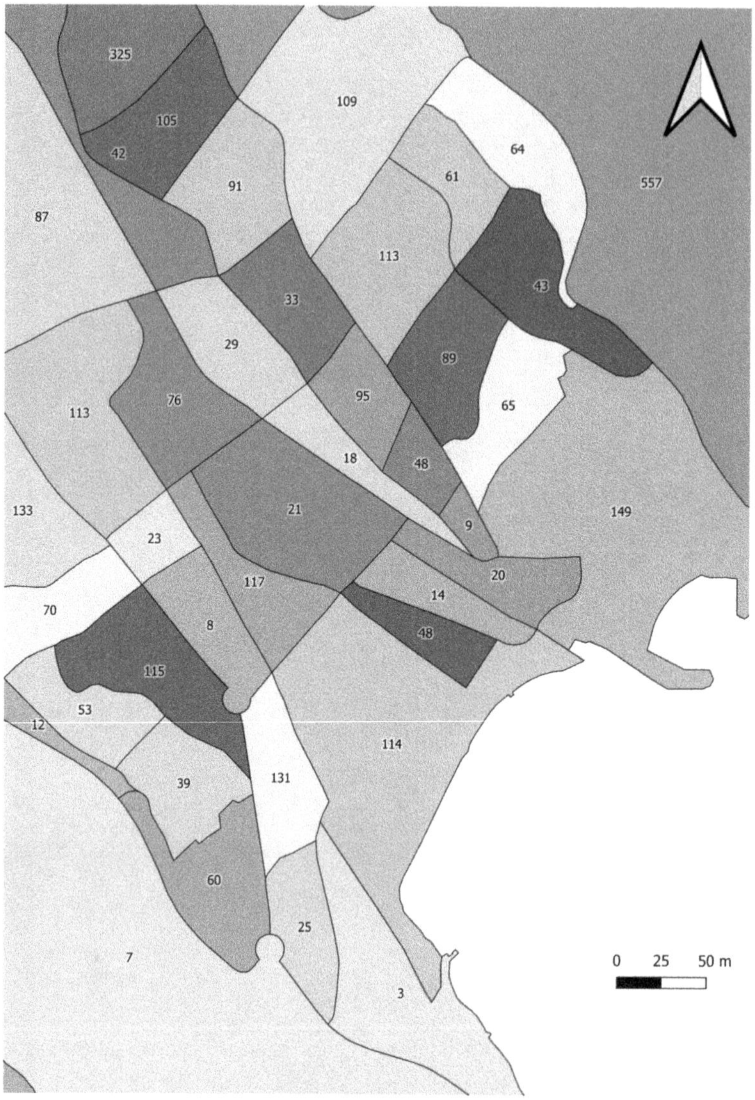

in red color) at two different time periods. To delimit the impermeable areas, the available cartography of the city of Machico was used, which included 1989 raster images at a scale of 1/1000, as well as orthophotomaps from 2018.

Therefore, it is possible to obtain the value of the increase in impermeable area and monitor the evolution of the urbanization process. With this, it is possible to identify which areas have become impermeable and the impact that this may have in case of a flood (visible on the map in gray color). These pieces of information can be visualized in Figures 23 and 24.

Table 8 shows that the delimitation of impermeable areas in Machico allowed us to verify that, in 1990, about 70.09 hectares of area were impermeable, a value that increased to 96.43 hectares in 2018, representing a 38% increase. When analyzing the variation by zones, it can be observed that Baixa de

Table 7. Evolution of the resident population and number of buildings in the municipality, parish, and study area of Machico variation of the number of buildings in the study area

Years	Study Area		Municipality		Parish	
	Population	Buildings	Population	Buildings	Population	Buildings
1991	4076	912	22016	6183	11916	3114
2001	3994	1280	21747	6637	11947	3494
2011	4132	1532	21828	8510	11256	4216
2021	3497	1332	19593	8464	9828	4193

Source: INE (1991, 2001, 2011, 2021)

Figure 16. Resident population in the study area in 1991
Source: Authors

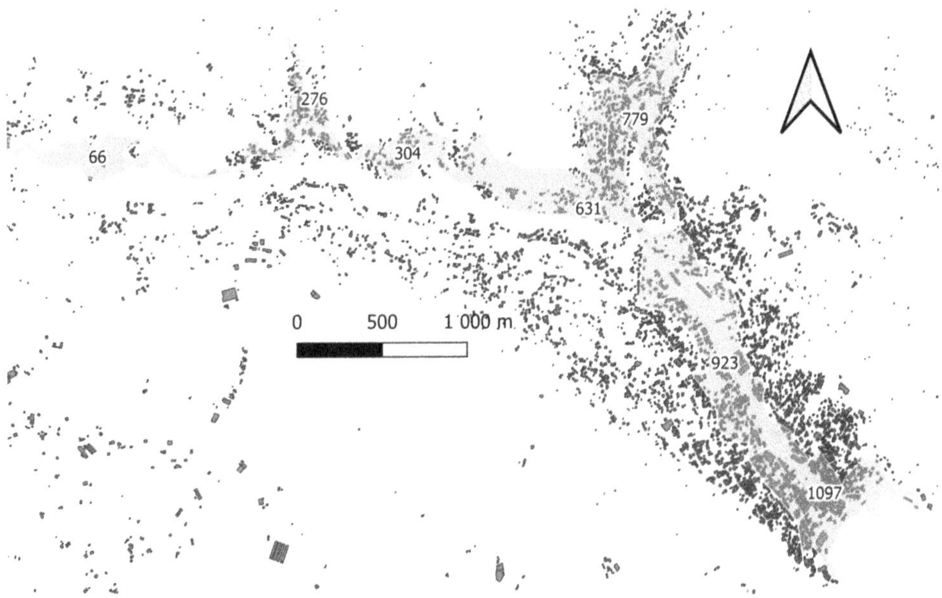

Machico had a 4% (0.72 ha) increase in impermeable areas, going from 17.87 ha in 1990 to 18.59 ha in 2018. In the Paraíso zone, a significant increase of 58% (7.75 ha) in impermeable areas was verified, going from 13.395 ha in 1990 to 21.14 ha in 2018. The Moinho da Serra zone showed a 56% (7.98 ha) increase, going from 14.34 ha in 1990 to 22.32 ha in 2018. In Ribeira Seca, there was a 28% (4.78 ha) increase, going from 17.15 ha in 1990 to 21.93 ha in 2018. The Ribeira Grande zone showed a 57% (2.17 ha) increase, going from 3.83 ha in 1990 to 6.00 ha in 2018. In the Maroços zone, there was a 58% (1.68 ha) increase, going from 2.91 ha in 1990 to 4.59 ha in 2018. In Madeira da Igreja, a significant increase of 215% (1.27 ha) was verified, going from 0.59 ha in 1990 to 1.86 ha in 2018. In total, the impermeable areas in Machico in 2018 corresponded to 56% of the total area of the study zone, representing a significant increase in urbanized areas in the region. This increase can have negative implications in case of floods, considering the reduction in soil infiltration capacity and the increase in surface runoff.

Figure 17. Number of buildings in the study area in 1991
Source: Authors

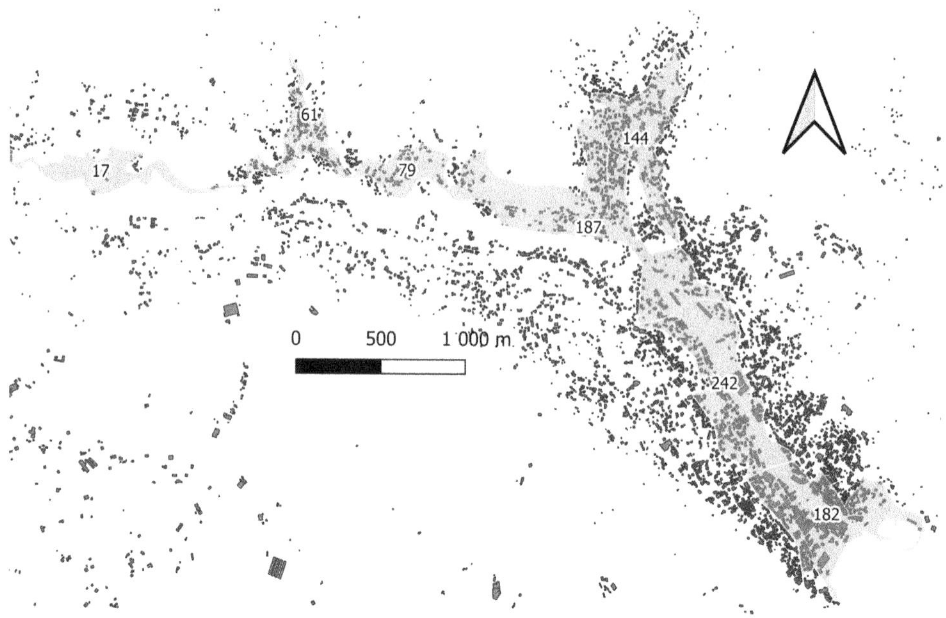

Figure 18. Resident population in the study area in 2001
Source: Authors

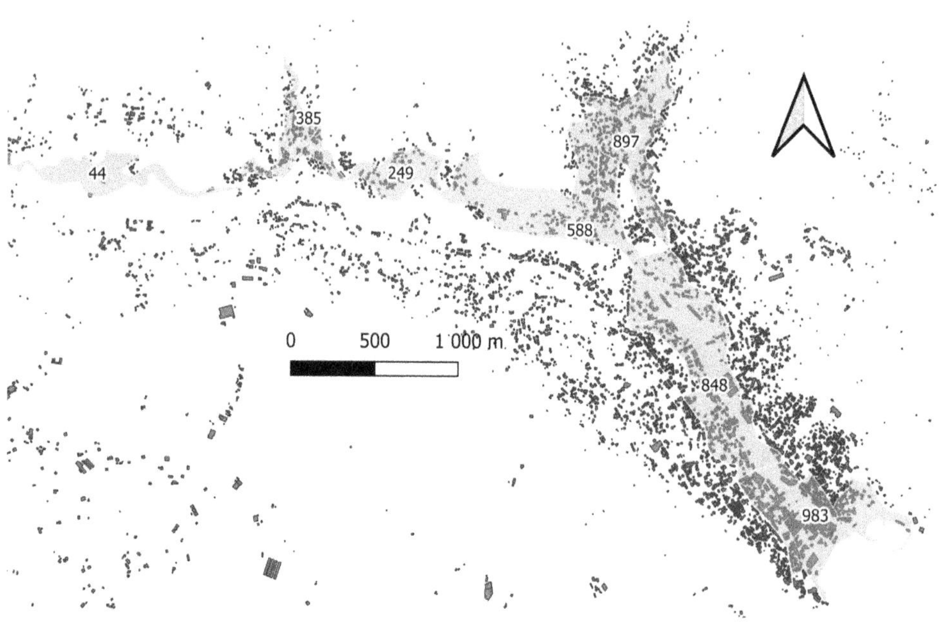

Figure 19. Number of buildings in the study area in 2001
Source: Authors

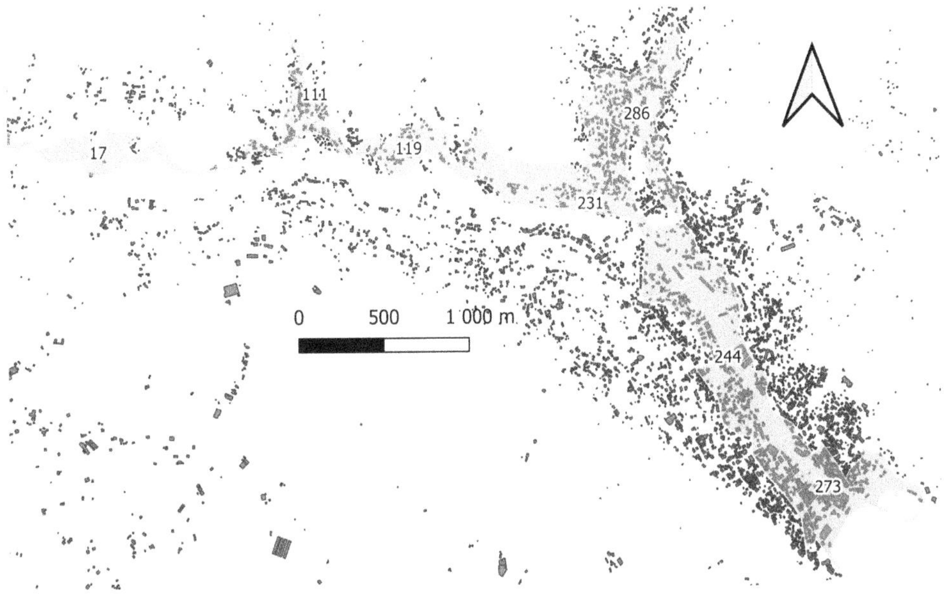

Figure 20. Resident population in the study area in 2011
Source: Authors

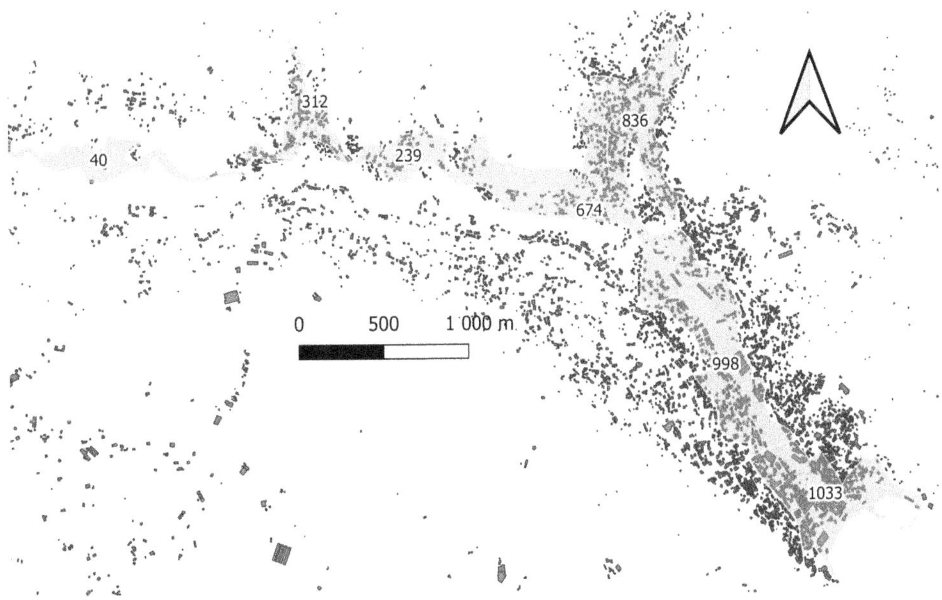

Figure 21. Number of buildings in the study area in 2011
Source: Authors

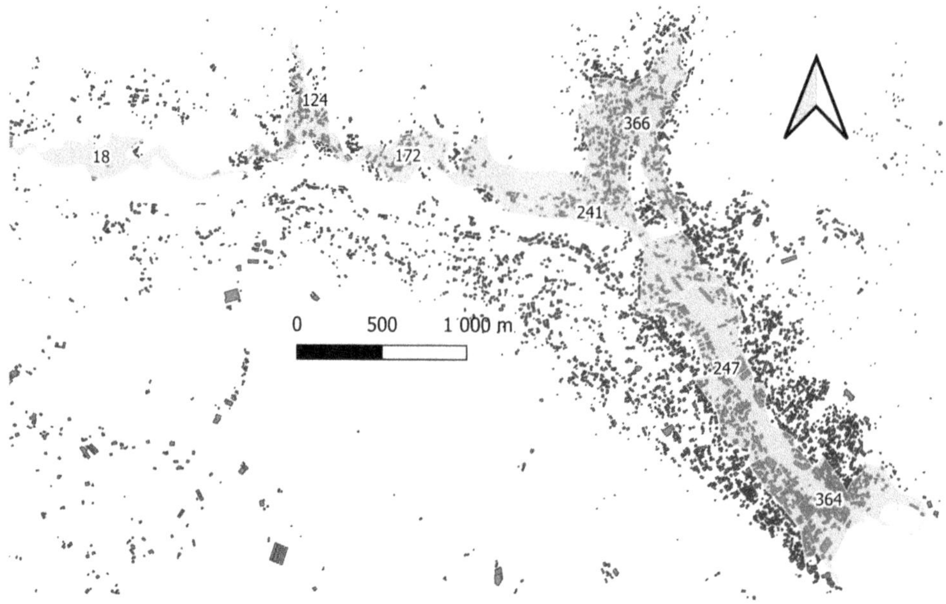

Figure 22. Resident population in the study area in 2021
Source: Authors

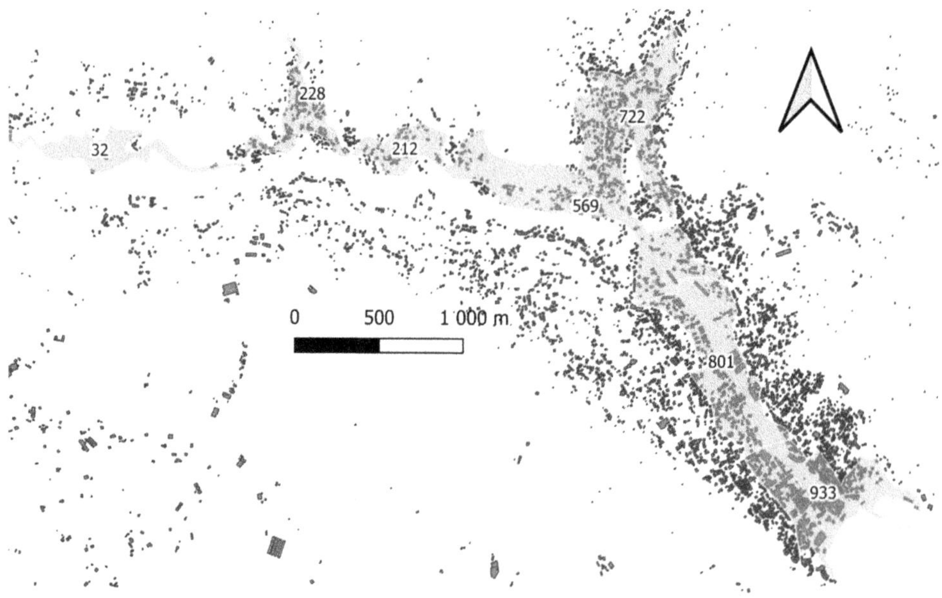

Figure 23. Number of buildings in the study area in 2021
Source: Authors

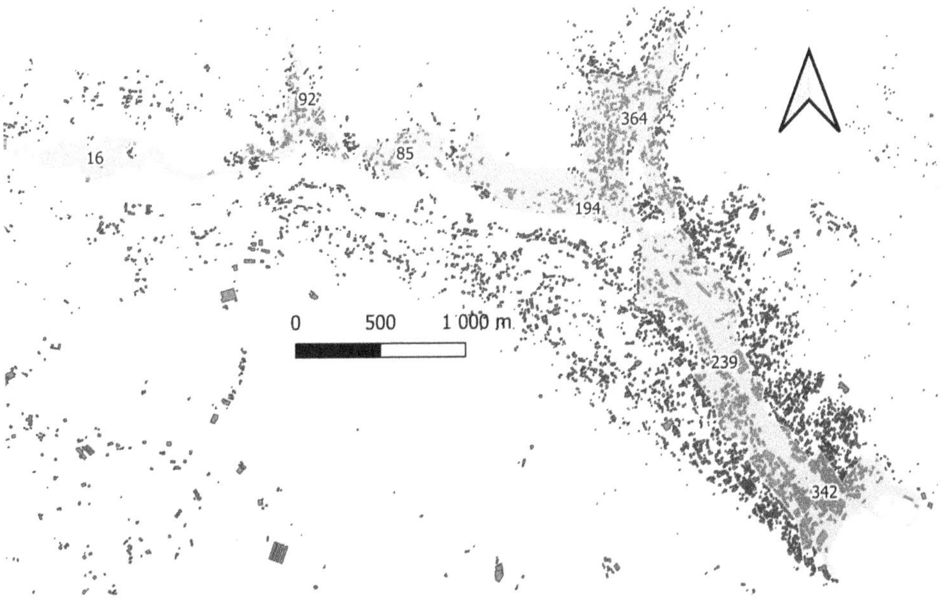

Figure 24. Impermeable areas in the study area in the 1990s

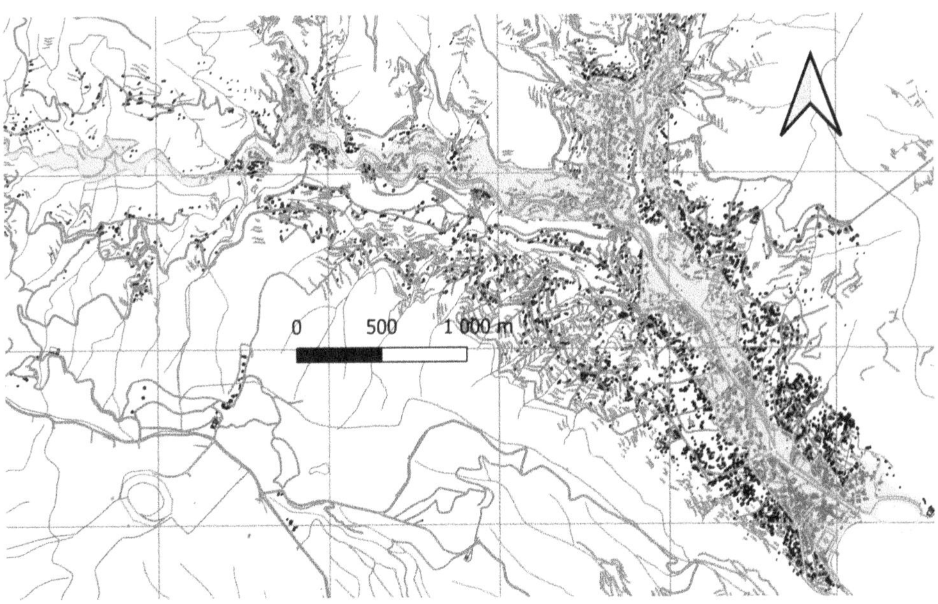

Figure 25. Impermeable areas in the study area in 2018

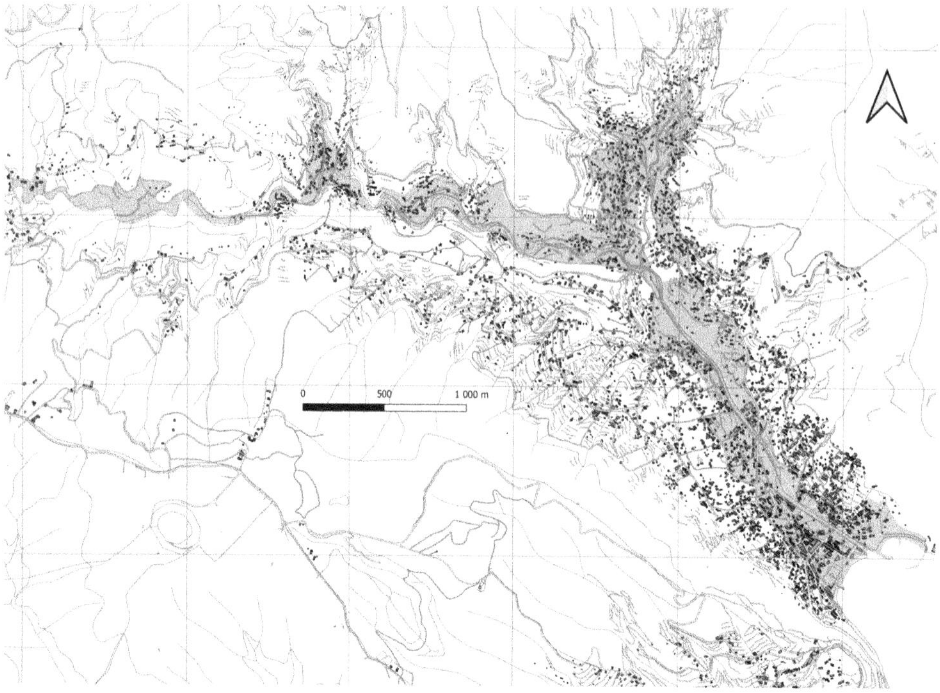

Table 8. Evolution of soil sealing in the study areas

Zones	Area (ha)	Sealed Area					
		Year1990		Year 2018		Difference	Variation
		Area (ha)	Area as a Proportion of the Zone (ha)	Area (ha)	Area as a Proportion of the Zone (ha)	1990-2018	1990-2018
Baixa de Machico	30,90	17,87	58%	18,59	60%	0,72	4%
Paraíso	31,56	13,395	42%	21,14	67%	7,75	58%
Moinho da Serra	43,77	14,34	33%	22,32	51%	7,98	56%
Ribeira Seca	29,01	17,15	59%	21,93	76%	4,78	28%
Ribeira Grande	12,81	3,83	30%	6,00	47%	2,17	57%
Maroços	8,47	2,91	34%	4,59	54%	1,68	58%
Madeira da Igreja	14,95	0,59	4%	1,86	12%	1,27	215%
Total	171,47	70,09	41%	96,43	56%	26,35	38%

Source: Authors

CONCLUSION

After analyzing the impact of urbanization on flood risk in the Machico River area, it is evident that the increase in impermeable surfaces, such as buildings and paved areas, has reduced the water infiltration capacity, leading to increasingly violent floods. Soil impermeability and obstacles such as buildings also worsen the situation by preventing rapid water runoff.

Although the local authorities have taken steps to address these issues, including canalization of the river and emergency planning, further measures are necessary to minimize the risks. It is recommended to adopt specific mitigation measures tailored to the unique characteristics of Machico, such as identifying retention capacity areas, informing stakeholders, implementing territorial interventions, and developing insurance plans. The implementation of these measures will contribute to more effective protection of people, animals, and heritage assets in the city of Machico.

Close collaboration between local authorities, stakeholders, and protection agencies is essential to developing comprehensive emergency and civil protection plans that include detailed risk mapping and the allocation of resources for flood management and mitigation. Furthermore, the implementation of land-use planning policies, such as PMOT and PDM, can regulate construction activities and limit the expansion of impermeable surfaces in flood-prone areas.

In addition to these measures, it is important to raise public awareness about flood risks and promote community engagement in emergency planning and response efforts to reduce the impact of floods in Machico. Investment in early warning systems for floods, including technology such as sensors, satellite imagery, and data analysis tools, is essential for providing timely and accurate information about flood events, allowing the authorities to take preventive measures and reduce the risk of casualties and damage to property.

Finally, it is crucial to recognize that flood risks are dynamic and can change over time due to climate change, urbanization, and other factors. Therefore, regular updates and reviews of flood risk assessments and emergency plans are necessary to ensure their continued relevance and effectiveness in addressing evolving risks in the Machico River area.

REFERENCES

Almada, J.A. (1865). Annaes of the Municipality of the Old Villa de Machico. *Flor do Oceano,* 247-286.

Barreto, V. H. (2013). *Study of the Hydrographic Basin of Ribeira de Machico – Vulnerability and Floods* [Master's dissertation]. University of Madeira.

Chow, V. T. (1956). Hydrologic Studies of Floods in the United States. *Inter. Assoc. Sci. Hydrol. Publ,* (42), 134–170.

Costa, F. S. (2009a). Flood risk in the city of Amarante (Northern Portugal): methodological contribution to its study. *Territorium, 16.* Available at: https://www.uc.pt/fluc/nicif/riscos/Documentacao/Territorium/T16_artg/T16art09.pdf

Costa, F. S. (2009b). Flood risk in the city of Amarante (Northern Portugal): methodological contribution to its study. *Territorium, 16.* Available at: https://repositorium.sdum.uminho.pt/handle/1822/22631

Cunha, S., & Taveira-Pinto, F. (2011). Application of a flood risk analysis methodology to the riverside area of Pesa da Régua. In *6th Hydraulic, Water Resources and Environment Conference*. FEUP.

Deltares. (2010). *Flood Risk Management*. Author.

Direção Regional do Ambiente e Alterações Climáticas - DRAAC. (2022). Flood Zone Maps and Flood Risk Maps. PGRI-RAM_2022-2027. Funchal.

EEA. (2005). Climate Change and Floods in Europe. European Environment Agency.

European Directive No. 2007/60/EC, of October 23, transposed into Decree-Law No. 115/2010, of October 22.

Freitas, M., & Andrade, C. (2007). *Climate Change and Impacts on the Coastline. XXIX Teacher Training Course in Geosciences*. Higher School of Education of Lisbon.

Geral dos Recursos, D., & Hidráulicos, A. (1984). International Course on Operational Hydrology (Vol. 2). Academic Press.

Gouveia, M., & Lourenço, L. (2017). Urban Flooding Risk in Mirandela City and Planning its Territory as a Form of Prevention. *Territorium, 24*, 249–256. doi:10.14195/1647-7723_24_18

Grothmann, T., & Reusswig, F. (2006). People at risk of flooding: Why some residents take precautionary action while others do not. *Natural Hazards, 38*(1-2), 101–120. doi:10.100711069-005-8604-6

Guedes, L. J. S. (2014). *Risks of Floods and Inundations. Case Study: Esposende. Paper written for the Environmental Geology course, BSc in Geology - UMinho, 2014/2015*. University of Minho.

INE - National Institute of Statistics, I. P. (1981). *Census 1981, Definitive Results for the Autonomous Region of Madeira*. Author.

INE - National Institute of Statistics, I. P. (1991). *Census 1991, Definitive Results for the Autonomous Region of Madeira*. Author.

INE - National Institute of Statistics, I. P. (2001). *Census 2001, Definitive Results for the Autonomous Region of Madeira*. Author.

INE - National Institute of Statistics, I. P. (2011). *Census 2011, Definitive Results for the Autonomous Region of Madeira*. Author.

INE - National Institute of Statistics, I. P. (2021). *Census 2021, Definitive Results for the Autonomous Region of Madeira*. Author.

Leandro, J., Chen, K. F., Wood, R. R., & Ludwig, R. (2020). A scalable flood-resilience-index for measuring climate change adaptation: Munich city. *Water Research, 173*, 115502. doi:10.1016/j.watres.2020.115502 PMID:32028251

Lousada, S., Cabezas, J., Castanho, R. A., & Gómez, J. M. N. (2021). Hydraulic Planning in Insular Urban Territories: The Case of Madeira Is-land—Ribeira Brava. *Water (Basel), 2021*(13), 2951. doi:10.3390/w13212951

Lousada, S., & Castanho, R. A. (2021). GIS-Based Assessment of Morphological and Hydrological Parameters of Ribeira Dos Socorridos and Ribeira Do Vigario Basins, Madeira Island, Portugal. *Current World Environment, 2021*(16), 408–426. doi:10.12944/CWE.16.2.08

Lousada, S., Gonçalves, L., & Atmaca, A. (2022). Hydraulic Planning in Insular Urban Territories: The Case of Madeira Island—São Vicente. *Water (Basel), 14*(1), 112. doi:10.3390/w14010112

Meyer, R. J., Baker, J., Broad, K., Czajkowski, J., & Orlove, B. (2014). The dynamics of hurricane risk perception: Realtime evidence from the 2012 Atlantic hurricane season. *Bulletin of the American Meteorological Society, 95*(9), 1389–1404. doi:10.1175/BAMS-D-12-00218.1

Miller, J. D., & Hutchins, M. (2017). The impacts of urbanization and climate change on urban flooding and urban water quality: A review of the evidence concerning the United Kingdom. *Journal of Hydrology. Regional Studies, 12*, 345–362. doi:10.1016/j.ejrh.2017.06.006

Mobini, S., Becker, P., Larsson, R., & Berndtsson, R. (2020). Systemic inequity in urban flood exposure and damage compensation. *Water (Basel), 12*(11), 3152. doi:10.3390/w12113152

Oulahen, G., & Doberstein, B. (2012). Citizen participation in post-disaster flood hazard mitigation planning in Peterborough, Ontario, Canada. *Risk, Hazards & Crisis in Public Policy, 3*(1), 1–26. doi:10.1515/1944-4079.1098

Owusu, S., Wright, G., & Arthur, S. (2015). Public attitudes towards flooding and property-level flood protection measures. *Natural Hazards, 77*(3), 1963–1978. doi:10.100711069-015-1686-x

Pagliacci, F., Defrancesco, E., Bettella, F., & D'Agostino, V. (2020). Mitigation of urban pluvial flooding: What drives residents' willingness to implement green or grey stormwater infrastructures on their property? *Water (Basel), 12*(11), 3069. doi:10.3390/w12113069

Public Journal. (2012). *Four dozen displaced people and roads closed in Madeira*. Author.

Quintal, R. (1999). Aluviões da Madeira. Séculos XIX e XX. *Territorium*, (6), 31–48. doi:10.14195/1647-7723_6_4

Radio News TSF. (2021). *Tree falls, landslides, and floods. Heavy rains promise not to give truce in Madeira*. Author.

Ramos, C. (2009). *Fluvial dynamics and land use planning*. CEG-UL.

Regional Secretariat for the Environment and Natural Resources – Regional Directorate for Spatial Planning and Environment. (2016). *Flood Risk Management Plan for the Autonomous Region of Madeira*. Author.

Sampson, N. R., Price, C. E., Kassem, J., Doan, J., & Hussein, J. (2019). "We're just sitting ducks": Recurrent household flooding as an underreported environmental health threat in Detroit's changing climate. *International Journal of Environmental Research and Public Health, 16*(1), 6. doi:10.3390/ijerph16010006 PMID:30577470

Sandink, D. (2013). Urban Flooding in Canada: Lot-Side Risk Reduction Through Voluntary Retrofit Programs, Code Interpretation and By-Laws. Institute for Catastrophic Loss Reduction, ICLR research paper series – number 52.

Silva, F. A., & Menezes, C. A. (1921). *Madeiran Elucidário. 1st Volume*. Esperança Typography.

Skougaard Kaspersen, P., Høegh Ravn, N., Arnbjerg-Nielsen, K., Madsen, H., & Drews, M. (2017). Comparison of the impacts of urban development and climate change on exposing European cities to pluvial flooding. *Hydrology and Earth System Sciences*, *21*(8), 4131–4147. doi:10.5194/hess-21-4131-2017

STI, ONE, & LREC. (2010). *Alluvial risk assessment study on the island of Madeira*. Academic Press.

Teles, V. M. B. (2010). *(Un)Awareness of Natural Disasters in Urban Areas. Case Study: the flood Risk in Braga's municipality*. Universidade do Minho.

Terpstra, T., Lindell, M. K., & Gutteling, J. M. (2009). Does communicating (flood) risk affect (flood) risk perceptions? Results of a quasi-experimental study. *Risk Analysis*, *29*(8), 1141–1155. doi:10.1111/j.1539-6924.2009.01252.x PMID:19558390

Zhu, Z., & Chen, X. (2017). Evaluating the effects of low impact development practices on urban flooding under different rainfall intensities. *Water (Basel)*, *9*(7), 548. doi:10.3390/w9070548

KEY TERMS AND DEFINITIONS

Floodings: Flooding is the act or process of water overflowing from its natural channels or boundaries, causing water to spread over usually dry land areas, often causing damage or inconvenience.

Floods: Floods refer to the overflow of water beyond its normal or natural confines, covering usually dry land areas.

Heritage: Heritage refers to cultural or natural assets that are considered to have significant value and importance for present and future generations. These assets can include sites, monuments, buildings, landscapes, or other tangible or intangible elements that represent a society's past and contribute to its identity.

Machico: Portuguese municipality on the island of Madeira, Autonomous Region of Madeira, where Tristão Vaz Teixeira and João Gonçalves Zarco landed in 1419 when they discovered the island of Madeira.

Madeira: Madeira - Madeira is a Portuguese archipelago located in the Atlantic Ocean, consisting of two inhabited islands, Madeira and Porto Santo.

Urbanization: Urbanization is the process by which an increasing proportion of a population lives in urban areas, characterized by high population density, infrastructure, and economic activity. It involves the growth and expansion of cities and urban centers, often accompanied by changes in land use, social structure, and cultural norms.

Chapter 4
Port of Machico, Madeira:
Assessment of the Consequences of Overtopping Using AHP Methodology

Raul Manuel Costa Alves
https://orcid.org/0000-0002-3680-6626
University of Extremadura, Spain

Sérgio António Neves Lousada
https://orcid.org/0000-0002-8429-2164
University of Madeira, Portugal

José Cabezas
https://orcid.org/0000-0002-7209-3731
University of Extremadura, Spain

José Manuel Naranjo Gómez
https://orcid.org/0000-0001-7998-9154
University of Extremadura, Spain

ABSTRACT

The Port of Machico, located on the Madeira Island, is an important infrastructure for the local economy and serves as a gateway for various commercial activities. However, the port is prone to overtopping events, which can result in damage to the infrastructure, disruption of activities, and harm to the environment. Therefore, it is necessary to evaluate the consequences of overtopping and identify measures that can reduce the impact of these events. The methodology proposed in this chapter uses the analytic hierarchy process (AHP) to evaluate the impact of overtopping events at the Port of Machico. The AHP is a multicriteria decision-making tool that allows the evaluation of alternatives based on a set of criteria and their respective weights. In this case, the criteria are related to the consequences of overtopping, such as economic losses, environmental impacts, and disruption of activities. The results obtained from the AHP can then be used to identify measures that can reduce the impact of overtopping, such as the construction of protective structures.

DOI: 10.4018/978-1-6684-6449-6.ch004

INTRODUCTION

A small sheltered port with a capacity for 80 small to medium-sized vessels, with a limit of twelve tons, was built in the bay of Machico city, located on the east coast of Madeira. It is situated between two Sites of Community Importance, namely, the Pico do Facho SCI and the Machico-Queimada SCI. There is also the more recent Madeira SCI that covers all coastal marine waters around the islands of the archipelago, the Cetaceans SCI (CMM, 2022). The port is highly relevant to Machico in terms of socio-economic importance due to the local fishing activity and its function as a shelter for the bay. According to Figure 1, the port of Machico is located in a highly exposed area of the Machico bay.

Figure 1. Location of the Port of Machico
Source: Authors

In situations of strong maritime agitation, the occurrence of overtopping in the port of Machico is very frequent, especially in the area of the main breakwater. In the port of Machico, according to data collected by the Regional Secretariat for Social Equipment (SRES), maritime agitation off Machico is observed (Table 1).

Among the values, it is noted that in the port of Machico the maritime agitation for a return period of 100 years, in the South direction, reaches the maximum value of 7 m. It is also known that southwest values are not very frequent, but when they appear, if the visible extension in Figure 2 did not exist, the sheltered area would be in vulnerable conditions (Freitas, 2020).

Table 1. Maritime agitation off Machico

Madeira Island (South Coast)			
Return Period (Years)	Seaward Maritime Agitation Extreme Hs Values in Meters		
	SW	S	SE
10	5,1	4,9	3
25	5,6	5,7	3,4
50	6,1	6,4	3,8
100	6,5	7	4,1

Source: SRES (2007)

Figure 2. Extension of the pier in Machico
Source: Authors

The occurrence of waves from the Southeast is not as problematic since there is an observable obstacle in Figure 3, a cable in the coastal zone to the east of Machico (Freitas, 2020).

Figure 3. Cape to the East: Queimada
Source: Authors

According to historical data, the main breakwater is frequently affected by overtopping, especially during periods of high maritime agitation, reaching its entire length, having already required four reinforcement works in recent years. Until almost 2000, there was only one masonry spur, visible in Figure 4, and there was no shelter in this area as the length of the pier was minimal compared to the current configuration (Freitas, 2020).

Later on, the structure of reinforced concrete caissons with protective layer was built, although it was not the current 150 meters long. Shortly thereafter, around 2001-2002, an extension was made, acquiring the current configuration (Freitas, 2020). Figure 5 shows the different phases of the cais construction with construction processes from different periods. Finally, the last intervention in May 2019 consisted of a reprofiling of the protective layer and localized rehabilitation and repair throughout the Machico port.

Although there is no systematic record of flooding events at the Machico port breakwater, there are recent reports of flooding during stormy weather in January 2022 (RTP, 2022), December 2020, the end of February 2018, December 2013 (RTP, 2013), and February 2010 (Alves et al., 2022). All of these events resulted in the inoperability of the Machico port, the destruction of some infrastructure and vessels, and the death of at least one person. In addition to these major flooding events, smaller events occur frequently that do not affect the operation of the Machico port.

The port is mainly used for fishing and recreational purposes, with several small boats and yachts docked. It is a popular port for tourists to enjoy panoramic views and boat trips along the coast. Machico itself is a historic city that played a significant role in the discovery and colonization of Madeira in the 15th century. It was the first capital of Madeira until it was replaced by Funchal, the current capital of the island.

Figure 4. Machico Pier in 1947
Source: Perestrellos Photographos

The importance of the activities associated with the breakwater and the port in general is undeniable, and the consequences of possible flooding events can affect both the port's operations and the safety of people and property in the surrounding area (Figure 6).

Figure 5. Different construction phases with different construction processes
Source: Authors

Figure 6. Machico Port
Source: Authors

Therefore, the assessment of the risk of occurrence of wave overtopping is crucial so that local authorities can take appropriate measures for planning and managing this area. The assessment may include the analysis of meteorological and oceanographic conditions that may contribute to the occurrence of wave overtopping, as well as the identification of areas that are more vulnerable to this phenomenon. Based on this assessment, preventive and mitigation measures can be taken to minimize risks and ensure the safety of people and property in the port and surrounding areas.

In addition, it is important that local authorities are prepared to deal with possible emergency situations, with clear and effective contingency plans to deal with the effects of wave overtopping and other adverse situations.

The pioneering studies Silva (2012) and Pereira (2013) for the Port of Ponta Delgada applied methodologies for wave overtopping calculation, analysis of its consequences, and risk assessment. These studies assessed the risk for five sections of the port structures, using a simple methodology to evaluate the consequences of wave overtopping exceeding the permissible limits.

Rodrigues (2014) extended this work, evaluating the risk of wave overtopping for twelve sections of the port structures, considering a 34-year period of sea agitation data (1979 to 2013) and the real tidal level.

The assessment of the risk of wave overtopping is essential to ensure the safety of the port and the surrounding area, and these pioneering studies are examples of how this assessment can be carried out using appropriate methodologies. Based on the results of these studies, preventive and mitigation measures can be taken to minimize risks and ensure the safety of port activities and people and property in the surrounding area.

For the case study presented, the same steps and methodologies adopted by the mentioned authors will be followed, adapting them to the specific situation of the Port of Machico. The assessment of the consequences of wave overtopping in the Port of Machico is based only on information provided by the responsible authorities and is a global and qualitative assessment for a single zone considered in the study, as shown in Figure 7. The spatial distribution or the existence of infrastructure was not directly considered, and no weights were assigned to the indicators to express the relative importance of each one for the studied zone. As a result, the consequence map does not simultaneously take into account all the indicators.

The present work aims to perform a more in-depth, quantitative, and detailed analysis of the consequences associated with the occurrence of excessive flooding in the port of Machico and to build a map of the consequences. To do so, a methodology based on a multicriteria analysis was used, which allows analyzing, classifying, and assigning weights to different spatial indicators (e.g., spatial distribution of the population, maritime and land infrastructure, built heritage, services and activities in the area, or the extent of the area affected by flooding, expressed in terms of flood levels) that characterize a given study area (Antunes, 2012; Poseiro et al., 2013b). This methodology consists of building a spatial index of anthropic pressure on the coast through the application of the Analytic Hierarchy Process (AHP), continuing the work already developed by, for example, (Carver, 1991; Malczewsky, 2006; Greene et al., 2011) on the application of this method together with GIS tools (J. Rodrigues, 2014). Thus, for the elaboration of the consequences map, a methodology based on multicriteria analysis was used, as described by (Antunes, 2012), which allows analyzing, classifying, and assigning weights to different spatial indicators that characterize a given study area, such as the spatial distribution of the population, maritime and land infrastructure, built heritage, services and activities in the area, or the extent of the area affected by flooding, expressed in terms of flood levels (Poseiro et al., 2013b).

Figure 7. Selected structure for consequences assessment
Source: Authors

The methodology used in this study aimed to quantify the anthropic pressure on the Portuguese coast by prioritizing each criterion or indicator considered. This approach was exemplified in the work of Craveiro et al. (2012) and applied to evaluate the consequences of excessive flooding in the port of Machico, using selected structures for the evaluation of consequences (Rodrigues, 2014).

The Analytic Hierarchy Process (AHP) methodology, which originated in the 1970s/1980s (Saaty, 1980), was initially developed to select alternative actions and investments in the business sector. However, its scope of application quickly expanded to other contexts, including environmental decision-making and public participation processes to weigh values and attitudes of stakeholders (Aragonés-Beltrás et al., 2015). AHP has been used in another project to value factors of trust or uncertainty concerning the adoption of new water production technologies for human consumption in the Algarve region (Freitas & Craveiro, 2015).

The AHP methodology has proven to be a robust tool with various possibilities for application, provided there are differentiated valuations between alternative options or conditions of analysis for decision-making. Therefore, AHP represents a rich tool for application and thematic scope, having been used in this study for the purposes stated above, and as a serial selection process on anthropic pressure in coastal areas.

The present study describes a methodology based on a Geographic Information System (GIS) to evaluate the consequences of flooding in the port of Machico. The methodology involves organizing data into different levels of information, represented by layers corresponding to various indicators such as Architectural Heritage, Special Places, Built, Altitude, Road Network, Land Use, and Population. The AHP method is applied to assign weights to each indicator, and the attributes of each indicator are classified into different weights. The Weighted Overlay function is then used to generate a consequences map that combines the spatial information of the indicators with their defined weights.

The methodology presented in the article is useful in generating a consequences map that can be overlaid with the probability map of flooding occurrence to obtain a realistic assessment of the flood risk in specific areas of the port of Machico. This approach facilitates effective preventive and mitigating measures to reduce the impact of excessive flooding in this area. The ModelBuilder tool available in ArcGIS is utilized to organize and process the data, and the methodology is based on the AHP method, which has been applied to a variety of contexts, including environmental decision-making and stakeholder participation processes.

Overall, the methodology presented in this study is a useful tool for assessing the consequences of flooding in the port of Machico and has potential for application in other coastal areas. By utilizing a combination of spatial information and the AHP method, the methodology provides a more comprehensive and quantitative analysis of flood risk, which can assist in making informed decisions to minimize the impact of flooding.

METHODOLOGY

AHP (Analytic Hierarchy Process) is a methodology used for making complex decisions that allows the analysis of different criteria and sub-criteria to define relative weights for each of them. In the case of the study at hand, the AHP methodology was applied to define the weight of each of the seven considered indicators: Architectural Heritage, Special Locations, Buildings, Altitude, Road Network, Land Use, and Population. Each indicator was subdivided into sub-indicators, and through the use of a questionnaire, a relative importance scale was established for each of them. This scale was used to define the relative weight of each indicator/sub-indicator. The assigned weight values were input into a Geographic Information System (GIS) along with the spatial data for each indicator, allowing the creation of a layer for each indicator. By using the Weighted Overlay tool, it was possible to combine the various layers and obtain the consequence map, indicating the intensity of consequences resulting from flooding in the Machico area. The ultimate goal of this methodology is to enable a more accurate and realistic evaluation of the risk of excessive flooding by combining the probability of flooding with the spatial characteristics of the studied area. This methodology is a common approach to assessing consequences in environmental impact and risk studies. It involves the collection and organization of relevant data in a GIS, applying the

Analytic Hierarchy Process (AHP) to determine the weight of each indicator, classifying the attributes of each indicator, the weighted overlay of indicators in the GIS according to the defined weights, and finally, implementing the process in the ModelBuilder tool. This approach enables a more systematic and objective assessment of the consequences of events such as flooding in specific locations, contributing to informed decision-making in terms of risk management and adaptation to climate change. The process of evaluating consequences begins with selecting the most relevant spatial indicators to characterize the location and integrating them into a GIS. For these indicators, the AHP process is applied, which is commonly used in decision-making situations where individuals work together to make decisions and where human perceptions, judgments, and consequences have long-term implications (Bhushan & Kanwal, 2004). This process allows the construction of a spatial index where different weights are assigned to various indicators. The comparison between indicators using AHP can be done in various ways, but the scale proposed by (Saaty, 2005) is the most commonly used. Saaty proposes a scale of relative importance between pairs of indicators, indicating the level of importance of indicator ‘a’ compared to indicator ‘b’. The assigned values range from 1 to 9, depending on whether ‘a’ is more important than ‘b’ or vice versa. The comparison matrix is constructed based on this scale and can be represented in a table for n indicators, as indicated in Table 2 (Saaty, 2005).

Table 2. Comparison matrix of n indicators according to

	Indicator 1	**Indicator 2**	**...**	**Indicator n**
Indicator 1	1	Numerical evaluation	Numerical evaluation	Numerical evaluation
Indicator 2	1/Numerical evaluation	1	Numerical evaluation	Numerical evaluation
...	1/Numerical evaluation	1/Numerical evaluation	Numerical evaluation	Numerical evaluation
Indicator n	1/Numerical evaluation	1/Avaliação numérica	1/Numerical evaluation	1

Source: Saaty (2005)

To determine the relative weights of each indicator in the comparison analysis using AHP, it is necessary to normalize the comparison matrix. This normalization is done by dividing each coefficient in the matrix by the sum of the coefficients in the corresponding column. After normalization, the arithmetic mean of the values in each row of the resulting matrix is considered as an approximate value of the eigenvector, which corresponds to the relative weight of each indicator (Vargas, 2010).

According to Vargas (2010), the AHP process has characteristics that ensure the consistency of its results. To verify the consistency of the weights assigned to the criteria and alternatives, the inconsistency index (λmax), the consistency index (CI), and the consistency ratio (CR) are calculated, according to equations 1, 2, and 3, respectively:

$\lambda max = \sum$(Eigenvector of indicator i × Total weight of each column of indicator i) i = 1,…,n; Equation 1

$CI = (\lambda max - n) / (n - 1)$; Equation 2

$CR = CI / RI < 0.1$. Equation 3

The inconsistency index (λmax) measures the difference between the largest eigenvalue and the number of indicators, while the consistency index (CI) measures the ratio between the inconsistency index and a consistent random index (RI), which varies between 0 and 1. The consistency ratio (CR) is the ratio between the consistency index (CI) and the consistent index for the number of indicators used (CI), which also varies between 0 and 1. CR values above 0.1 indicate acceptable consistency, while values below 0.1 indicate inconsistency in the results (Vargas, 2010). The value of RI is a fixed value for a given number of evaluated indicators, n, according to Table 3 (generic).

Table 3. Values of the random consistency index, RI, as a function of the number of indicators considered, n

n	1	2	3	4	5	6	7
RI	0	0	0,58	0,9	1,12	1,24	1,32

Source: Saaty (2005)

Each indicator, in the process, is characterized by a set of spatial attributes, to which different weights are assigned. As an example, in the "Land Use" indicator, the following attributes can exist: natural soils, urbanized soils, and special use soils. The process is finalized with the weighted overlay of the indicators in the Geographic Information System (GIS), according to the different weights defined from AHP, and their attributes, also with different weights, resulting in a map of consequences (Brito, 2012).

To apply this methodology, a diagram is created with the ModelBuilder tool, available in the commercial software ArcGIS. It is a tool that allows creating a flowchart that shows the necessary operations to obtain the map of consequences, using the tools provided by this software (Rodrigues, 2014).

This new method of assessing the consequences of overtopping has numerous advantages over the expedited method. According to the references, the expedited method (Poseiro et al., 2013a) assesses the consequences of overtopping through degrees ranging from "insignificant" to "catastrophic". However, in the new method, indicators with spatial distribution are used, allowing obtaining a map of the degree of consequences that takes into account the weight of all indicators simultaneously. In addition, the indicators are composed of various attributes that characterize them spatially, with different weights assigned to each of these attributes, allowing a more detailed categorization of each level of information according to its degree of importance (Ramos et al., 2019).

PORTO AND ITS SURROUNDINGS

The Machico pier-quay, located in the Autonomous Region of Madeira, is a structure of great socio-economic relevance, both for local fishing activity and for sheltering a wide area of the bay. The quay is also used as a leisure area for local residents. According to Pita (1986), tidal conditions in the region are a crucial factor in the design of port structures, such as the crest level and the sizing of protective elements. The agitation coming from the south and southwest is especially problematic, as it can leave the sheltered area vulnerable if there is no extension of the structure. Offshore agitation data obtained by PLANEGE for the Machico Bathing Zone project in 2007 show that although southwest values are not very frequent, they are important to consider. The Machico quay was built in three phases over the last 20 years, with the first stretch being made of masonry, the second of voussoirs, and the rest of reinforced concrete caissons with a protective layer. Originally, the masonry spur served mainly to support local fishing activity (Figure 8), but over time, the structure evolved to provide shelter and protection for coastal areas, including housing the local marina. With the occurrence of south and southwest wave direction, the sheltered area became unprotected, which made it necessary to construct an extension of the structure to improve its protection capacity.

Figure 8. Plan of the Primitive Machico Pier
Source: Municipal Archive

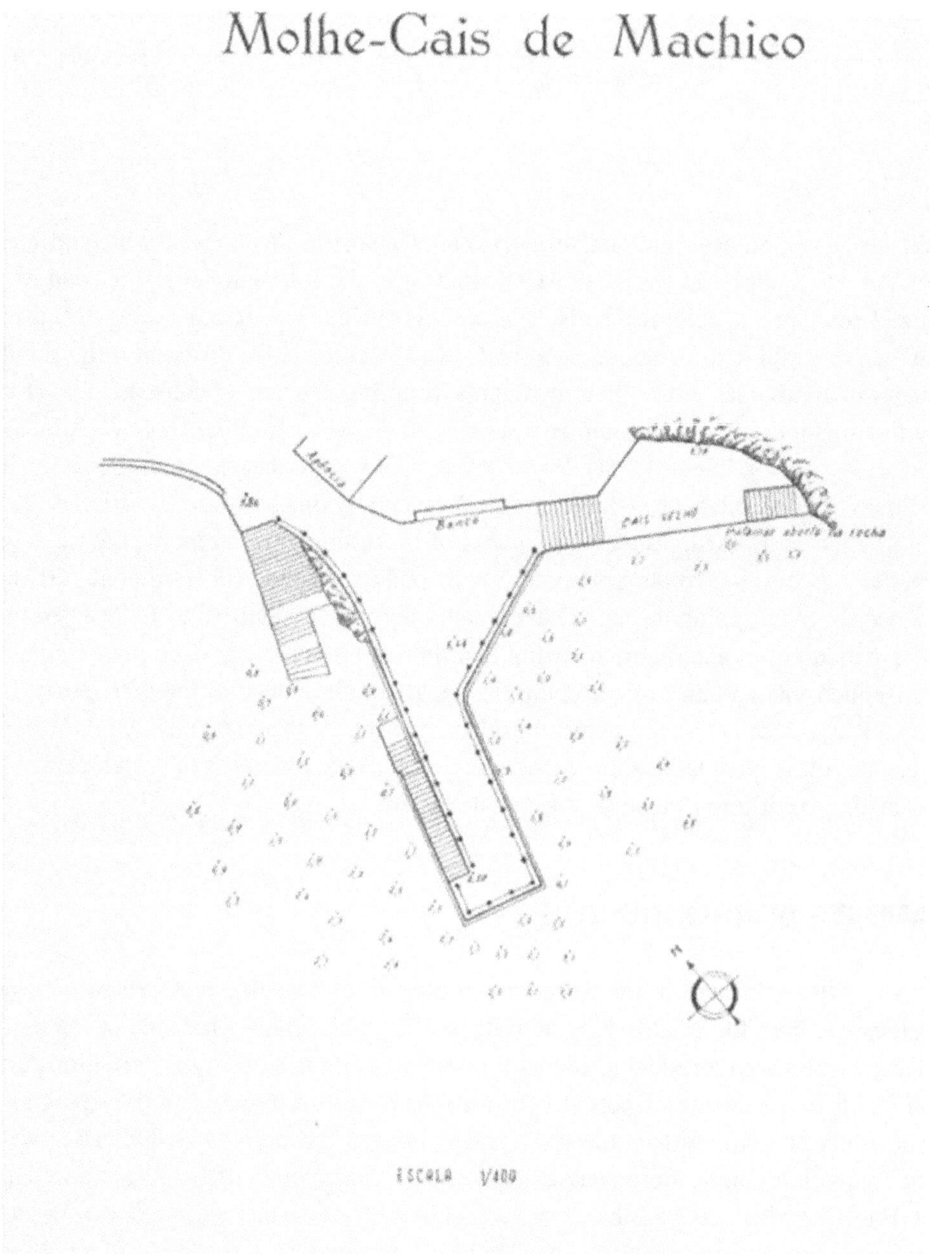

Figure 9. Plan of the Machico Pier-Jetty
Source: Freitas (2020)

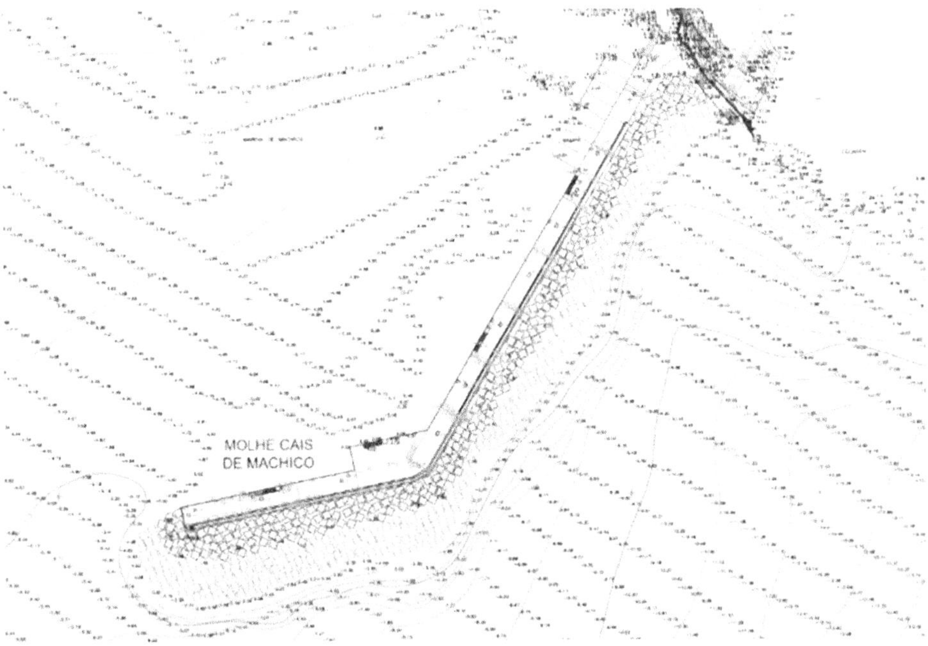

According to Freitas (2020), the pier-jetty has a length of about 150 meters and a variable width, with an average of 5.75 meters. As shown in Figure 9, the pier-jetty consists of three identical caissons at the head, an intermediate caisson with a different geometric configuration, and regular-sized caissons upstream. The caissons are made of reinforced concrete and filled with all-size riprap, and the function of the curtain wall is to minimize overtopping. The foundation is composed of a regularized riprap prism, and the culverts are reinforced concrete elements filled with sand and/or riprap.

The cais at Machico is made of reinforced concrete caissons filled with TOT (all-size rock) and has a foundation composed of a regularized rock prism, on which the caissons rest (Freitas, 2020). In 2020, the Regional Government of Madeira intervened to ensure the functionality and integrity of the cais at Machico, due to its deteriorating condition and frequent wave overtopping. The proposal was to reprofile the caisson armor layer by increasing the weight of the blocks in the upper layer. The aim was to strengthen and rehabilitate the structure to avoid more intrusive and costly interventions in the future. The interventions carried out included reprofiling the armor layer and localized rehabilitation and repair (Freitas, 2020). The plan for the implementation of the new protection armor layer is shown in Figure 10 and generally coincides with the existing structure.

After the interventions carried out on the dock, several new characteristics can be observed. Starting with the banquette, now two blocks are used horizontally between the curtain wall and the beginning of the slope, resulting in a width of 4.60 meters along the entire length of the dock (Figure 11). The solution to increase the size of the Antifer blocks in the upper layer involved replacing 7.5-ton elements with 15-ton elements (Freitas, 2020).

According to Freitas (2020), regarding the sub-mantle, the existing 7.5-ton blocks were rearranged as a base for the new layer, to guide it to the desired inclination. The remaining 7.5-ton blocks were

Figure 10. Plan for the installation of the new protective mantle
Source: Freitas (2020)

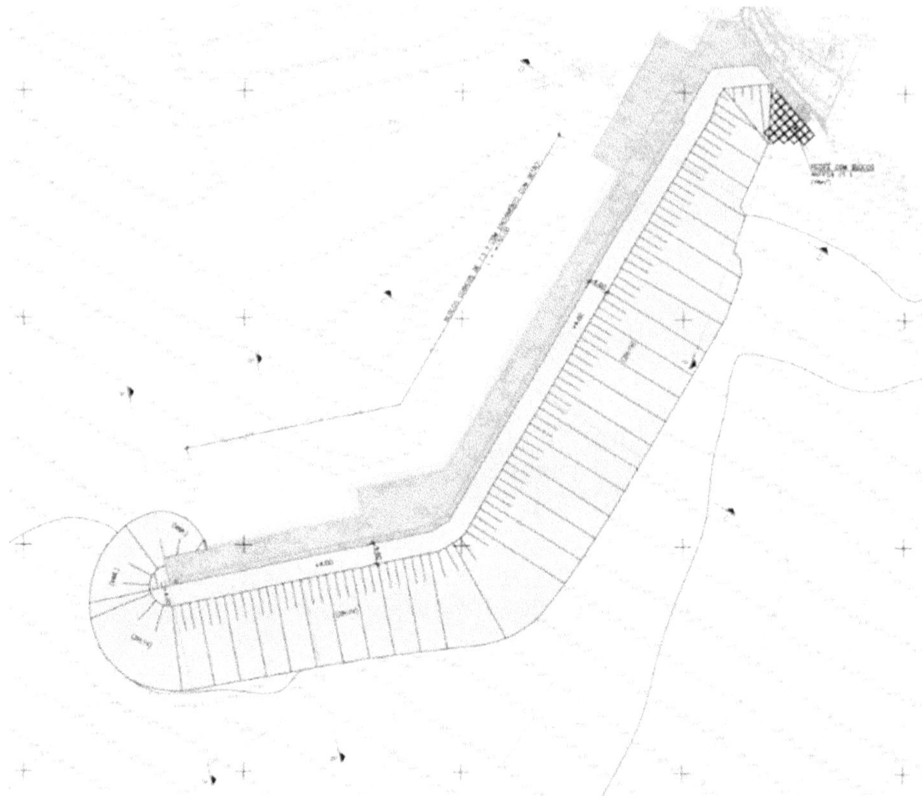

Figure 11. Typical profile of the protective mantle after reshaping intervention
Source: Freitas (2020)

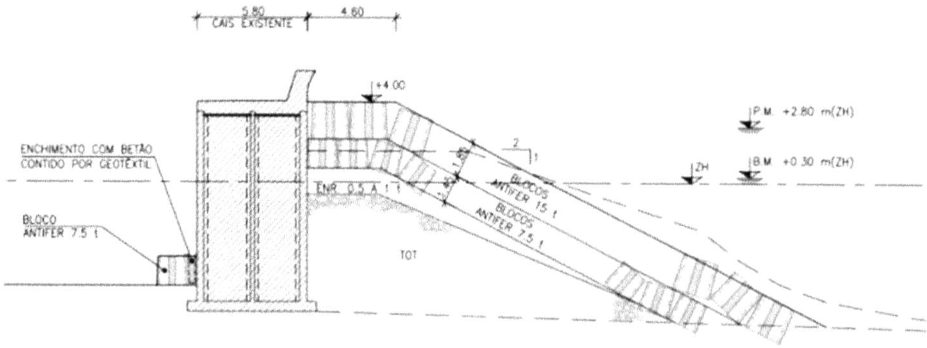

used for various purposes, such as solving the absence of a protection rock prism on the sheltered side, being sent to the Contractor's yard for reuse in other works, helping the crawler crane works, and being placed next to the base of the caisson, on the inside.

On the sheltered side of the caisson, next to the base, a row of 7.5-ton Antifer blocks was created to increase stability against caisson sliding and overturning. The construction process involved creating this row in such a way as to serve as a lost formwork and later filling it with concrete (Freitas, 2020).

Regarding the old caisson, located at the base of the cliff in the eastern zone of the Machico Bay, near the initial stretch of the exposed side trunk, a reinforcement solution was carried out, which involves the construction of a submerged reef in the vicinity of this "old caisson." The aim is to minimize the damage that may result from its existence due to the reflection of waves on the caisson's slope (Freitas, 2020).

The Fort of São João, also known as the Fort of São João Baptista or the Fort of São João do Desembarcadouro de Machico, is a historical fortress located in the city of Machico on the island of Madeira, Portugal. The fortress was built in the sixteenth century by the portuguese to defend against pirate attacks and protect the city's port.

The Fort of São João (Figure 12) is a small stone fortress, with a circular shape and a central tower. It was strategically placed on a hill overlooking the Machico Bay, allowing the portuguese to have a clear view of any approaching enemy ships. The fort underwent several renovations and changes over the centuries, including the addition of a chapel in the seventeenth century. It is located in the northern area of the study area, maintaining a connection with the port and starting the promenade.

Figure 12. Cardinal Tolentino Mendonça Street
Source: Author

Cardinal Tolentino Mendonça Street (Figure 12), formerly known as Leiria Street, is an important road that connects Machico Marina and Fort São João Baptista. Due to the constant flow of pedestrians and vehicles, it is important to ensure their safety against the sea agitation. Along much of the avenue's length, it is possible to identify two levels of circulation: one for pedestrians and cyclists at lower levels, where there are only two commercial spaces, and another for vehicles at higher levels with heavy traffic.

Machico Marina (Figure 13) is a marina located in the city of Machico. Inaugurated in 2004, the marina has 80 berths for recreational boats and offers a wide range of services, including electricity, water supply, and maintenance services. In addition, the marina offers several leisure options such as restaurants, bars, shops, and a nautical club. The marina's privileged location in front of Machico Bay provides visitors with a stunning view of the sea and surrounding mountains. The marina is frequently used as a starting point for boat excursions along the coast of Madeira Island, as well as for water sports such as diving, fishing, and jet skiing (CMM, 2022).

Figure 13. Machico Marina
Source: Authors

EVALUATION OF CONSEQUENCES USING AHP

In the case of the Machico port, the indicators presented in Figure 14 were considered.

The indicators were characterized by data from various sources, including the BGRI provided by the National Institute of Statistics, the Municipal Master Plan, and the Zoning Plan. These sources provided a variety of information, ranging from flood elevations to the distribution of buildings and transportation routes. The authors chose the indicators based on elements they deemed relevant to characterize the site and that have a significant impact on the flooding phenomenon. Other indicators may be considered with the opinion of local authorities, but the methodology to be applied will not be altered. Table 4 presents the attributes considered for the mentioned indicators.

Figure 14. Indicators and data sources for the case study
Source: Rodrigues et al. (2017)

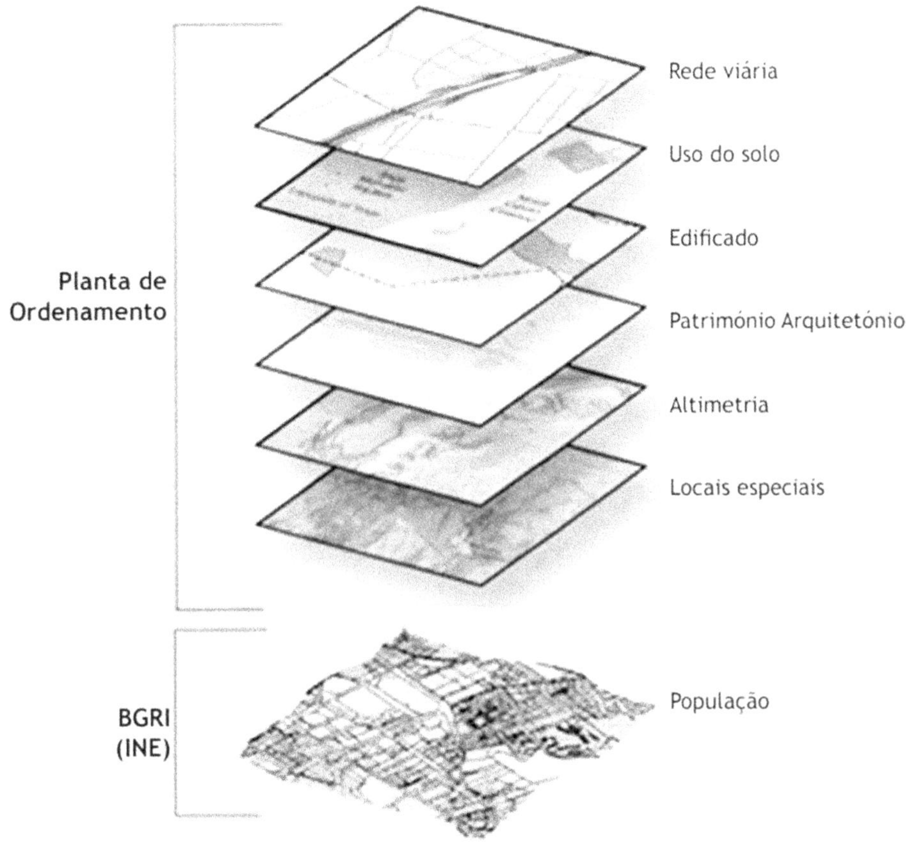

Table 4. Considered indicators and attributes

Indicators	Attributes
Architectural Heritage	Public Interest Property
	Municipal Interest Property
Special Sites	Wastewater Treatment Plant
	Pumping Station
Built Environment	Buildings (of any nature)
Altitude	Elevation varies between 0 m and 21 m
Road Network	Municipal roads and paths
Land Use	Urbanized land
	Natural land
	Land with special use
Population	Resident population

Table 5. Comparative matrix of selected indicator group

Indicators			A	B	C	D	E	F
A	Architectural Heritage	1,00	1,00	2,00	6,00	6,00	6,00	0,50
B	Special Sites	1,00	1,00	4,00	8,00	6,00	6,00	0,50
C	Built Environment	0,50	0,25	1,00	4,00	4,00	5,00	0,25
D	Altitude	0,17	0,13	0,25	1,00	0,50	0,50	0,25
E	Road Network	0,17	0,17	0,25	2,00	1,00	1,00	0,25
F	Land Use	0,17	0,17	0,25	2,00	1,00	1,00	0,25
G	Population	1,00	1,00	3,00	8,00	8,00	8,00	0,50
	Total	4,00	3,71	10,75	31,00	26,50	27,50	2,50

Table 6. Normalized matrix

Indicators			A	B	C	D	E	F
A	Architectural Heritage	0,250	0,270	0,186	0,194	0,226	0,218	0,200
B	Special Sites	0,250	0,270	0,372	0,258	0,226	0,218	0,200
C	Built Environment	0,125	0,067	0,093	0,129	0,151	0,182	0,100
D	Altitude	0,042	0,034	0,023	0,032	0,019	0,018	0,100
E	Road Network	0,042	0,045	0,023	0,065	0,038	0,036	0,100
F	Land Use	0,042	0,045	0,023	0,065	0,038	0,036	0,100
G	Population	0,250	0,270	0,279	0,258	0,302	0,291	0,200

Table 7. Calculation of the eigenvector

Indicators	Weight	Weight (%)
Architectural Heritage	0,221	22,06
Special Sites	0,256	25,63
Built Environment	0,121	12,10
Altitude	0,038	3,83
Road Network	0,050	4,98
Land Use	0,050	4,98
Population	0,264	26,42
Summation		100,00

Table 8. Inconsistency index, consistency index (CI), and consistency ratio (CR)

λ_{Max}	**7,67**
CI	0,11
CR	0,08

ASSIGNING WEIGHT TO INDICATORS

Through the comparison matrix, which was constructed as presented in Table 2, Table 5 was generated for the case study of Machico, based on a consultation with local authorities. To obtain the relative weights of each indicator, it is necessary to normalize this matrix. To do this, for each column, the value of each coefficient is divided by the sum of all coefficients in that column (Table 6).

The value representing the weight of each indicator is the eigenvector, and to obtain this value, it is common to normalize the comparison matrix and then calculate the arithmetic mean of the values obtained in each row of the normalized matrix. Vargas (2010) proposes this method as an approximate value for the eigenvector. This value is presented in Table 7 of the Machico study as the average of the normalized values for each indicator.

To ensure the validity of the weights assigned to the indicators, the inconsistency index (λMax, equation 1) is calculated. The consistency index (CI, equation 2) and the consistency ratio (CR, equation 3) are also calculated, as shown in Table 8, considering the value of Ri=1.32 and the fact that 7 indicators were evaluated (Table 3).

According to (Saaty, 2005) and as mentioned earlier, the matrix can be considered consistent since the consistency ratio (CR) is less than 0.1 (equation 3). Therefore, the weights assigned to the indicators can be used in evaluating the consequences of flooding and in creating the corresponding map.

DISTRIBUTION OF ATTRIBUTE WEIGHTS

Each of the indicators was incorporated into a Geographic Information System (GIS) as a layer, each with a set of attributes. In this case, the classification of these attributes did not require the AHP method and therefore, different weights were arbitrarily assigned to each one, on a scale of importance from 1 to 9. Table 9 shows the different weights assigned to each indicator and attribute. It is important to mention that the weights assigned to the indicators were defined in collaboration with the Municipality of Machico - CMM, and the application of the AHP method served to analyze the consistency of this distribution. In general, the weights obtained by AHP (Table 7) validate the weights assigned to the various indicators according to the CMM, since, as can be seen in Table 9, the values are of the same order of magnitude.

Where:

1 Values resulting from CMM.
2 Values obtained from AHP, Table 7.

For the attributes, although it is possible to apply the AHP method, the assignment of weights was made arbitrarily on a scale of 1 to 9, where 1 corresponds to the least important attribute and 9 to the most important (Poseiro et al., 2013b).

To work with the data indicated in Table 9, the ArcGIS software was used, and a model was created with the ModelBuilder tool that included all the necessary steps to reach the final result (Rodrigues, 2014; Rodrigues et al., 2014).

Table 9. Weights assigned to the various considered indicators and attributes

Indicators	Attributes (Weight)
Classification according to AHP	Arbitrary classification from 1 to 9
Architectural Heritage	Public Interest Property (9)
18%[1] (22,06%[2])	Municipal Interest Property (9)
Special Sites	Pumping Station (8)
30%[1] (25,63%[2])	
Built Environment	Buildings (of any nature) (9)
12%[1] (12,10%[2])	
Altitude	0 - 21 m (9)
2%[1] (3,83%[2])	
Road Network	Municipal roads and paths (9)
7%[1] (4,98%[2])	
Land Use	Urbanized land (5)
6%[1] (4,98%[2])	Natural land (8)
	Land with special use (9)
Population	Resident population (9)
25%[1] (26,42%[2])	

Regarding the “Land Use” indicator, it was considered relevant to consider three scales of different attributes, with different scales of importance. The “urbanized soils” attribute was given a weight of 5 as it was considered the least relevant of the three. For the “natural soils” attribute, a weight of 8 was assigned, and for the “soils with special use” attribute, a weight of 9 was considered, reflecting the greater importance of this attribute.

Regarding the “Architectural Heritage” (Figure 15), the data considered deserving of the same classification were gathered in a single layer (for example, the “classified or about to be classified properties” and the “properties with architectural value”). These properties were classified from 1 to 9, depending on whether they were of little or great interest.

Regarding “Built Environment” (Figure 16), which considers constructions of any nature, the classification is 1 for areas where there are no buildings (less severe) and 9 for areas where there are buildings (more severe).

Regarding “Altitude,” it is possible to represent a three-dimensional model indicating the elevation of each point using contour lines. Thus, a layer with levels of severity was obtained that decreases as the terrain elevation increases. In the “Road Network” layer (Figure 17), the roads were classified with a severity level of 9, while the rest were classified with a severity level of 1.

Regarding the “Population” indicator, it was possible to obtain information on the resident population in the city of Machico from the 2011 Census data provided by INE, at the level of statistical subsection. A correspondence (Join) was established in ArcGIS between the shapefile with the polygons corresponding to each statistical subsection and the table containing the 2021 Census information corresponding to the resident population. Since these results refer to the resident population in each subsection, it was

Figure 15. Representative layers of architectural heritage
Source: Authors

Figure 16. Representative layers of built environment
Source: Authors

Figure 17. Layers representing the road network
Source: Authors

necessary to calculate the population distribution per square meter. Thus, a new attribute was created resulting from the division between the resident population in each subsection and its área (Figure 18).

DEVELOPMENT OF THE CONSEQUENCES MAP

Finally, as mentioned in (Poseiro et al. 2013b), the flowchart developed in ModelBuilder was used to obtain the consequences map shown in Figure 19. This map represents the area with the most severe consequences.

Figure 18. Layers representing population distribution per area (statistical subsection)
Source: Authors

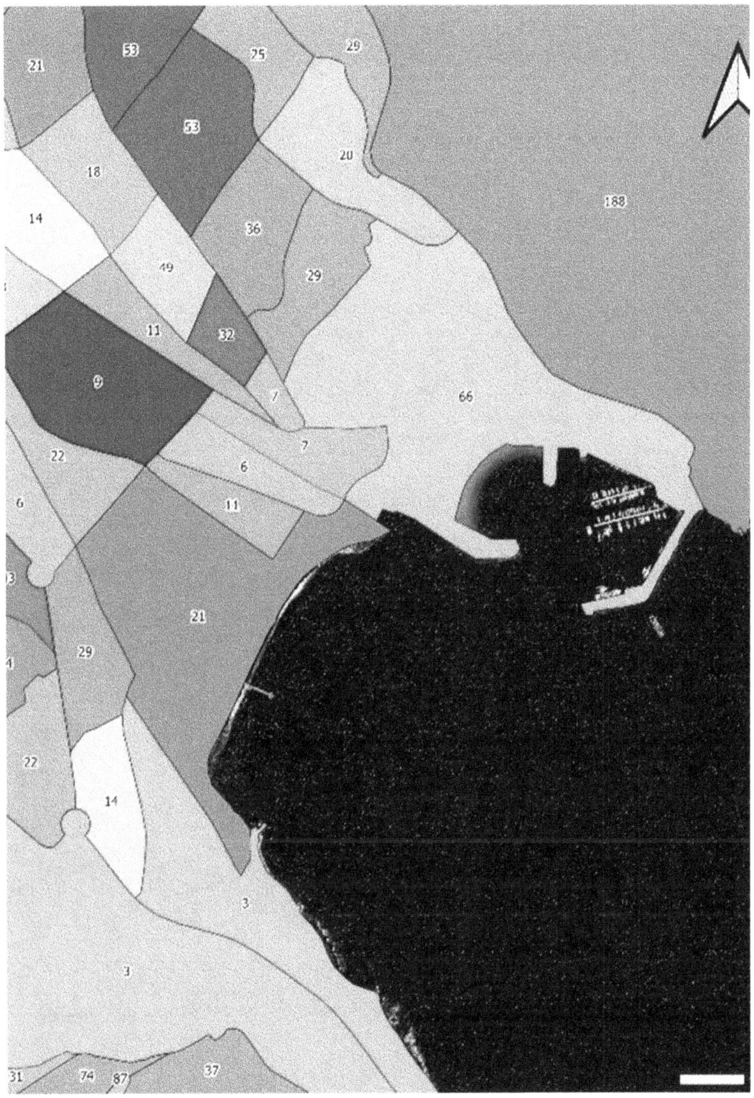

Figure 19. Consequence map resulting from the application of AHP
Source: Authors

CONCLUSION

In conclusion, the application of the Analytic Hierarchy Process (AHP) and GIS tools allowed for a more detailed and precise analysis of the consequences of overtopping at the port of Machico, in comparison to the simplistic method used previously. The AHP enabled the assignment of weights to each indicator/attribute, resulting in a map of consequences that took into account all relevant indicators simultaneously. This methodology can be applied to improve the spatial characterization of elements exposed to various hazards and to develop a flood risk map for the area.

In the future, it is necessary to continue refining the weights assigned to each indicator/attribute based on information provided by responsible authorities and consider expanding the use of AHP to other attributes. In summary, the use of AHP and GIS tools was a useful and advantageous approach for the analysis of flood risk at the port of Machico.

REFERENCES

Alves, R., Lousada, S., Cabezas, J., & Naranjo Gómez, J. (2022). *Flood Risk and Vulnerability: An Approach to Urban Flooding in the Machico River of the Autonomous Region of Madeira*. Geoinformatics in Support of Urban Politics and the Development of Civil Engineering.

Antunes, O. (2012). *Multi-criteria analysis in GIS for determining a spatialized index of coastal anthropic pressure. Cases of Espinho, Caparica and Faro* [Master's thesis]. Faculty of Social and Human Sciences, Nova University of Lisbon.

Aragonés-Beltrás, P., García-Melón, M., & Estruch-Guitart, V. (2015). Analysis of the participation of stakeholders in environmental management based on AHP: Application to a Spanish natural park. *International Journal of the Analytic Hierarchy Process*, *7*, 1.

Bhushan, N., & Kanwal, R. (2004). *Strategic Decision Making: Applying the Analytic Hierarchy Process*. Springer-Verlag. doi:10.1007/b97668

Brito, L. (2012). *Multicriteria analysis for selecting areas for the implementation of agroforestry systems using geoprocessing and fuzzy logic* [Doctoral dissertation]. Federal University of Lavras.

Carver, S. (1991). Integrating multi-criteria evaluation with geographical information systems. *International Journal of Geographical Information Systems*, *5*(3), 321–339. doi:10.1080/02693799108927858

CMM. (2022). *Machico Marina*. https://www.cm-machico.pt/investir/espacos/porto-comercial-e-marinas

Freitas, J. (2020). *Rehabilitation of Machico's Dock. Internship Report for the degree of Master in Civil Engineering in the area of Buildings. Departmental Area of Civil Engineering. Higher Institute of Engineering of Lisbon*. ISEL.

Freitas, J. (2020). *Rehabilitation of Machico's Dock. Internship Report for the degree of Master in Civil Engineering in the area of Buildings. Higher Institute of Engineering of Lisbon*. ISEL.

Freitas, M.; Craveiro, J. (2015). Report 1 of Workshop 1 of the Stakeholders Panel. LIFE/HYMEMB, Tailoring Hybrid Membrane Processes for Sustainable Drinking Water Production [LIFE12 ENV/PT/001154]. Rep.1, t.B5, 31st May 2015.

Greene, R., Devillers, R., Luther, J., & Eddy, B. (2011). GIS-Based Multiple-Criteria Decision Analysis. *Geography Compass, 5/6*(6), 412–432. doi:10.1111/j.1749-8198.2011.00431.x

Malczewski, J. (2006). GiS-based multicriteria decision analysis: A survey of the literature. *International Journal of Geographical Information Science, 20*(7), 703–726. doi:10.1080/13658810600661508

Pereira, D. (2013). *Evaluation of the Flood Risk of Port Structures: The Case of the Port of Ponta Delgada (Azores, Portugal)* [Master's thesis]. University of Madeira, Funchal.

Poseiro, P., Fortes, C., Reis, M., Santos, J., Simões, A., Rodrigues, C., & Azevedo, E. (2013a). A methodology for overtopping risk assessment in port areas: Application to the Port of Praia da Vitória (Azores, Portugal). *Proceedings of the 6th International Conference on Applied Coastal Research (SCACR 2013).*

Poseiro, P., Fortes, C., Santos, J., Reis, M., & Craveiro, J. (2013b). Application of the Analytic Hierarchy Process (AHP) to the analysis of the consequences of flooding. The case of Praia da Vitória Bay. *Proceedings of the 8th Portuguese Conference on Coastal and Port Engineering.*

Ramos, S. (2019). *Assessment of flooding risk in coastal areas - Case study of Praia da Rocha. 12th National Congress of Geotechnics*, Braga, Portugal.

Rodrigues, J., Fortes, C., Reis, M., Poseiro, P., & Taveira-Pinto, F. (2017). Evaluation of the consequences of flooding at the Port of Ponta Delgada, S. Miguel, Azores, using the AHP method. *Territorium Journal, 24*(April), 127–139. doi:10.14195/1647-7723_24_9

Rodrigues, J., Poseiro, P., Reis, M., Fortes, C., & Pinto, F. (2014). Application of the Analytic Hierarchy Process (AHP) to the evaluation of flooding consequences in port areas - the case of Ponta Delgada, Azores. *Proceedings of the 3rd International Congress; 1st Ibero-American Symposium; 8th National Meeting on Risks; November 5-7, Department of Geography, University of Minho, Guimarães, 2014 - Multidimensionality and Risk Territories.* 10.14195/978-989-96253-3-4_71

Rodrigues, M. (2014). *Selection of priority areas for ecological restoration in the Rio das Velhas-MG watershed* [Master's thesis]. Universidade Federal de Minas Gerais.

RTP. (2013). *Bad weather in Madeira causes one death.* https://www.rtp.pt/noticias/pais/mau-tempo-na-madeira-faz-um-morto_v702388

RTP. (2022). *Machico dock closed* [video]. https://www.rtp.pt/madeira/sociedade/cais-de-machico-interdito-video_84415

Saaty, T. (1980). *The Analytic Hierarchy Process.* McGraw Hill.

Silva, D. (2012). *Evaluation of the Overtopping of Port Structures* [Master's thesis]. Civil Engineering, Faculty of Sciences and Technology, University of Coimbra.

Vargas, R. (2010). *Using Multi-Criteria Programming (Analytic Hierarchy Process - AHP) to Select and Prioritize Projects in Portfolio Management.* PMI Global Congress.

KEY TERMS AND DEFINITIONS

Analytic Hierarchy Process: The Analytic Hierarchy Process (AHP) is a decision-making method developed by Thomas Saaty in the 1970s. It is a structured approach to making complex decisions, based on breaking down a problem into smaller parts and comparing them pairwise to determine their relative importance. AHP involves creating a hierarchy of decision criteria and alternatives, and then comparing them using a series of pairwise comparisons to determine their relative priorities. This process results in a numerical weight or priority for each criterion or alternative, which can be used to make a final decision. AHP has been widely used in fields such as engineering, management, and environmental sciences for a variety of decision-making applications.

Machico: Machico is a city located on the eastern coast of the island of Madeira, which is an autonomous region of Portugal. It is known for its historical importance as the landing site of the discoverers of Madeira in the 15th century, and for its beautiful beaches and natural landscapes. The city has a population of around 20,000 people and is an important commercial and tourist center on the island.

Madeira: Madeira is an autonomous region of Portugal located in the Atlantic Ocean, southwest of the portuguese mainland. It is an archipelago composed of two inhabited islands, Madeira and Porto Santo, and several uninhabited islands and islets. The region has a subtropical climate, with mild temperatures throughout the year, and is known for its beautiful landscapes, including mountains, forests, and coastlines. The economy of Madeira relies heavily on tourism, as well as agriculture and fisheries.

Overtopping: Overtopping is a coastal engineering term that refers to the phenomenon of waves breaking over a barrier, such as a seawall, and flowing onto the landward side. This can lead to flooding, erosion, and damage to infrastructure and property. Overtopping is typically quantified in terms of volume or discharge rate, and is influenced by factors such as wave height, wave period, and the elevation and roughness of the barrier. Overtopping can be managed through the design of coastal structures, such as breakwaters and revetments, that are intended to reduce wave energy and prevent or minimize overtopping.

Port: A port is a facility where ships can load and unload cargo or passengers. Ports can be found in coastal areas, along rivers or lakes, and they play a crucial role in global trade and transportation. Ports can have various types of infrastructure, including docks, wharves, piers, and terminals, as well as cranes, containers, and other equipment for handling cargo. They can also provide services such as customs clearance, storage, and maintenance for ships.

Chapter 5
Flood Risk and Vulnerability:
An Approach to Urban Floods in Ribeira de Machico in the Autonomous Region of Madeira

Raul Manuel Costa Alves
https://orcid.org/0000-0002-3680-6626
University of Extremadura, Spain

Sérgio António Neves Lousada
University of Madeira, Portugal

José Cabezas
https://orcid.org/0000-0002-7209-3731
University of Extremadura, Spain

José Manuel Naranjo Gómez
https://orcid.org/0000-0001-7998-9154
University of Extremadura, Spain

ABSTRACT

The city of Machico grew in an accelerated and unbalanced way. This unplanned lack of control generated urban flooding problems with some damage. Currently, efforts are being made to avoid and mitigate these floods or alluviums, as flash floods are known on the island of Madeira, by investing in studies. The aim of these studies is to understand the problem and its application in the management of prevention and control of this type of risk. In this way and based on the Flood Risk Management Plan for the Autonomous Region of Madeira, the mitigation of alluvial problems for the Ribeira de Machico is idealized.

DOI: 10.4018/978-1-6684-6449-6.ch005

INTRODUCTION

Among all-natural disasters, flooding affects the most people in the world (Moore, 2005). In the Autonomous Region of Madeira – RAM, due to geographical specificities, floods have already caused substantial material damage and a high number of fatalities. It is estimated that public investment in the repair and reconstruction of public infrastructure damaged by floods, since 2020, will exceed 250 million euros. It appears that urban floods have been increasing considerably in RAM, due to pressures exerted by human occupation, weatherproofing sealing, increased population density and disorderly urban occupation. With the increase of this problem at a global level and, mainly in tropical regions, it became necessary to develop measures and devices to mitigate the impacts of floods (Lousada & Castanho, 2021).

Since the beginning of the 19th century, around 30 alluviums of significant intensity have occurred on Madeira Island, characterized by rapid and violent floods that transport high concentrations of solid material. It is worth mentioning the occurrence of 1803, the year in which about 1000 people died, victims of the catastrophic floods that occurred throughout the island, but with greater intensity in Funchal, Machico and Santa Cruz (Quintal, 1999). Figure 1 shows the most serious recent event that occurred on February 20, 2010.

Figure 1. Photographs of the alluvium of February 20, 2010
Source: Diário de Notícias

In 2010, an extremely well-known event regionally was the alluvium, which caused damage and losses in several locations on Madeira Island. It took place on February 20, 2010, resulting in the death of 47 people on the island, 600 displaced and 250 injured (RTP, 2016).

Vulnerability represents the degree of loss of component or components exposed to floods, this aspect being different from the concept of risk, being strongly linked to the economic, age, cultural, social characteristics of a society, as well as the occupation of the territory (Ferreira, 2010). Urban flooding in the city of Machico arises since there is a large occupation, somewhat manageable, of potentially floodable areas, where there is a great waterproofing of the soil, associated with the obstruction of several devices responsible for the drainage of the city, namely, culverts and gutters, the high-water table, physical aspects of the city somewhat unfavorable and the bad design of drainage systems.

Associated with the high daily maximum annual rainfall (>200 mm) and the fact that a large part of the city's population is located by the sea (high tides), the risk of flooding in Machico increases, especially during the months of October and November. This phenomenon causes several inconveniences to the lives of many people, especially those who live along the stream and by the sea. Thus, it is extremely important to mitigate these disasters, not only in Machico but throughout the island.

Therefore, this article proposes to describe the indicators of flood vulnerability, also identifying, through Directive 2007/60CE, the Flood Area, the Flood Risk, the Potentially Affected Inhabitants (> 400 inhabitants), the Affected Cultural Heritage (5 buildings), Potentially Affected Roads (> 9 km), Sensitive Potentially Affected Buildings (Machico City Hall, Machico Parish Council, Machico Municipal Fire Department and Machico Health Center) and Potentially Affected Infrastructure.

RIBEIRA DE MACHICO HYDROGRAPHIC BASIN: GEOMETRIC AND PHYSIOGRAPHIC CHARACTERISTICS

In this stage, will be addressed issues related to:

- Geometric characteristics of Ribeira de Machico, namely its location, area, perimeter, and shape of the basin;
- Physiographic characteristics such as geology, soil and vegetation, characteristics that may or may not be favorable to the occurrence of alluvium.

The Municipality of Machico is located on the westernmost tip of the island, extending from the sea to the northern mountains and ending at Ponta de São Lourenço, with a total area of 68.31 km^2 (Alves et al, 2022). The city of Machico corresponds to the parish of Machico, one of the five parishes in the Municipality. According to the latest data from the National Statistics Institute (INE) in 2020 the resident population was around 19,926 inhabitants. With around 287 inhabitants per km2, the municipality of Machico has the fourth highest population density in the Autonomous Region of Madeira. In Machico's economy, trade, hotel services, fishing, catering, civil construction, agriculture, and tourism, among others, predominate as main activities (CMM, 2022).

In the city of Machico it is possible to observe a single stream, the Ribeira de Machico (Figure 2) and the corresponding hydrographic basin (Antunes, 2014).

The U-shaped Ribeira de Machico basin, seen in Figure 3, is in the municipality of Machico, on the east side of Madeira Island. It is an open valley, which extends from altitudes of 1000 m to the sea. This is ramified into a huge number of secondary valleys where lower courses flow.

Regarding the shape of the watershed, it is important to consider different aspects, such as the compactness index, shape factor and elongation index (Barreto, 2013).

Analyzing in a first phase, we have the basin compactness index (KC) of Ribeira de Machico that relates the basin perimeter with its area. Figure 4 shows that Ribeira de Machico has a watershed with a drainage area of 24.58 km^2 and a perimeter of 24.97 km, which is one of the seven largest watersheds in RAM (Castro, 2011).

The drainage area is the area of the horizontal projection of the surface of the hydrographic basin, being normally determined by planimetry or using GIS (geographical information system) (Hipólito, 2011). It should be noted that the larger the area of the hydrographic basin, the greater the peak flood

Figure 2. Map with the location of Ribeira de Machico
Source: Google Earth (2022)

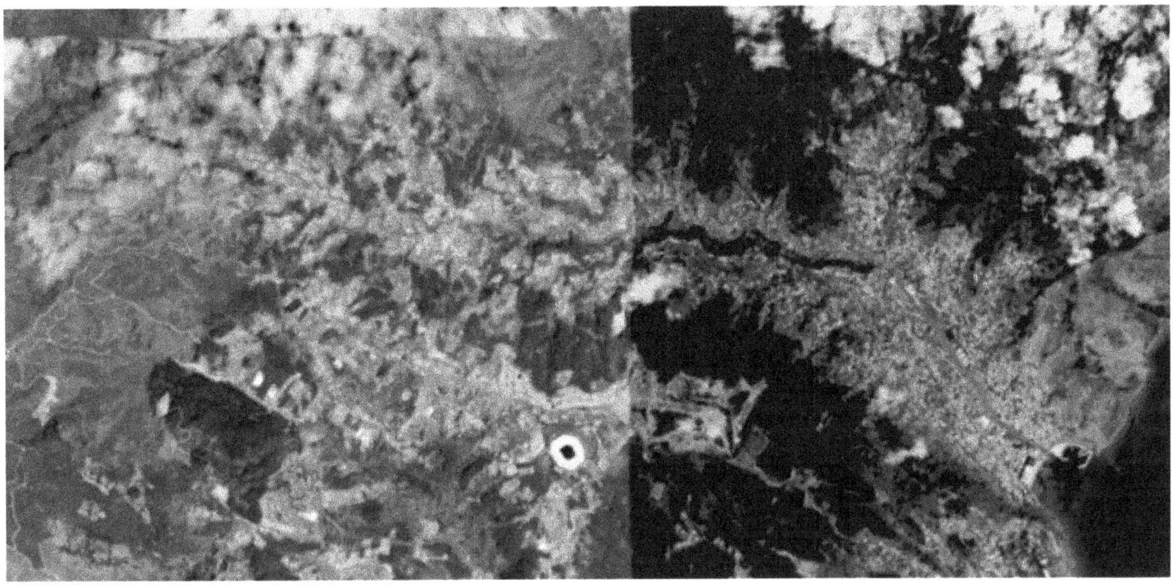

Figure 3. Map with the Ribeira de Machico watershed
Source: Google Earth (2022)

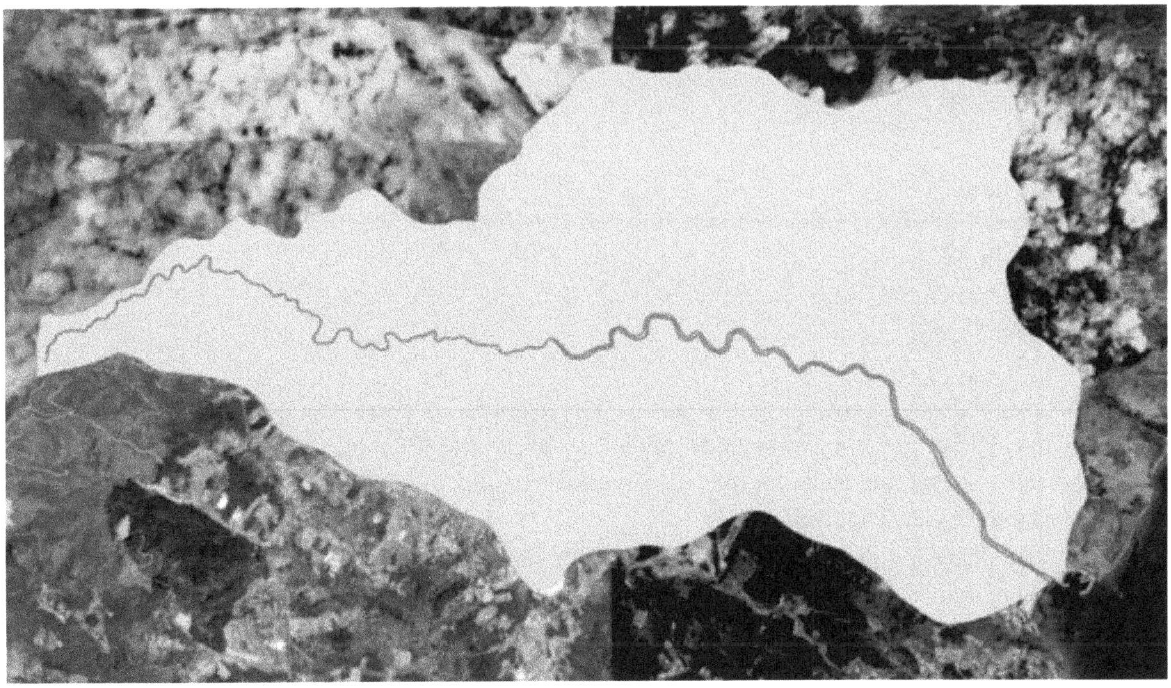

Figure 4. Dimensions of the largest hydrographic basins in Madeira Island
Source: Author

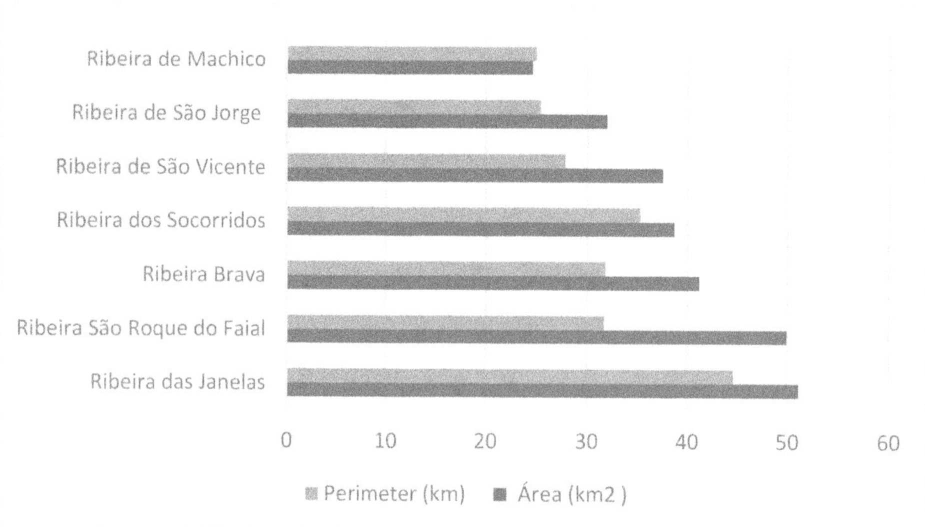

flow will be, since the formula used for its calculation, that of the peak flood flow by the rational method, refers to the area being entirely related (Barrett, 2013). The perimeter is the length of the boundary line that delimits the watershed, which is represented in the horizontal plane (Barreto, 2013).

The shape of the basin will influence or not the susceptibility of flooding. In the case of the Ribeira de Machico basin, the KC value is 1.42 (Silva et al, 2021). This value was obtained by the following formula:

$$KC = 0,282 \times \frac{P}{\sqrt{A}} \tag{1}$$

Where we have:

P = Perimeter (km);

A = Basin Area (km^2).

The determined value indicates that we have a Gravelius Index greater than 1.128, that is, it will be necessary to calculate the elongation index, otherwise, and if the KC indicated a value close to unity, it would indicate a greater tendency for the occurrence of flooding imagining precipitation instantaneous and regular in the basin (Hipólito, 2011).

In a second phase it will be necessary to determine the form factor (KF). It is important here to determine the length of the basin "L" and its average width "b". The length of the basin is measured by the longest watercourse from the furthest headwater of the basin to the mouth, on the other hand, the average width is obtained by dividing the basin area by the basin length (Barreto, 2013). Basins with low KF will tend to flood with smaller flood points than others with high form factors (Hipólito, 2011). In the case of the Ribeira de Machico basin, the KF value is 0.16. In the Machico Hydrographic Basin there is a low form factor, that is, conducive to smaller flood points (Barreto, 2013). This value was obtained by the following formula:

$$KF = \frac{A}{L^2} \tag{2}$$

Where:

A = Basin area (km^2);

L = Length of the main watercourse of the basin (km).

In this third and final phase, the elongation index (KL) was determined, which relates the length to the width of the basin (Barreto, 2013; Lousada et al, 2021). In the case of the Ribeira de Machico basin, the KL value is 4.10. There is an elongated basin, which implies a lower risk of flooding: longer routes, longer runoff times, therefore longer response time (Barreto, 2013). This value was obtained by the following formulas:

$$KL = \frac{L}{b} \tag{3}$$

Where we have:

L = Length (km);

b = Width (km).

Therefore, the variable "L" and "b" are presented in the following equations and the following values:

$$L = \frac{KC \times \sqrt{A}}{1{,}128} \times \left| 1 + \sqrt{1 - \left(\frac{1{,}128}{KC}\right)^2} \right| = 10{,}04 \text{ km} \tag{4}$$

$$b = \frac{KC \times \sqrt{A}}{1{,}128} \times \left| 1 - \sqrt{1 - \left(\frac{1{,}128}{KC}\right)^2} \right| = 2{,}45 \text{ km} \tag{5}$$

Viewing Figure 5, the island of Madeira is made up of several streams and streams (water resources) that are important for the population (source of subsidence) but also a source of risk.

Observing Figure 6, we can see the different hydrographic basins, which are narrow in the main watercourses, elongated and with relatively small areas.

Figure 5. Water resources on the island of Madeira
Source: Google Earth in the Classroom

Figure 6. Watersheds on the island of Madeira
Source: Google Earth in the Classroom

Figure 7. Geological map of the island of Madeira adapted to the basin under study
Source: Geological Chart of the island of Madeira (2010)

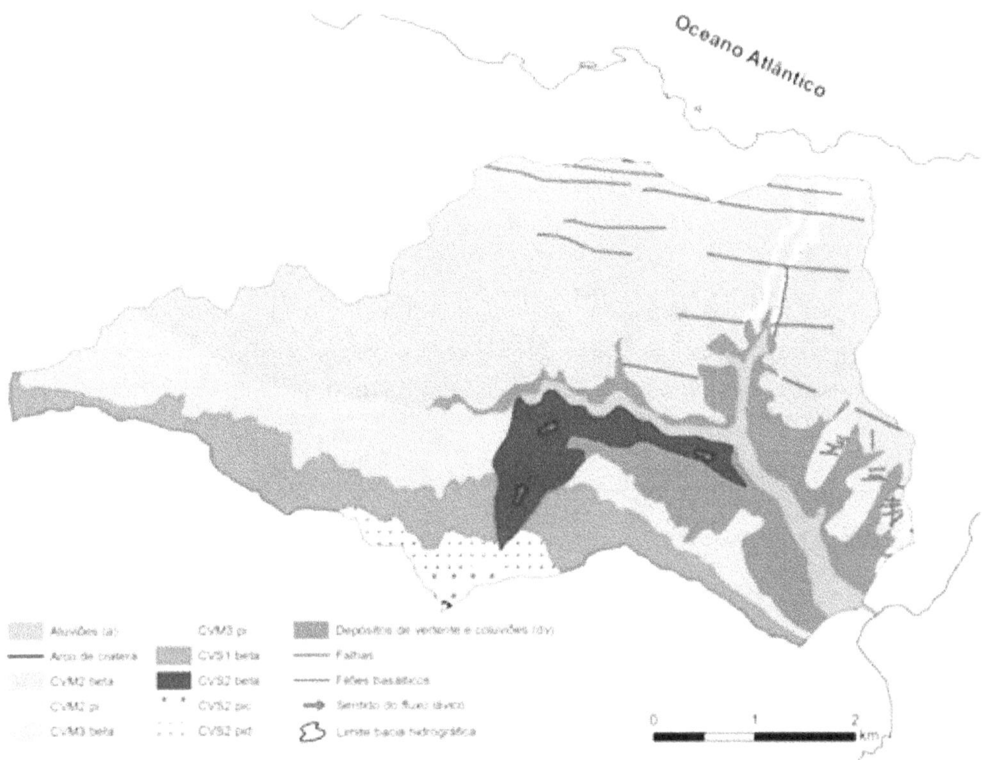

Regarding the geology, through Figure 7, the composition of the Machico watershed can be observed. The Machico watershed is composed of intermediate (CVM) and Upper (CVS) volcanic complexes, which have medium to good permeability (Caetano, 2014).

Visualizing Figure 8, the city of Machico has a single watershed, called Ribeira de Machico, which has a high area of Andosols (Caetano, 2014). According to the Madeira Island Soil Chart (Ricardo et al, 1992), the vegetation present in the Machico watershed is very diverse.

LAND USE, URBAN GROWTH AND FLOODING

The accelerated growth of cities and disordered consumption requires the elaboration of studies that allow the adoption of public policies that can build a socio-economic and environmentally sustainable environment. One of the studies refers to the problems of water resources on:

- **Surface Runoff.** From the moment that the precipitation falling on the ground exceeds the capacities related to the processes of interception and infiltration, the volume of excess water, in obedience to the laws of gravity, drains from the surface of the land to the nearest water line, giving origin of surface runoff (ISEL, 2015). The water lines of smaller section (furrows, ravines, streams, streams, and streams) are associated with others of successively larger section (rivers), which will eventually, with rare exceptions (endorheic basins), communicate with the sea (Lencastre, 1972).

Figure 8. Extract from the soil map of the island of Madeira for the Ribeira de Machico basin
Source: Soil Chart of Madeira Island (1992)

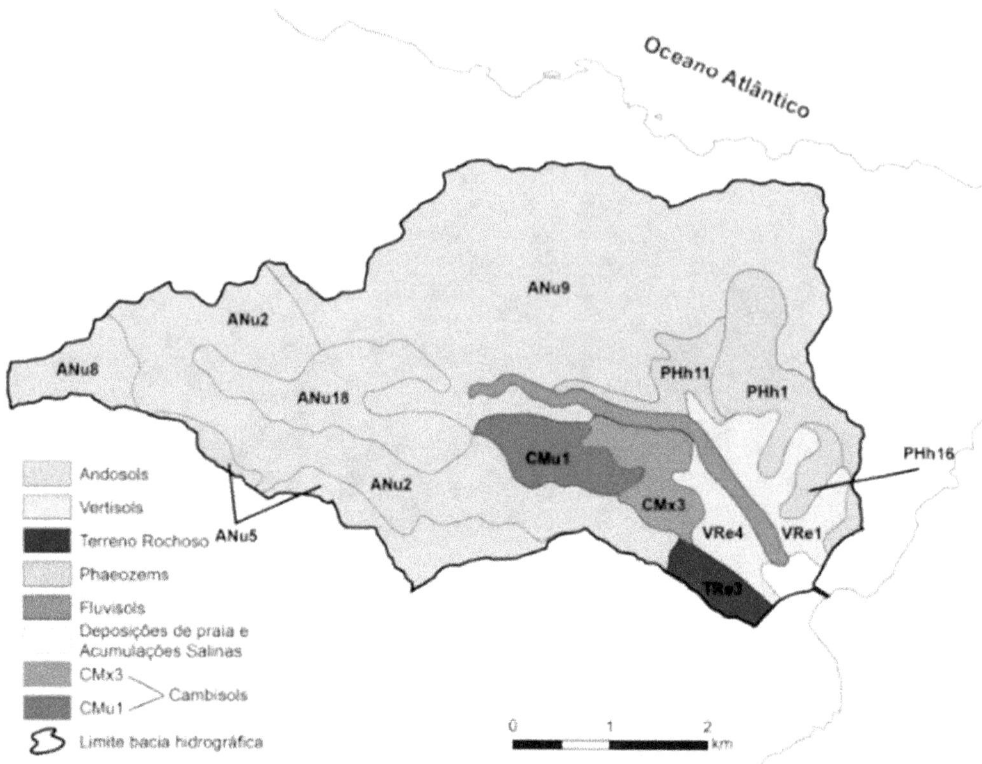

- **Population Growth.** It results in the waterproofing of the soil and subsoil and the pollution of water resources, since rivers, streams and coastlines are sought after for the construction of buildings and the expansion of cities. There is also the strangulation of these water sources that are altered to respond to the needs of the population. Flooding in urban areas through soil impermeability and overloading of artificial water drainage systems, reduces the infiltration of rainwater and increases surface runoff (Tucci, 2005; Ferreira et al, 1996). The river channeling works, using concrete, cause an increase in the peak of the floods and a decrease in the response time of the drainage basins (Tucci, 2005; Ferreira et al, 1996).

In the city of Machico, the first buildings appeared in the east, on the left bank of the stream, around the primitive Chapel of Christ, currently known as Capela dos Milagres (Leite, 1989). Thus, the idea is that the genesis of the settlement of Machico occurred in the eastern part of the Ribeira de Machico, which occasionally threatened the local community during the rainy season (Leite, 1989). It is not by chance that the Sítio da Banda D'Além is often referred to as the oldest "neighborhood" in the City of Machico (Azevedo, 1866).

Machico grew up with the phenomenon of alluvium as a constant, making this part of its history. With the passage of time and with technological advances, people continue to inhabit the floodplains and the problems of alluvium continue. Among the areas most affected by the alluvium, Figure 9 shows the area designated in the Regulation of the Municipal Master Plan of Machico - PDMM, as the old or historic area of Machico (RAM, 2005).

Figure 9. Extract from the PDMM with the U1 zone designated as ancient or historic Machico
Source: Machico PDM

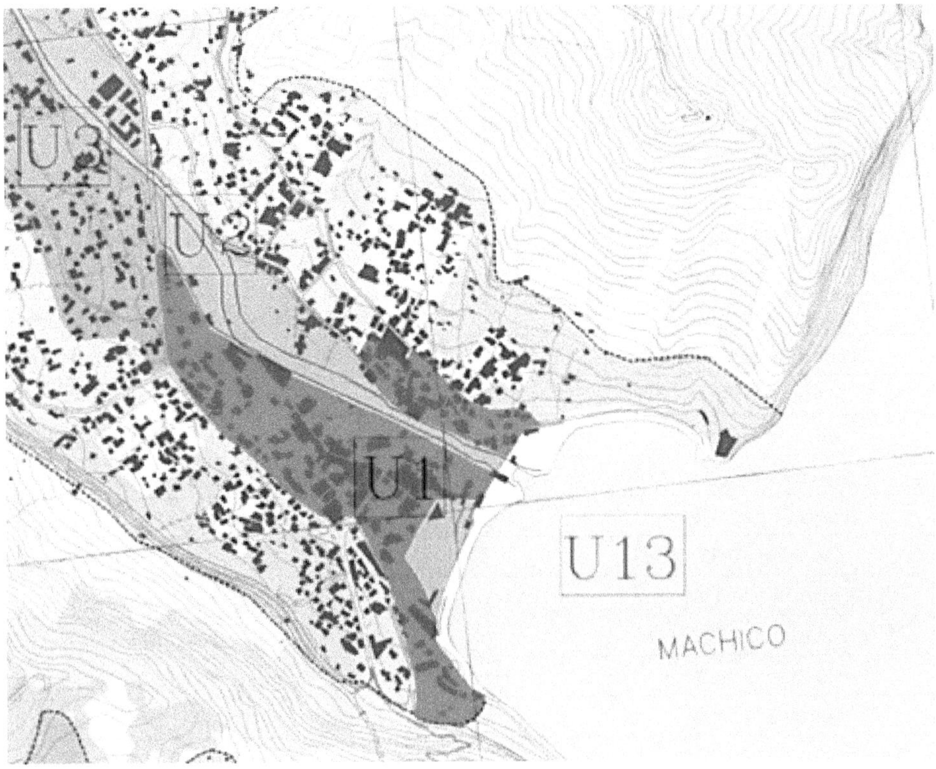

In Table 1 it is possible to observe the history of the most significant alluviums recorded in the city of Machico since its settlement until the present day. The first record in the city of Machico dates to 1467, causing the death of Franciscan friars who had a hermitage by the stream, in front of the Misericórdia chapel (Almada, 1865).

Of all the catastrophes that martyred the Madeirans, the greatest was the flood of October 9, 1803. According to reports at the time, about 1000 people must have died, due to the brute force of the waters, most of them in Funchal (Quintal, 1999). Every year, on October 9, Madeirans remember the tragic date, participating, with wide adherence and deep feeling, in the solemn procession of Senhor dos Milagres in Machico, another of the martyred places (Quintal, 1999).

On November 29, 1901, there were floods and landslides in Funchal and Machico. "In several streets of the city, the water rose to a height of 3 palms, entering some establishments and bringing part of its content to the street. Torrential rains have fallen in the village of Machico. Moinho da Serra and heading to Banda de Além, causing great panic among the inhabitants of that place. A ravine landed a house on the Ribeira de Machico site, killing nine people. The village was in great danger because it continues to rain" (A Cruz, 1901).

March 5th and 6th, 1921. It rained heavily throughout the island, with flooding and damage to agriculture in Machico, Santana, Faial, and São Jorge. In Machico the waters rose in some places to almost the first floor of the houses. In Ribeira Brava, three children died due to the collapse of a house and another child succumbed, dragged by the waters (Quintal, 1999).

Table 1. Historical summary of the main alluviums on the island of Madeira

Date	Source
Winter of 1467	(Almada, 1865)
November 9, 1678	(Almada, 1865)
November 18, 1724	(Almada, 1865)
October 9, 1803	(Almada, 1865)
October 23, 1815	(Almada, 1865)
October 23, 1842	(Almada, 1865)
February 22, 1853	(Almada, 1865)
November 29, 1901	(Silva & Menezes, 1921; Quintal, 1999)
February 25 e 26, 1920	(Silva & Menezes, 1921; Quintal, 1999)
March 5 e 6, 1921	(Quintal, 1999)
November 3, 1956	(Quintal, 1999)
September 27, 1989	(Quintal, 1999)
October 24 e 29, 1991	(Quintal, 1999)
October 29, 1993	(Quintal, 1999)
October 19 e 20, 1997	(Quintal, 1999)
February 20, 2010	(IST & UMA & LREC, 2010)
November 6, 2012	(Jornal Público, 2012)
November 29, 2013	(DROTA, 2016)
March 27, 2021	(Radio TSF, 2021)

November 3, 1956. Machico, Santa Cruz, Água de Pena, Porto da Cruz and Santo da Serra were the localities affected by the violence of the waters. Roads, bridges, and houses destroyed; cultivated land and animals washed out to sea; 6 dead (Quintal, 1999). "The parishes of Porto da Cruz, Machico, Água de Pena and Santa Cruz were yesterday the theater of a horrible tragedy. The floods of October 9, 1803, and October 24, 1842, were repeated there. of real anguish and natural panic, waiting for the moment, which they thought would arrive, to be dragged by the impetuosity of the waters. It was around 10:30 am. . . . This is the only way to explain the immediate increase in the flow of streams and the sudden floods in this region." (Diário de Notícias, 1956). In Porto da Cruz, four bridges were destroyed. The flood washed away fruit trees, willows, and vines. Many cows, pigs, goats, and sheep ended up in the sea. In Machico, 3 people were killed and several people were injured (Quintal, 1999). "The bridge that connected the Piquinho site to Paraíso was destroyed; the same happened with the one at Ribeira Grande. As for the Ponte Nova, on the Estrada to Caniçal, only the central part exists, as the ends were washed away by the flood. Banda d' Além site, the waters that crossed the walls invaded the chapel of Milagres. the terrible flood, suffice it to say that the waters reached the height of the roofs of some buildings. sea, and the destruction of the Seixo bridge. In the parish of Santo da Serra, the Ribeira de Machico site was the most affected. There a woman died, two houses were destroyed, several haystacks disappeared with the flood, the same happening to seven cows" (Diário de Notícias, 1956).

January 20 to 24, 1979. Southwest wind with speed between 40 and 50 km/hour, with gusts that reached 70 km/hour; rough sea; heavy downpours. Floods, landslides, houses, bridges, and paths destroyed throughout the island (Quintal, 1999). The "diluvian" rains began to fall around one and a half hours yesterday morning - the 20th - and with greater intensity between three and four hours. the stream of Ribeira de Machico" (Diário de Notícias, 1987), which ran one meter below the deck of the access bridge to Banda d' Além (Quintal, 1999). Several municipal roads and bridges were destroyed in Ribeira Grande and Maroços due to the rising flow of the stream. In that place, a flood destroyed two houses and killed two people (Quintal, 1999). "The paths and streams in the areas overlooking the county seat became authentic streams, dragging boulders and other debris that cut short the paths and main roads, in particular the E.N. 101 between Machico and Porto da Cruz, where several landslides occurred. and collapses" (Diário de Notícias, 1987).

The alluvial problem is not something current, given the various reports and records shown in Table 1 and observable in Figures 10, 11, and 12.

Figure 10. Alluvium in the municipality of Machico in 1956
Source: Museum – Vicentes

Figure 11. Alluvium in the municipality of Machico in 2010
Source: Facebook – António Nóbrega

Figure 12. Alluvium in the municipality of Machico in 2012
Source: Facebook – Foto Mundo

Figure 13. Flooding of the road next to Ribeira de Machico
Source: Diário de Notícias de 2021

Years later, even carrying out containment and drainage works, the problem of alluvium was not resolved, as can be seen in Figure 21, in recent records on March 27, 2021 (Diário de Notícias, 2021).

There is a need to develop solutions for floods, and it is essential to consider technical, engineering and planning issues, including the political actors responsible for critical decisions in the management and mitigation of these problems, as well as the population itself.

On the island of Madeira, after the 20th of February 2010, there was a change in the way the Regional Government approached natural disasters, in particular, alluvium. There is a greater public investment in mitigating the effects of natural disasters, to reduce the damage caused by these phenomena, which are mostly associated with poor urban planning and planning. Anthropogenic interventions associated with unplanned urbanization, changes in river channels, added to the orographic effect and hydrogeomorphological susceptibility of the basins make alluviums recurrent and cause significant damage and losses

Figure 14. Alluvium in the municipality of Machico in 2010
Source: Facebook – António Nóbrega

in Madeira Island (Freitas et al, 2020). An example of this situation was the 2010 Alluvium, where the severity of the data was related to natural and anthropic factors, where the urban occupation of the city of Funchal occurred without planning (Aragão, 1987). In parallel with human actions, the geological formation of the island and its geomorphological evolution resulted in sloping basins with a short time of concentration, making them naturally susceptible to the occurrence of extreme hydrogeomorphological events (Almeida & Lopes, 2014).

In Machico, what is currently verified is that the existence of high walls, helped to control the force of the waters that circulate in Ribeira de Machico. Adjacent to these walls are the communication routes and paved areas that, whenever an alluvium occurs, transform into authentic rivers (Figure 14).

If the floodable area remained free, some of these situations would be avoided (Figure 15).

In this context, the risk of flooding produced in the city will depend on the form and degree of occupation of the areas close to Ribeira de Machico (Riverside Zones) and on the frequency of occurrences of alluvium. In the city of Machico, there is an increase in the incidence of small floods that affect the population, even when the rain is less intense, revealing the vulnerability of those who live here (Diário de Noticias, 2022).

Vulnerability to natural disasters is not only associated with "human action" and the "climate framework". There are other physical aspects, namely geometry, drainage network, relief, geological substrate, soils, vegetation, etc. (Lousada et al., 2022). There are also those who argue that, in addition to this set of factors, the socioeconomic situation of the population will play an important role in the vulnerability to alluvium (Monteiro, 1991).

Figure 15. Alluvium in the municipality of Machico in 2012
Source: Facebook – Foto Mundo

Figure 16 shows an excerpt from the Flood Risks Management Plan – PGRI, of the Autonomous Region of Madeira (DROTA, 2016). It is possible to visualize the extension of the flooded area in the downtown area of the village of Machico, with a large part of the urban area conquered by the blue stain. This is since it is not possible for this water to flow, since this area of Machico is next to the Ribeira de Machico bank, at a low elevation associated with the significant rise in sea level, during high tide.

It is important to identify the floodable zones based on the definition of flood risk in these areas at different topographic levels. Afterwards, regulation of the type of occupation of these areas must be carried out, indicating those that present greater and lesser risk of flooding. Finally, these areas must be introduced in the Municipal Master Plan of Machico. This combination of prevention measures makes it possible to reduce the effects of floods and improves planning for the occupation of the area subject to flooding (Pegado et al, 2014). Regional policies and societal behaviors are changing often in these insular territories leading to the necessity of close monitoring and new analysis over the directions and dynamics of the land-use changes as well as the management of sustainable development strategies (Naranjo Gómez, 2020).

Figure 16. Map of flooded zones (extension): Ribeira de Machico
Source: PGRI (2016)

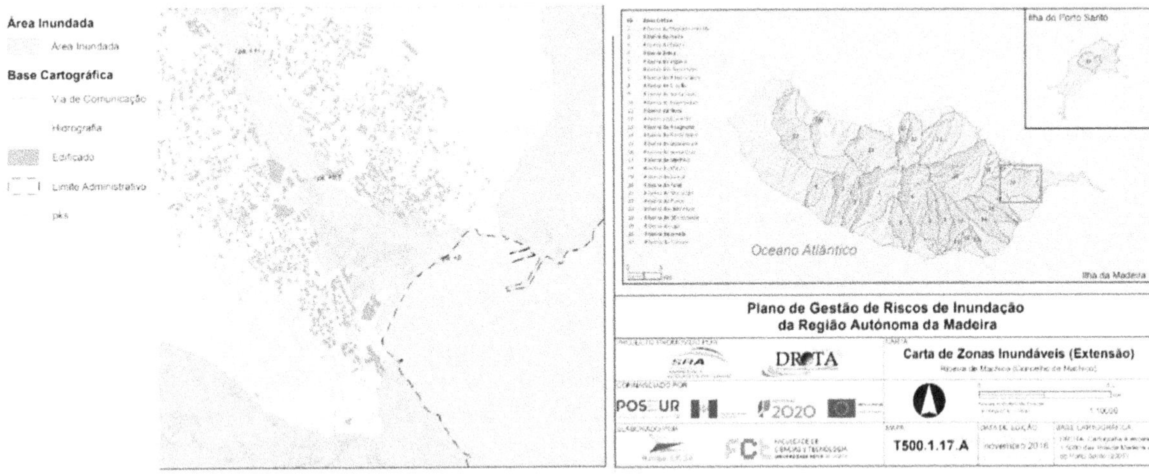

PROMOTE FLOOD RISK MANAGEMENT IN THE CITY OF MACHICO

Over time, it appears that in the growth of the urban core of Machico, there were inadequate occupations of flood zones, most of the time legal and consolidated. This type of urbanization quickly destroys the environment, aggravating urban drainage problems.

To mitigate the risks associated with alluvium, the zoning of floodable areas is defined to analyze the records in terms of time and space, always leaving the floodable areas unobstructed (Pegado et al, 2014). The use of land use planning as a flood risk mitigation strategy is one of the most appropriate in the task of managing the expansion of urban areas and new constructions (DROTA, 2016).

All streams in the Autonomous Region of Madeira have their natural flooding area. Its rugged orography determined a progressive urban occupation of the floodplains (Lousada & Loures, 2020). When the population crosses the limits of the natural environment where they live, the degree of exposure to these natural phenomena increases, becoming a serious social, environmental, and economic problem. Along the coast is where there is a more concentrated pattern of land occupation, so this is where the problem is aggravated, because in periods of high tide, floods and flooding become frequent. In general, it appears that about 40% of the flooded areas are urban areas or areas with equipment or infrastructure (DROTA, 2016). These floods are characterized by the extravasation of water from the stream bed to the floodplain.

Figure 17 shows an image taken from Tucci (Tucci, 2009) where the urban planning analysis is visible, showing the regulation of the floodplain, subdividing the flow section into three lanes. In number 1, there is the flood crossing lane, in number 2 the lane with restrictions and in number 3 the low-risk lane.

It is clearly noted that in the case of the City of Machico, the walls (Figure 18) that confine the Ribeira de Machico put out the three strips described above. The walls, when being built, should be between strip two and three, since strip three has a small probability of flooding, being reached in exceptional years by small water depths at low speeds (Pegado et al, 2014).

Figure 17. Flood zone regulation
Source: U.S. Water Resources Council (1971) and Tucci (2009)

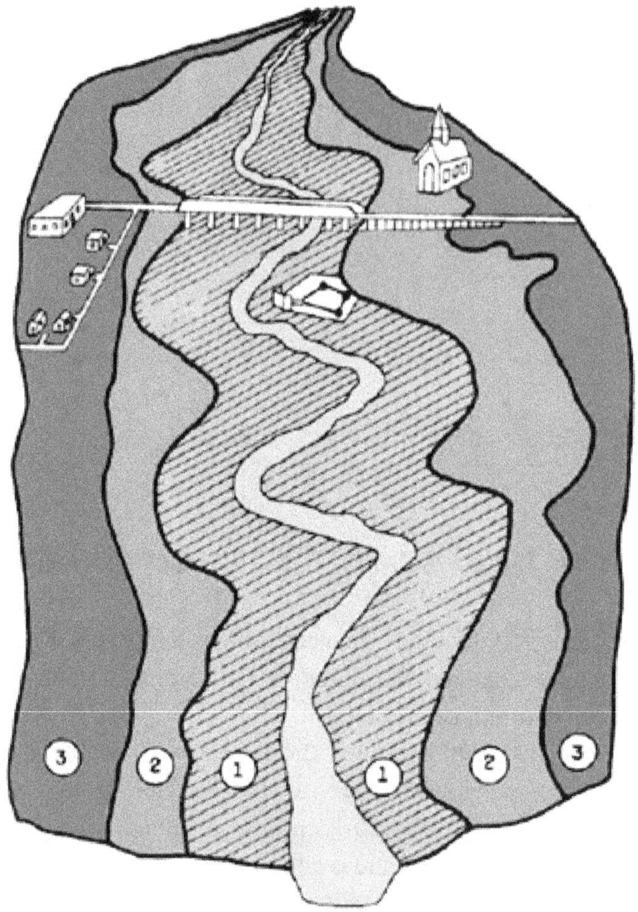

The growth of the city along the stream, resulting from the occupation of the floodplains with various types of infrastructure such as farms, housing, warehouses, industrial units and even power stations, left a narrow path for the water to go to the sea. This growth is largely due to the proximity to the city center, the easy access to public services (Schools, Health Center, Machico City Council, Tax Authority, Social Security, Bus Central), private buildings (Conservatory of Music, Forum Machico, Church of Nossa Senhora da Conceição, Capela dos Milagres, Several hotels and commercial spaces) among others. These zones have favorable conditions for the occurrence of alluvium (Pegado et al, 2014).

In this way, it is imperative that all Emergency Plans are carried out/reviewed, that river rehabilitation and revitalization programs are defined, with the aim of mitigating the problems that may arise in the face of these alluviums, that the reorganization of the urban land takes place, using the displacement of sensitive infrastructure and buildings located in floodable areas, that longitudinal structures (walls) are built and reinforced, that the population is involved (for example, in the participation of simulations of flood situations), that municipal supervision is more active in order to prevent the occupation of the Ribeira de Machico bed and even if monitoring systems are installed/reinforced.

Figure 18. Walls in Ribeira de Machico
Source: Author

FLOOD RISK FOR RIBEIRA DE MACHICO

The determination of the Flood Risk consists of the combination of the probability of flooding, considering its magnitude, and its potential harmful consequences for human health, the environment, cultural heritage, infrastructure, and economic activities. Its adverse consequences are assessed by identifying the number and type of activity affected, which can sometimes be supported by a quantitative analysis (Decree-Law no. 115/2010).

Based on the Flood Risk Management Plan – PGRI, of the Autonomous Region of Madeira (DROTA, 2016), this chapter presents several tables showing various elements exposed to the possible occurrence of alluvium. The global risk analysis recommended in the article is identical to that elaborated in the PGRI (DROTA, 2016). This relates the degree of threat of the flood to the consequences. The flowchart (Figure 19) summarizes the analysis that leads to the flood risk mapping (DROTA, 2016).

As a result of this analysis, Table 2 shows the population exposed to the possible occurrence of floods, identified for the three return periods (20, 100 and 500 years), regardless of the associated risk level.

Table 3 shows the exposed elements related to economic activity, cultural heritage, and sensitive buildings. Regarding tourist establishments, the List of Projects of Tourist Enterprises was also considered, favorably appreciated by the Regional Secretariat for Economy, Tourism and Culture (list sent by protocol of 11/10/2016), noting that no enterprise is in an area risk of flooding (DROTA, 2016).

In the agricultural or livestock areas that exist in risk zones, specific measures for areas with a return period of 20 years, associated with high to very high risk, which cover about 2 ha, fundamentally located in Ribeira de Machico. Of the 12 tourist developments present in the municipality, for the return period of 500 years, 3 are affected by floods. Regarding the sensitive buildings affected by the floods, there are three Day Care Centers, Nurseries and Schools, a church, and the Municipal Civil Protection Service, located in the Machico City Council Building. Cultural Heritage is also exposed to flooding in

Figure 19. Modeling and risk analysis flowchart
Source: PGRI (2016)

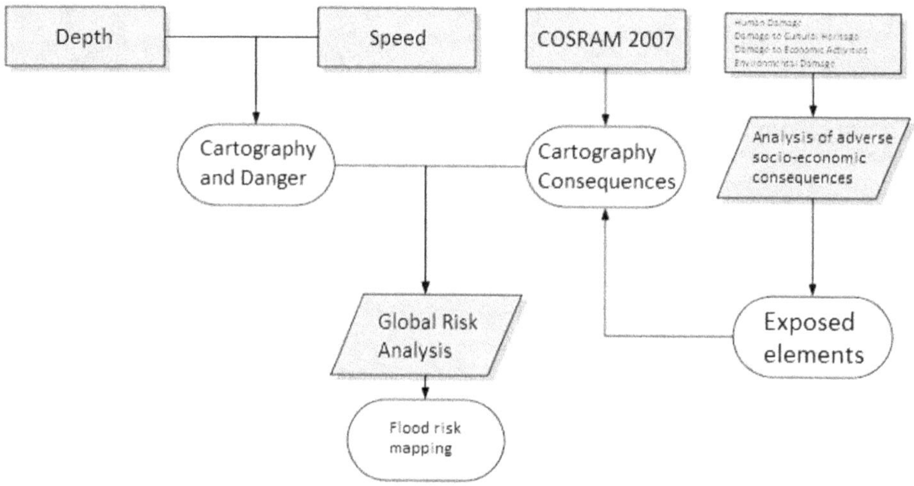

Table 2. Exposed elements related to the population

	Payback Period, T (Years)	No. of Potentially Affected Inhabitants
Ribeira de Machico	T20	496
	T100	657
	T500	714

Table 3. Exhibited elements related to economic activity, cultural heritage, and sensitive buildings

	Payback Period, T (Years)	Agricultural Activity (ha)	No of Tourism Establishments	No of Cultural Heritage	No of Sensitive Buildings
Ribeira de Machico	T20	5,853	-	5	2
	T100	6,637	-	5	4
	T500	7,201	3	5	5

Table 4. Supply and basic sanitation infrastructure and road infrastructure exposed to floods for T=100 years

Ribeira de Machico	Water Supply System Infrastructure	Potentially Affected Extent (m)	
		ERP	ERC
	1	208	8

critical areas, with a significant number of exposed elements (two buildings of municipal interest and three buildings of public interest), with a vulnerability, in many cases, relatively frequent (20 years).

Table 4 shows the number of sanitation and water supply infrastructure exposed to floods and road infrastructure located in critical areas (main regional roads - ERP and complementary regional roads - ERC) for a return period of 100 years.

Considering the data presented, it is important that the Municipality of Machico anticipates alluvial prevention and mitigation measures. In this way, it is important to mention that a good intervention strategy in spatial planning involves:

- Keep the natural inundation beds of watercourses free of constructions. According to the WATER LAW (Law no. 58/2005), given the importance of floods and inundations, Article 40 provides for protection measures against floods;
- Raise awareness of the population for more sustainable land use with the aim of mitigating the negative aspects;
- Pay attention to the licensing of works, where alluvial risks are foreseeable. Define non edificant zones in the Machico Municipal Master Plan, based on risk maps.

FLOODED AREA OF RIBEIRA DE MACHICO IN THE CONTEXT OF RAM

Through directive nº 2007/60/EC, of 23 October, which established the framework for the assessment and management of flood risks, with the aim of mitigating the consequences associated with harmful floods for human health, the environment, the heritage cultural and economic activities, other critical areas were identified in RAM besides Ribeira de Machico. In 2010, this directive was transposed into national law, through Decree-Law nº 115/2010, of 22 October, defining that the state must proceed with the elaboration of the following instruments: maps of floodable areas for risk areas; flood risk charts and flood risk management plans, which resulted in the PGRI (DROTA, 2016).

Through the Plan, it was possible to carry out an analysis of the different streams and streams of RAM, defined as critical areas. Table 5 shows the twenty-seven critical areas analyzed by municipality.

Figure 20 shows, for the return period of 500 years, the number of inhabitants potentially affected. Among the twenty-seven critical areas, Ribeira de Machico is the second critical area in RAM with the highest number of potentially affected inhabitants.

Figure 21, for a return period of 500 years, represents the exposed elements related to economic activity, cultural heritage, and sensitive buildings. Regarding the exposed elements related to economic activity, it is observed that Ribeira de Machico has the second largest agricultural area exposed to alluvium in RAM, after Ribeira da Abegoaria, however the largest number of tourist establishments exposed to flooding belongs to Ribeira from Machico. The number of buildings classified as Cultural Heritage exposed to floods in the case of Ribeira de Machico is five, more than any other critical area in RAM. In the case of sensitive buildings, for Ribeira de Machico there are only four, the same as for Ribeira de Santa Luzia, less than Ribeira Brava which has fifteen and less than Ribeira de São Vicente with five.

The number of sanitation and water supply infrastructures exposed to floods located in critical areas, for a return period of 500 years, are shown in Figure 22. Of the twenty-seven critical areas, only seven have sanitation and supply infrastructures of water, making a total of fourteen sanitation and water supply infrastructures.

Table 5. Streams in RAM by municipality

County	Designation of the Critical Zone
Ponta do Sol	Ribeira da Madalena do Mar
	Ribeira da Ponta do Sol
Ribeira Brava	Ribeira da Tabua
	Ribeira Brava
Câmara de Lobos	Ribeira do Vigário
Funchal / Câmara de Lobos	Ribeira dos Socorridos
Funchal	Ribeira do Ribeiro Seco
	Ribeira de São João
	Ribeira de Santa Luzia
	Ribeira de João Gomes
	Ribeiro da Nora
	Ribeiro do Lazareto
Porto Moniz	Ribeira da Laje
	Ribeira da Janela
Santa Cruz	Ribeira do Porto Novo
	Ribeira da Abegoaria
	Ribeira da Boa Ventura
	Ribeira de Santa Cruz
Machico	Ribeira de Machico
	Ribeira da Maiata
	Ribeira do Junçal
Santana	Ribeira do Faial
	Ribeira de São Jorge
São Vicente	Ribeira do Porco
	Ribeira dos Moinhos
	Ribeira de São Vicente
Porto Santo	Ribeira do Tanque

Finally, in Figure 23, the extension, in meters, of the Road Infrastructures exposed to floods for T=100 years, for the twenty-seven critical areas of RAM. In total, approximately 11.5 km of roads are potentially affected for the 100-year return period. The most critical situations are found in Ribeira dos Socorridos, Ribeira Brava, Ribeira de Santa Luzia, Ribeira de João Gomes and Ribeira de São Vicente (DROTA, 2016).

In addition to these situations, it is worth mentioning, in terms of communications and electricity supply to the populations, the historical occurrence resulting from floods, service supply interruptions for several hours and covering a significant population (DROTA, 2016).

Figure 20. Comparison between the 27 critical areas of RAM: Number of inhabitants potentially affected

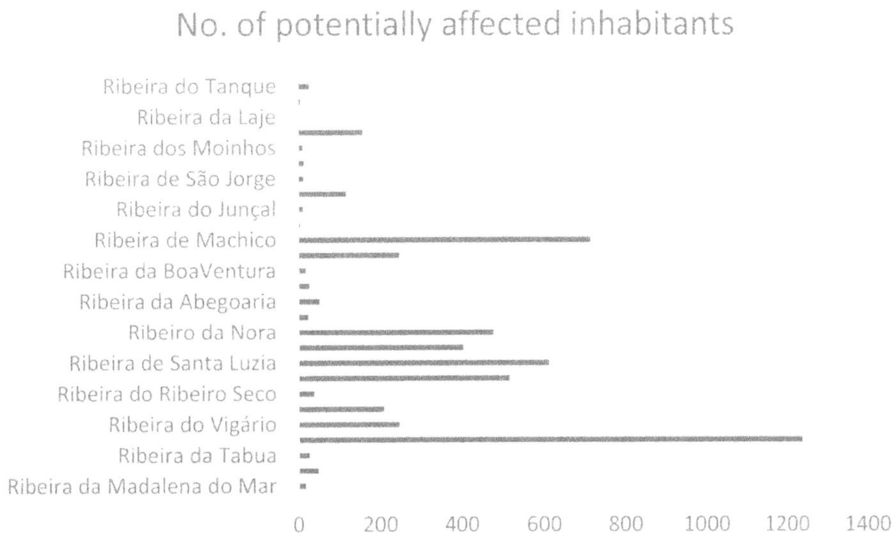

Figure 21. Comparison between the 27 critical areas of RAM: Economic activity, cultural heritage, and sensitive buildings

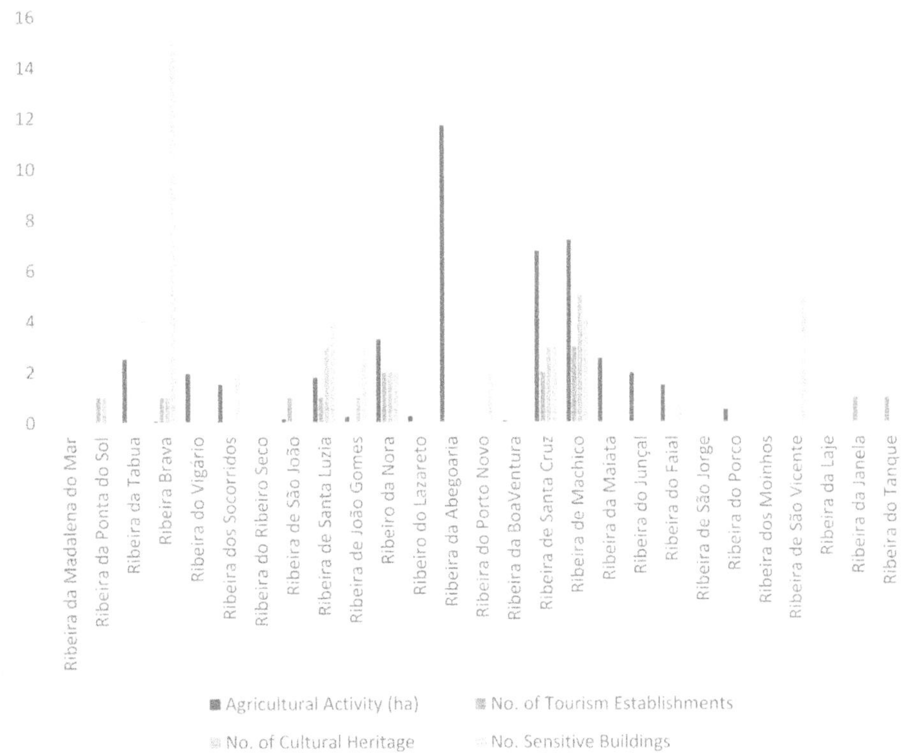

Figure 22. Supply and sanitation infrastructure exposed to floods

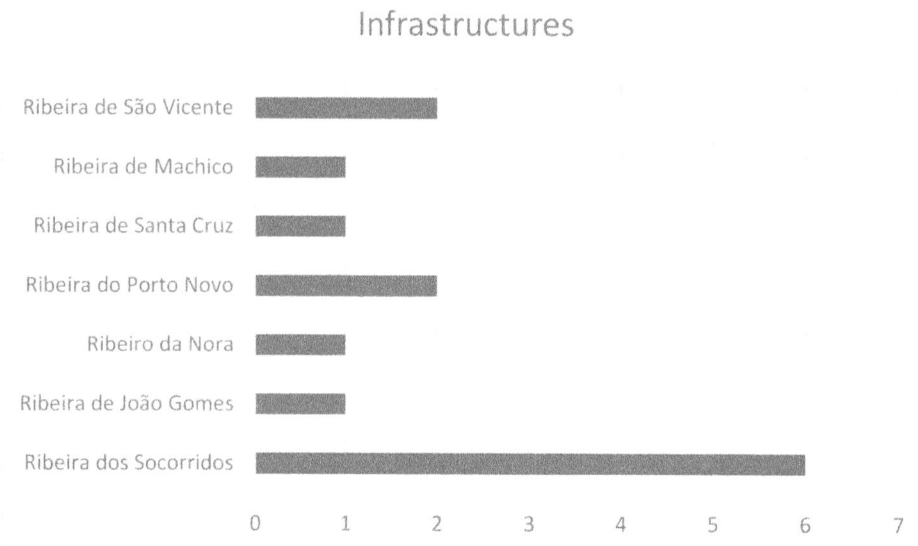

Figure 23. Road infrastructure exposed to flooding for T=100 years

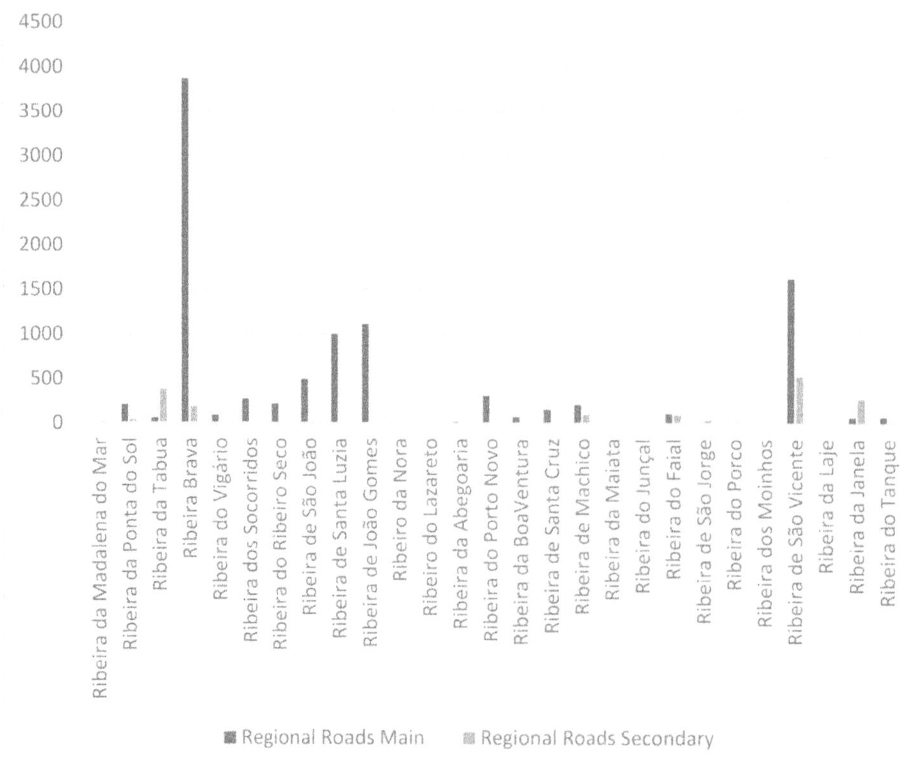

CONCLUSION AND FUTURE CONSIDERATIONS

Based on the information presented in this article, it is concluded that the island of Madeira presents itself as a region favorable to the occurrence of alluvium. These are due to several factors, such as the longitudinal and transversal slope of the hydrographic basins associated with high precipitation.

Alluvium problems are exacerbated due to poor public management of the territory. The village of Machico does not have a municipal flood risk management plan that mitigates the problems caused by the alluvium. There is indeed a RAM PGRI that includes some measures that serve to reduce the damage caused by a flood event. However, it would be important, on the part of the municipality, to integrate some measures and rules, namely minimum distances / distances on the banks of the Ribeira de Machico and even non edificant areas in the Machico PDM.

The need arises to implement a system for monitoring the stream, to prevent the risk of flooding. This forecasting system will minimize the negative consequences arising from the occurrence of floods. It will be necessary to sensitize the population about the risks and how to act in the face of an alluvium (simulacra). The municipality of Machico will be able to carry out monitoring using video cameras installed in critical sections along the Ribeira de Machico. Using automatic image recognition technology, the system will be able to determine the flow rate and speed, among others.

Knowing how to live with floods by mitigating their impacts is a measure that must be adopted by civil society and public bodies (Silveira, 2008). Therefore, in addition to chambers in critical sections, it will be important to place them near weirs, analyzing their behavior in retaining solid material. which will constitute an important instrument to contribute to a possible improvement and adaptation of these infrastructures in the performance of their retention functions. Installation of other technological equipment such as sensors (water level) and geophones (solid material that circulates in the stream) assisting the video cameras. Installation of udographers (rainfall monitoring), contributing to the correct sizing of the drainage network.

The Municipality of Machico should launch a municipal system for monitoring and warning of floods, as a measure to mitigate the risk of floods for populations, infrastructure, and the environment, which should be part of a specific PGRI for the City of Machico, with management objectives risk assessment, risk assessment, definition of the level of protection and implementation of prevention measures that may result in the availability of online hydrometeorological data to the population.

Risk prevention is a concern progressively rooted in RAM and in the Municipality of Machico. The elaboration of this municipal plan would be very important for the planning of the city, increasing the population's awareness of this theme.

Regarding future studies, it would be of great importance to add and deepen more information about other characteristics of the Basin under analysis, namely characteristics on the relief and on the drainage system. On the relief, different parameters could be determined, such as the hypsometric curve (which makes it possible to obtain the average height of the watershed), the dimensionless hypsometric curve (determines the maturity state of the basin under analysis), longitudinal profile, slope, and relief index (allows you to obtain the flow velocity). About the drainage system, some parameters could be determined, such as the drainage network (ease with which water flows), bifurcation ratio, drainage density, average path over the ground and water density.

It would also be relevant, in the analysis of the flood risk, to consider Lagoa da Portela. Since this artificial lagoon may rupture associated with the phenomenon of floods. It is an embankment dam (homogeneous earth) with a maximum height of 19.5 m above the thalweg of the downstream watercourse

(22 m above the lowest level of the foundation), with faces inclined at 1V:1.5H and crowning, implanted at elevation 740.0 with 6 m wide and about 140 m of development and a total water storage capacity of 87,800 m^3 (Teixeira Duarte, 2012).

REFERENCES

Almada, J.A. (1865). Annaes of the Municipality of the Old Villa de Machico. *Flor do Oceano*, 247-286.

Almeida, A. B., & Lopes, S. (2014). *New Structural Measures to Reduce Hydrological Risk in Funchal.* In *VIII National Risk Meeting*, Guimaraes.

Alves, R., Silva, A., & Lousada, S. (2022). The impact of local accommodation in the sustainable development of Machico. In T. R. Aranzadi (Ed.), Planeamiento Ecológico En Las Iniciativas de Desarrollo (pp. 305–322). Academic Press.

Antunes, L. C. (2013). *Google Earth in the Classroom.* https://www.mapasnasaladeaula.org/mapas-do-projeto/portugal/regi%C3%A3o-aut%C3%B3noma-da- madeira/ribeiras-principais-pt-ram

Aragon, A. (1987). *For the history of Funchal.* Regional Secretariat for Tourism and Culture.

Autonomous Region of Madeira - Presidency of the Government. (2005). *Resolution No. 4/2005/M. Diário da República 2nd series, 219 (November): 16018 – 16030.* https://dre.pt/dre/detalhe/resolucao/4-2005-3022266

Azevedo, A. R. (1866). The Villa de Machico and the Capella de Machim. *Archivo Pittoresco*, *9*, 353.

Barreto, V. H. (2013). *Study of the Ribeira de Machico Hydrographic Basin – Vulnerability and Floods* [Master's dissertation]. Madeira University.

Caetano, C. (2014). *Risk assessment of alluvium in the streams of the island of Madeira* [Master's dissertation]. STI.

Castro, J. (2011). *Hydraulic, hydrological, and solid transport characterization of the February 20, 2010, event in Ribeira de João Gomes Madeira Island* [Master's dissertation]. Madeira University.

CMM. (2022). *Characterization of the Municipality of Machico.* Author.

Cruz, A. (1901). Funchal, December 1, 1901. Academic Press.

da Silveira, A. B., Madeira, J., Ramalho, R., Fonseca, P., & Prada, S. (2010). *Geological Map of the Island of Madeira at Scale 1:50,000, Sheets A and B.* Regional Secretariat for the Environment and Natural Resources of the Autonomous Region of Madeira and University of Madeira.

Decreto-Lei n. ° 115/2010, of 22 October of the Ministry of the Environment and Spatial Planning. Diário da República: I série, n. ° 206. Available at https://dre.pt/dre/detalhe/decreto-lei/115-2010-307873

Diário de Notícias. (1956). Funchal, November 4, 1956. Academic Press.

Diário de Notícias. (1979). Funchal, January 21, 1979. Academic Press.

Diário de Notícias. (2021). *Falling rocks and flooding in Machico*. Academic Press.

Diário de Notícias. (2022). *Bad Weather causes 13 occurrences in Machico*. Academic Press.

Directive Nº. 2007/60/EC, of 23 October of the European Parliament and the Council of the European Union. Available at https://eur-lex.europa.eu/LexUriServ/LexUriServ.do?uri=OJ:L:2007:288:0027:0034:PT:PDF

Ferreira, A. (2010). *Natural and Technological Risks* [Master's dissertation]. Academic Press.

Ferreira, F.P.M., & Cunha, S.B. (1996). *Urban Hydrology: Floods in Rio de Janeiro – Jacarepaguá*. Federal University of Rio de Janeiro/CNPq. Rio de Janeiro.

Freitas, H., Goerl, R., & Michel, G. (2020). Alluvium in Funchal, Madeira Island: Historical and conceptual aspects and damage potential. *Geosaberes: Journal of Geoeducational Studies*, *11*, 608–622. doi:10.26895/geosaberes.v11i0.971

Hipólito, J. R., & Vaz, A. C. (2011). *Hydrology and Water Resources* (1st ed.). IST Press.

ISEL. (2015). *Hydrology*. Lisbon: ISEL. Obtained from https://issuu.com/joanadecastrocortesao/docs/hidrologia_resumo_luis_mende s.

IST, UMA, & LREC. (2010). *Alluvial risk assessment study on the island of Madeira*. Academic Press.

Jornal Público. (2012). *Four dozen displaced people and roads closed in Madeira*. Academic Press.

Lei n. º 58/2005, of 29 December of the Assembly of the Republic. Diário da República: I-A series, no. 249. Available at https://data.dre.pt/eli/lei/58/2005/12/29/p/dre/pt/html

Leite, J. D. (1989). *Discovery of Madeira Island*. Alpha Publications.

Lencastre, A. (1972). *General Hydraulics Manual*. Academic Press.

Lousada, S., Cabezas, J., Castanho, R. A., & Gómez, J. M. N. (2021). Hydraulic Planning in Insular Urban Territories: The Case of Madeira Island—Ribeira Brava. *Water (Basel)*, *2021*(13), 2951. doi:10.3390/w13212951

Lousada, S., & Castanho, R. A. (2021). GIS-based Assessment of Morphological and Hydrological Parameters of Ribeira dos Socorridos and Ribeira do Vigário Basins, Madeira Island, Portugal. *Current World Environment*, *16*(2), 408–426. Advance online publication. doi:10.12944/CWE.16.2.08

Lousada, S., Gonçalves, L., & Atmaca, A. (2022). Hydraulic Planning in Insular Urban Territories: The Case of Madeira Island—São Vicente. *Water (Basel)*, *14*(1), 112. doi:10.3390/w14010112

Lousada, S., & Loures, L. (2020). Modelling Torrential Rain Flows in Urban Territories: Floods - Natural Channels (The Case Study of Madeira Island). *American Journal of Water Science and Engineering.*, *6*(1), 17–30. doi:10.11648/j.ajwse.20200601.13

Monteiro, C. A. F. (1991). Climate and exceptionalism – conjunctures on the performance of the atmosphere as a geographical phenomenon. Florinópolis: UFSC.

Moore, R. J., Bell, V. A., & Jones, D. A. (2005). Forecasting for flood warning. Comptes Rendus Geoscience – External Geophysics. *Climate and Environment*, *337*, 203–2017.

Naranjo Gómez, J. M., Lousada, S., Velarde, J. G., Castanho, R. A., & Loures, L. (2020). Land-Use Changes in the Canary Archipelago Using the CORINE Data: A Retrospective Analysis. *Land (Basel)*, *9*(7), 232. doi:10.3390/land9070232

Pegado, R. S., Blanco, C. J. C., Roehrig, J., Caroça, C., & Costa, F. (2014). Flood Risk and Vulnerability: An Approach to Urban Flooding in Béle/Pará/ in Brazil. *Territory*, *21*, 71–76. doi:10.14195/1647-7723_21_6

Quintal, R. (1999). *Alluviums of Madeira. 19th and 20th centuries*. Territorium Magazine.

Radio TSF. (2021). *Tree falls, landslides, and floods. Heavy rains promise not to give truce in Madeira*. Author.

Regional Secretariat for the Environment and Natural Resources – Regional Directorate for Spatial Planning and Environment. (2016). *Flood Risk Management Plan for the Autonomous Region of Madeira*. Author.

Ricardo, R. P., da Câmara, E. M. S., & Ferreira, M. A. M. (1992). *Madeira Island Soil Chart*. RAM.

RTP. (2016). https://www.rtp.pt/programa/tv/p32494/e4

Silva, A., Gonçalves, L., Moura, A., Lousada, S., Gouveia, J., & Camacho, R. (2021). Flood Hazard Assessment in Madeira (Portugal) – The Case Study of Machico. *American Journal of Water Science and Engineering.*, *7*(1), 1–13. doi:10.11648/j.ajwse.20210701.11

Silva, F. A., & Menezes, C. A. (1921). *Madeiran Elucidário. 1st Volume*. Esperança Typography.

Silveira, W. N. (2008). *Historical analysis of flooding in the city of Joinville - SC, focusing on the Cubatão do Norte river basin, Florianópolis* [Dissertation]. Technological Center. Graduate Program in Environmental Engineering.

Teixeira Duarte. (2012). https://teixeiraduarteconstrucao.com/projetos/lagoa-portela/

Tucci, C. (2005). *Urban Flood Management*. Urban Flood Management Manual.

Tucci, C. E. M. (2009). *Hydrology: Science and application*. ABRH.

KEY TERMS AND DEFINITIONS

Flooding: Result of the concentration of excess rainwater that cannot be absorbed by already saturated soil and other forms of runoff.

Machico: Portuguese municipality on the island of Madeira, Autonomous Region of Madeira, where Tristão Vaz Teixeira and João Gonçalves Zarco landed in 1419 when they discovered the island of Madeira.

Risk Prevention: Preparation of something in advance for a certain purpose, the fact of foreseeing damage or anticipating possible barriers, conflicts, or problems.

Urban Flooding: Floods caused by floods due to the overflow of water courses that cross urban areas.

Urban Planning: Is a technical and political process concerned with the control of the use of land and design of the urban environment, including transportation networks, to guide and ensure the orderly development of settlements and communities.

Chapter 6
Water Policies:
Management and Planning of Water Resources in Spain

María Sotelo Pérez
King Juan Carlos University, Spain

Ignacio Sotelo Pérez
University Institute of Environmental Sciences, Complutense University of Madrid, Spain

ABSTRACT

In a country, Spain, in which drought and desertification mark the environmental reality, water policies acquire a fundamental role when dealing with the management and planning of water resources. From the perspective of what has come to be called "postmodernity," which predicts the end of the territory and the "no place," the refutation of the region concept is established, at a time when no subspace can be separated from reality, complex and disparate, what happens to it, where the processes of globalization and fragmentation mark the future of its activities, relationships, developments, etc., and, therefore, are marked by regionalization and individualization, taking into consideration that, at the present time, the regional ones are shown as the basis and condition of the set of global relations—from the local to the global. In this way, the knowledge and study of the water reality of a country like Spain has as a reference the region and its problems from the territorial perspective, thus laying down what should be the bases of the water policies.

INTRODUCTION

The so-called "market economy", based on the principles of the neoclassical school, continues to be the fundamental basis of current economies. Accepted and assumed by most Western geographers, economists, politicians and jurists, they would show that we are facing a pure, perfect economy, without complexities or problems, Although, in practice, these parameters are certainly not linked to reality, therefore, the imperfections or "faults" become more than palpable, especially from an environmental perspective, in general, and of water as a fundamental element for life, in particular. Thus, the intervention of the State

DOI: 10.4018/978-1-6684-6449-6.ch006

in the economy -inexact, where the defects or imperfections of the market become visible-, through policies that try to mitigate, stop or eradicate the negative externalities produced by the lack of conservation and preservation, or by the deterioration or contamination of the environment, it becomes essential -as well as to promote the so-called "positive" externalities-, which will favor the use of new instruments or mechanisms that are not very closely linked –where the symbiosis between law and economy become essential-, at the present time, with the environmental sphere, in general, and with water in particular; in addition to deepening a structural change in the current economy.

Thus, the preservation, conservation, regeneration and improvement of quality are the fundamental bases on which the so-called "environment" gravitates[1], and, therefore, they must be constituted as the main objectives of the environmental analysis[2], as well as the primary finalist policies of the Economic Policy, in general, and of the Sectoral Policies, such as that of Water, in particular.

Following this dynamic, in 1974, José Jané Solá (based on Thedjeck, 1963), established a primary scheme in which the objectives of economic policy were related to specific policies (both sectoral and instrumental), incorporating the environment into said structure, through preservation and conservation as objectives of the so-called finalist policies -deterSolmined by their economic and social characteristics, and granting them a rank similar to economic growth, sustainable development, price stability, balanced balance of payments, equitable distribution of income, full employment, etc.-; although, preservation and conservation show characteristic features that denote their own nature, from economic policy-which, we must not forget, is linked by jurisprudence, as a binding "entity" of the processes of regulation, action, management, planning, etc. As stated by professors Armando Villamil Serrano and Joan Maties García (1998), preservation is a static objective, which reflects the state of an ecosystem, at a given moment . . . and aims to "correct attacks on the environment, preferably at the source itself"; while "conservation is, on the contrary, a dynamic objective. Here an economic policy is needed that constantly avoids or minimizes ecological deterioration. For this reason, it is connected, mainly, to the principles[3] of "caution and preventive action"[4], "correction of attacks on the environment", and "whoever pollutes pays"[5].

In this regard, it should be noted that the implication of economic policy (whether through economic planning, finalist or specific -sectoral or instrumental- policies) in environmental prevention and conservation, is marked by the decision-making and management nature of the interdisciplinary nature of environmental matter. Therefore, economic policy -based on the previously analyzed objectives and instruments- is the central axis in decision-making, as well as in the specification and delimitation of intervention methods that control the actions of the "market" -being also necessary in making production and consumption decisions-. In this way, the real importance of promoting the involvement of economic policies in the set of disciplines -interdisciplinary nature- that make up the study of the environment, in general, and its natural resources, where water is consolidated as a central axis of action -along with the air-, since it is an essential element for life, as well as promoting the control of the so-called "market failures"[6] such as the so-called "externalities". Let us therefore proceed to their analysis.

MARKET FAILURES AND ENVIRONMENTAL POLICIES IN THE FIELD OF WATER RESOURCES

Despite the fact that all environmental problems have dimensions of an extra-economic nature, as a result of the incidence of the ecological crisis that does not allow Science Economy itself to provide total solutions to said problems, Legal Economy and, more especially, the Policies –instrumented and regulated-, allow the design of "non-conventional" situations or strategies that need to redistribute scarce natural resources among diverse alternative uses. Thus, the failure of the market to generate and promote an optimal allocation, in Pareto terms, of natural resources, is clearly evident.

All this becomes more important when we bear in mind that the environment falls within the sphere of public goods and services[7] -for being, in principle, indivisible, and not exclusive-, where "water" is constituted, in the beginning, as a pure public good, considered as a preferential good that satisfies the preferential needs that are not seasoned to exhibit and expose adequately through the unique preferences of each individual because they do not properly value, due to ignorance and not acting rationally, the benefits produced by said good and, therefore, the real preferences of the same. In this way, the main objective of the neoclassical model stems from the urgent need to show how the economy manages to reach, in a "natural" way, without intervention, a situation of equilibrium with full employment; in such a way that, in terms of Economic Policy, in general, and, in Water Policies, in particular, the ideal would be "non-interventionism" in the economy[8]. Although, the main complexity of this situation arises from the fundamental idea that these hypotheses are merely theoretical, and reality exceeds such pretexts. Thus, it can be stated that, in practice, these adjustment mechanisms are extremely complex, since the level of information is transferred to the agents in a gradual, slow and discontinuous manner. And, therefore, a real adaptation in the adjustments of the market economy is not generated, fundamentally, because the agents develop an optimizing behavior[9] (how it is logical to think, from these approaches, that water and its problems must be dealt with in the field of the market economy).

From the perspective of traditional economic analysis, individuals try to maximize their income levels within existing constraints, such as property rights, transaction costs, diminishing returns, or resource scarcity. Thus, each person will act rationally according to their private costs or benefits, but without taking into account external effects that they may generate for society, with which, according to Vilfredo Pareto, an optimal allocation can only be achieved when the Social Marginal Benefict is equal to the Social Marginal Cost, these are issues of notable interest when studying a subject as complex as it is attractive, water, and its Policies.

Thus, in the original analysis of the relationship between the economy and the environment -and, more specifically, that linked to environmental degradation, and the problems dealt with here-, various principles are established on which environmental assimilation is based, the capacity of load and the capacity of recovery of the environment, always since it subject water[10]; and, only when the levels of degradation of the natural environment are exceeded -water pollution, overexploitation of aquifers, degradation of fresh water . . . economic relevance is given to the processes of pollution, always bearing in mind that this can never be null (because the opportunity cost would be very high), and, therefore, an optimal level of contamination must be reached –where costs and benefits are equal. Therefore, it is possible to know -in theory- the optimal level of environmental degradation of water resources that must be assessed when putting water policies into practice, according to the function of maximizing social welfare in profit and cost function.

Taking into consideration that, in general terms, the costs of controlling said environmental degradation or loss increase as environmental quality increases -that is, when pollution or degradation is reduced, as well as the pressure on the environment-, and that the costs of environmental degradation or damage intensify as pollution levels increase, the so-called optimum of degradation or contamination will be reached when it and the control and management costs[11] are the same.

There is, therefore, technological evidence that the neoclassical solution of the optimal environmental degradation, certainly optimistic, fails to perceive. In short, we have that, as a consequence of dynamic ecological phenomena (dynamics of accumulation and dispersion of pollutants) tending to reduce the assimilative capacity of the natural environment, the neoclassical solution of the Paretian optimum disregards the ecological risks and external costs in the future. In short, there is a certain incompatibility between the economic optimum and the ecological balance (ecological optimum) that cannot be resolved immediately with the formulas of "appropriate technology" and "zero growth". . . . "Appropriate technologies" do not provide a definitive solution; if it means a reduction in profits and a contraction in production, it can reduce the "ecological gap", but if it is more profitable and stimulates production, it only temporarily displaces the problems. "Zero growth" does not imply "zero pollution" nor does it mean "zero production", but that production would remain at constant levels and as long as the level of non-assimilated pollution continues to increase, the ecological gap continues to be produced, although at a slower pace. (Jiménez Herrero, 1998)

It is from this "simple" theoretical analysis that the foundations of Economic Policy are laid, through which to indicate what are the "desirable" levels of environmental degradation of water resources, as well as the instruments and mechanisms necessary for their control and management, and through the application of the aforementioned Water Policies. Although, it must be taken into consideration that the complexity of said analyzes and theoretical concretions are, at least, more numerous and complex, especially if we take into consideration that it must be taken into account that there are ecological impacts or damages that are complex or that cannot be estimated in the long term, as well as a notable complexity for the real concretion of a Paretian optimum in the environmental field due to the difficulty, as well as inaccuracy, of foreseeing and assessing the social and environmental impacts.

All these issues have been studied by economists called "new Keynesians"[12] by developing the principle of approximate rationality[13] and the analysis of externalities (Buchanan & Stubblebine,1962)–fundamentally, applied to the "negative"; considering that economic agents will carry out their activity acting rationally –that is, not reacting-, in the face of possible imbalances or changes -based on self-regulation due to changes in the adjustment variables-, and, therefore, the so-called "approximate rationality" will give rise to these adjustment variables being rigid, generating costs or benefits for the bulk of individuals that constitute the community, through the so-called externalities. Professor Jose Luis Martin Simon (1983), established that, in the economic literature, the concept of externalities has been the subject of inadequate treatment. Sometimes, they have been analyzed in a somewhat confused and partial way, which has caused errors when using them as an argument, sometimes basic, in certain theories.

At the time when studies were dedicated to achieving a balance referring to greater well-being for the community, which would be automatically given by the hidden forces of the free competition market, external economies, caused by the interdependencies of the subjects of the economic activity, would show how there could be a dissociation between the competitive equilibrium and the welfare optimum of society. In this context, a first definition of external economies appears that is restricted and different from the one followed in the analysis of economic development. Consequently, we find two types of definitions: strict ones, referring to the microeconomic study of equilibrium and, therefore, of a static nature; and others, with a broader sense and of a dynamic nature, related to the theory of industrialization". . . . "Marshall introduces the concept of external economies to safeguard the notion of competitive equilibrium, when long-run supply curves are downward". . . . "The essential role played by external economies in the Marshallian treatment: being able to reconcile decreasing cost and competitive balance" (Simon, 1983)

In this way, externalities can be defined as any economic effect that occurs outside the market (uncompensated benefits or costs), that is, as the benefit or harm that an agent or group of economic agents causes to another agent or group of economic agents –that is, when the costs[14] and private benefits differ from social ones[15]-; establishing a dilemma between the external effects that affect or fall on the production or consumption of goods and services, as well as a differentiation between actions that can be described as "positive" ("external economies") and those considered as "negative" ("external diseconomies"). Emphasizing, later, the need[16] or not[17] -according to the established precepts-, of the intervention of the State in the regulation and control of said situations.

In the specific case that concerns us, outsourcing processes can affect water resources positively or negatively, through the general characteristics of the different processes, levels or models of production and/or consumption developed by the population, both resident or not, in our national territory, thus conferring considerable importance on the economic valuation of water.

In the case of externalities characterized as positive, the impacts vary depending on whether they originate in production -in the specific case of water resources we mention the processes of water supply, distribution or treatment- or in consumption -whether direct or indirect-. In this way, in the face of a positive externality on consumption, the consumer demand curve will show a displacement -giving rise to the so-called social demand curve-, producing a situation in which the marginal social benefit to be consumed will be greater than the marginal benefit private –keeping the supply curve in equilibrium, where the private and social marginal costs are identical-, giving rise to a new equilibrium situation –in this case, social-, where the utilities of the set of individuals that make up a society will be greater than the private ones. Framed as positive externalities on water consumption, we find, for example, those produced by improvements in environmental education which, in turn, lead to a reduction in water demands for domestic or industrial uses, industrial processes in which substitute elements of water resources, decrease in the use of non-recycled water resources for public maintenance of cleaning or irrigation of gardens, protection of natural spaces when leisure activities are carried out -avoiding the degradation of groundwater, surface water, coastal or coastal waters,…-, use of cooling systems that do not require the use of water, etc.

On the other hand, the positive externalities on the production in which water resources are used, are marked by a displacement of the supply curve where the private marginal costs will be greater than the social ones, in this way, a new equilibrium is produced social, that favors the environmental protection of water resources. Examples of these can be found in the use of new cultivation techniques where the lowest possible level of water is used and that can be reused, research and technological innovation that involves an improvement in the use of water resources, carrying out studies of environmental impact assessment and strategic environmental assessment, industrial processes that prevent water degradation or contamination –avoiding the use of outfalls, wastewater evacuation processes or the discharge of chemical products into water, both underground and surface-, companies in charge of water purification and desalination, improvements in channeling infrastructures and distribution of water resources –for example, through investment in the reconstruction and maintenance of canals, siphons and beacons-, among others.

Similarly, external processes are generated, whether by demand or supply that negatively affect the production or consumption of water resources. A negative externality on consumption supposes a displacement of the demand curve, breaking with the initial equilibrium situation, where the social marginal benefit to be consumed is less than or less than the private benefit -keeping the supply curve in equilibrium, where the marginal costs private and social are identical-, giving rise to a new situation of equilibrium -in this case, social-, where the utilities of the group of individuals that make up a society will be less than the private ones.

Perhaps, these are the most visible externalities, as well as recurring, given the character of a public good[18] such as "water", as has been analyzed previously, and, in the same way, the social and institutional concern -at least theoretical- to eradicate or mitigate the damage to it is relevant. In this way, negative externalities on the consumption of water resources stand out as those in which drinking water is wasted, intentional discharges into rivers or streams, outfalls from homes or private buildings into surface waters, overexploitation of aquifers -where there is no optimal natural regeneration of the same-, greater demands for water in cities as well as less efficiency in transport, increased demands for water by households, etc. As well as due to the existence of negative externalities on production, where there are shifts in the supply curve where the private marginal costs will be lower than the social ones, in this way, a new -social- equilibrium is produced, which significantly harms, environmental protection of water resources. Highlighting, among the many conditions, the contamination of groundwater due to the concentration of nitrates and worsening of the levels of Biological Oxygen Demand and Chemical Oxygen Demand, changes in the physical conditions of the water -over heating or cooling of the same, even affecting even ecosystems, etc.

Unquestionably, the measurement processes, that is, monetarily quantifying the effects or repercussions of the aforementioned externalities on water resources, have become essential in recent decades, conferring the value of an economic good on a public resource[19] and that, in theory, it should have no price. This has been feasible thanks to the potential of the water policies carried out in a country like Spain, where the original bases of action were aimed at favoring supply policies and, therefore, the price associated with water has been marked by the costs of storage, distribution, sanitation and recycling of the same.

With all this, taking into account that the market is imperfect and "fails" in the optimal allocation of environmental resources, in general, and of water resources, in particular -because they are public and preferential goods- as well as in the I proceed from the generation of externalities, the Public Sector is "placed" to act and intervene to achieve an optimal allocation of them, so the management of water resources, as an essential element for life, must be developed through the relevant water policies. For its part, the contributions of the aforementioned economic theory and its link with the environment, from the neoclassical perspective, have tended to be oriented, fundamentally, in thc correction of the so-called "external diseconomies" -negative externalities-, through different instruments[20] and mechanisms to "internalize"[21] the deterioration of the environment until these are harmonized with the satisfaction of the needs of individuals.

WATER POLICIES AND THE 2030 AGENDA

The fundamental reference point of the Agenda is the achievement of the so-called Sustainable Development. In fact, of the seventeen objectives that are set, two are relatively important in the subject studied-, "11. Sustainable Cities and Communities" and "6. Clean Water and Sanitation". As the 2030 Agenda becomes a political instrument of relative importance, it deserves to be studied in this section. We are facing the continuation of the so-called Sustainable Development Goals (2030 Agenda), whose point of The 2030 Agenda is the interrelation between the environmental and international development, generating the so-called "lever policies", that is, "transversal" instruments that allow overcoming sectoral compartments, contributing to the intersection of policies, sectors and actors (especially important in the case of Water Policies, from a sustainability perspective). The approval and application of the 2030 Agenda and its 17 Sustainable Development Goals constitute the broadest and most hopeful civic process of our era. Wide because its adoption extends among governments, administrations, companies, social organizations and citizens from all over the world. Hopeful because the SDGs are demonstrating their ability to become a common language for the collective interpretation of the great challenges of the planet, especially in everything related to access to water (Mataix et al., 2019).

We must not forget that, at the present time, the two catastrophes we are suffering -credit and environmental- are totally linked. In the early years of the 21st century, the world is suffering the consequences of a declining economic model that is giving rise to a plague of incalculable dimensions: poverty. The inequality gap between rich and poor is widening, along with a continuous environmental degradation of the planet. Surprisingly, the model of economic growth based on cheap labor, an undervalued currency, excessive exports . . . is beginning to seem unsustainable, to say the least. Even so, it underlies the decision of many States not to recognize the inalienable right of access to water by human beings, of all human beings.

This must be one of the fundamental principles of a necessary change in the development model and new economic schemes capable of replacing the current ones. These have shown predators with the planet, socially unfair and economically unviable. Water is, therefore, a good that constitutes a fundamental right of human.

ENVIRONMENT, LEGISLATION, AND INSTRUMENTS FOR THE PROTECTION, CONSERVATION, AND REGENERATION OF WATER RESOURCES

After the previously mentioned, we can start from the fundamental idea that, in Spain, the collection efficiency of the exactions stipulated in the Water Law 29/1985 is very low, as a direct consequence of the economic-financial regime and due to the scarce effectiveness of the system collection of levies. "All this prevents, in the first place, the recovery of the financial resources necessary for adequate surveillance, control, administration, maintenance of hydraulic infrastructures and protection of the public hydraulic domain. Secondly, it shows the scarce internalization of the costs generated in the process by the users" (White Paper on Water, 1998). In the same way, the distribution of said collections in the Spanish territory is diverse, complex and not very equitable at the present time, since it is the Autonomous Communities themselves that set the prices of water, taking into account the bioclimatic and territorial conditions of the region itself, hydrographic basin or basins within which it is framed; this is a fact that becomes very important at the point in some areas of Spain water is wasted or is not given the value that this good requires, while in other Communities the price exceeds the national average as direct consequence of the lack of water.

And this is how we find the first of 1866. Although it did not enter into force due to the political and social conditions of the time, it served as the basis for the later Water Law of 1879, which included in its text all and each of the public domain waters existing in Spain, whether they were rivers, lakes, natural streams . . . although groundwater was not included (matter of great importance that if it was included in the Water Law of 1985). In this way, the new legal regime of the hydraulic public domain was established, around the constitutional bases of the 1978 Constitution, which includes the division of powers between the State itself and the delegation of certain powers to the Autonomous Communities, thereby establishing a new regulatory system for the different competent Public Administrations.

With all this, we must bear in mind that, when dealing with the complex management of water resources in Spain, there is a great variety of legislation that regulates and manages water supply. For this, we can make a distinction between the International Legal Framework, the National Legal Framework of Spain, through the Hydrographic Confederations and the Basin Organizations, although what interests us in this study is, only, the National Legal Framework of Spain, in which the Spanish Water Law of 1985 itself is framed, which has been progressively modified in recent years, presenting great interest in the different economic instruments linked, fundamentally, to the protection of the Environment. Thanks to this Water Law and its corresponding modifications, it has been possible to verify both the existence of various practical problems in water management at the national level, as well as the absence of effective instruments in it to face the new demands in relation to said resource, both in quantity, given that its consumption is increasing exponentially, and in quality, and, more specifically, in terms of environmental protection.

The Water Law (art. 38. Law 29/1985) has the purpose of achieving a good ecological state of the hydraulic public domain, as well as achieving the satisfaction of water demands that allow a balance in regional and sectoral development, as well as increasing the availability of the quality resource, in relation and harmony with the environment and other natural resources. At the same time, we must take into account that the best reallocation of available resources, through planning mechanisms, they allow to increase water production through the use of new technologies, granting legal status to legal regime for desalination or reuse procedures; to promote efficiency in the use of water, for which the required flexibility of the current concession regime is necessary through the introduction of the new contract for the transfer of rights to the use of water, which will make it possible to socially optimize the uses of such a scarce resource, and, lastly, to introduce policies to save said resource, either by establishing the general obligation to measure water consumption through approved control systems or through the administrative setting of Reference consumption for irrigation (explanatory statement of the Water Law 29/1985). This is why it is necessary to carry out new modifications and reforms in the current Water Law, with special emphasis on economic instruments, based not only on the homogenization of the pricing of water resources, but also in pursuit of "preventing to cure", in environmental matters. As the Council of State already pointed out in 1962 on the Water Law of 1879, the 1985 Law must subsist, although with the tweaks and modifications necessary to update its technique and solve the most pressing problems that administrative experience has come accusing, but without altering its principles.

SPANISH WATER LAW AND ITS ECONOMIC INSTRUMENTS IN ENVIRONMENTAL MATTERS

To understand the repercussion that the Spanish Water Law has had and still has, we have dealt with the different instruments that the Law has had to regulate the different uses of water; for this reason, we break down all the economic instruments of said Law, attending to the basic tax concepts and, later, comparing the evolution of these in the Water Law 29/1985, the Water Law 46/1999 and the Consolidated Text of the Law of Waters, Royal Decree 1/2001.

Regarding the 1985 Water Law, it introduced three major innovations interwoven with its purpose: "The regulation of the hydraulic public domain, the use of water, and the exercise of the powers attributed to the State in matters related to said domain within the framework of the powers defined in article 149 of the Constitution" (Art. 1, Water Law 29/1985). In the first aspect, the great novelty was the consideration as public domain of all the waters that are part of the hydrological cycle, including the renewable groundwater. In the second, the regulation of hydrological planning as a key instrument for resource management. In the third, the adaptation of state powers to constitutional and statutory statements, not easy to interpret (Embid Irujo, 2016).

And, it is that, the Water Law of 1985, and the Consolidated Text of the Water Law of 2001, have tried to solve the historical problems that have arisen with the supply policy generated by the regenerationist ideals of Joaquín Costa, adapting to a new culture of water by modernizing its management, although the dispossession of the State at the hands of the Autonomous Communities, in many cases, is holding it back; all this together with the large amount of money that is being dedicated to the modernization of infrastructures whose benefits are not tangible in the short term, and, due to the so-called user culture, far removed from sustainable development, which leads Professor Embid to affirm that "the distinctive sign of Spanish water law today is its lack of definition" (Embid Irujo, 2002). All this coupled with the unsustainability of having a hydraulic policy in which the idea of generality, of union, of unity practically does not exist.

We live in a stage marked by particularisms, and by the fragmentation generated by territorial interests that have segmented water management, with the only exception, in our opinion, of the aforementioned Water Law. Likewise, highlighting the breakdown of the principle of integrated management and basin unit, the breakdown of the principle of regional cooperation, and the inability of the State to exercise its authority in a field in which it is essential.

For all these reasons, analyzing the main economic mechanisms leads us to approach Title VI of the economic-financial regime for the use of the hydraulic public domain and, specifically, the "occupation canon" (art. 104 of the Water Law of 29/1985)[22], modified by article 40 of Law 46/1999 of December 13, where it is called "fee for the use of assets in the hydraulic public domain" and, later, regulated by article 112 of the Consolidated Text of the Water Law, Royal Legislative Decree 1/2001.

The Water Law entrusts the economic regime referring to the availability of water to the rates called regulation canon and water use rates. Thus, we are faced with a canon which regulates the occupation or use of the hydraulic public domain. Since the text sets out the protection and improvement of said domain as an objective, clearly, we are dealing with an economic instrument, which we could call "extra-fiscal". The Law establishes the importance of the occupation or authorized use of the public hydraulic domain, of the channels of natural currents and the beds of lakes, lagoons and surface reservoirs in public channels. In fact, as stated in article 104 of said Water Law; the taxable base of this levy will be the value of the good used, taking into account the yield it reports. The taxpayers will be all those holders of the concessions or the authorizations of occupation or use, while the annual tax rate will be four per cent on the value of the taxable base, with a tax quota that in our case, the result of apply the aforementioned tax rate to the tax base. We must take into account that said tax quota will be determined by each Basin Organization, taking into account various intrinsic conditions of the territory in which said "canon" is applied. For its part, the criterion of temporal imputation of said canon will be determined by the exact moment in which the concession of the use of the hydraulic public domain is proceeded, taking into account the specific conditions set forth in said authorizations.

On the other hand, we can highlight the importance of Title VI of the economic-financial regime for the use of the hydraulic public domain, the "discharge canon" (art. 105 of the Water Law of 29/1985)[23], was modified by article 41 of Law 46/1999 of December 13, where it is called "discharge control canon" and, subsequently, it has been regulated by article 113 of the Consolidated Text of the Water Law, Real Legislative Decree 1/2001. Similarly, we must bear in mind Royal Decree 849/1986.

From the attached figure it can be deduced that we are facing a Canon, which regulates authorized discharges, exclusively, made directly by the use or employment of the hydraulic public domain. The fact that gives rise to the tax obligation are the authorized discharges, that is, those that are carried out directly or indirectly in the channels, whatever their nature, as well as those that are carried out in the subsoil or on the land, rafts or excavations, through evacuation, injection or deposit (in accordance with art. 92 Law 29/1985). The pollution load of the discharge, expressed in pollution units, that is, the total volume of authorized discharges, together with the taxpayers, will be all those obliged to bear other economic charges, produced by the authorized discharges in the hydraulic public domain. For its part, the tax rate is determined by the price of control of authorized discharges that will be determined and reviewed, where appropriate, in accordance with the provisions of the Hydrological Plans regarding the quality of continental waters. The tax rate, on the other hand, will be obtained by multiplying the tax base (since there are no reductions), by the type of tax, that is, multiplying the volume of authorized discharge by the unit price of discharge control. All this without forgetting that article 105 of the Water Law 29/1985 does not stipulate the exact moment of the accrual period.

Likewise, it is necessary to highlight Title VI of the economic-financial regime for the use of the public hydraulic domain, the "regulation canon and the pricing of water use" (art. 106 of the Water Law of 29/1985)[24], was modified by article 42 of Law 46/1999 of December 13 and, subsequently, it has been regulated by article 113 of the Consolidated Text of the Water Law, Royal Legislative Decree 1/2001.

From the above information it can be inferred that we are faced with two taxes, which are the "regulation canon" and, on the other hand, what we can describe as "special contributions", which is the "pricing of water use". The use or enjoyment, direct or indirect, of surface or groundwater regulation works carried out totally or partially, at the expense of the State, and the use and enjoyment of other specific hydraulic works carried out entirely at the State's expense. The expected amount of operating, conservation and administration expenses by the organizer, of the works carried out; taking into account that all those who benefit directly or indirectly from surface or groundwater regulation works carried out totally or partially by the State, and the use and enjoyment of other specific hydraulic works carried out entirely by the State. For its part, the tax rate is 4% of the value of the investments made by the State, duly updated, taking into account the technical amortization of the works and installations and the depreciation of the currency, in the manner determined by regulation. The tax rate is the result of applying the tax rate to the tax base, that is, applying the tax rate to the amount of operating, conservation and administration expenses by the organizer, of the works carried out. The temporary imputation criterion of said canon will be determined by the exact moment in which the use or enjoyment of said hydraulic works occurs (all this has been expanded with the modifications made later).

LEGISLATIVE CHANGES AND ECONOMIC INSTRUMENTS IN THE SPANISH WATER LAW

The Water Law 46/1999 practically does not alter the existing legislation of the Water Law 29/1985, although it does make certain modifications, with the aim of perfecting and renewing it so that, as stated in the legislation itself, "to respond to its insufficiencies, to the new challenges that water management requires at the gates of the 21st century, in accordance with our full integration into the European Union and the need to grant maximum protection to said natural resource as an environmental asset of first order". For these purposes, Law 46/1999 proceeds to the modification and restructuring of the different economic instruments and incorporates new ones, as we are going to analyze, with the intention of introducing technical improvements, at the same time that it regulates some matters that did not appear included in it, such as desalination, transfer of water use rights or public works. In it, Title VI of the economic-financial regime for the use of the public hydraulic domain stands out for its importance, and, specifically, the "canon for the use of assets in the public hydraulic domain" (art. 40 of the Water Law of 46/1999)[25], by which article 104 of Law 29/1985 is modified, where it is called "occupation canon" and, subsequently, it has been regulated by article 112 of the Consolidated Text of the Water Law, Royal Legislative Decree 1/2001. It should be noted that the wording of article 104 of Law 29/1985 is modified and sections 2 and 3 are incorporated.

The above information shows us that, as in the original Water Law 29/1985, we are faced with a Canon, which regulates the occupation or use of the hydraulic public domain. Since the text sets out the protection and improvement of said domain as an objective, we are clearly faced with an economic instrument that we could call extrafiscal. It is not modified, the taxable event continues to be the occupation or authorized use of the public hydraulic domain, only, of the channels of natural currents and the beds of lakes, lagoons and surface reservoirs in public channels; There is a modification in which it is stipulated that the tax base will depend on three key factors: the first of them will be the market value of the occupied land (at market prices), taking as a reference the bordering land in the case of occupation of hydraulic public domain lands; the second is determined by the value of occupation or use of the land; and, finally, the value of the goods and services of the use of land in the hydraulic public domain. In this amendment to the Water Law, it is stipulated who will be the taxpayer, defining this as the concessionaires or authorized persons or, as the case may be, those who are subrogated in their place (and the fact is that, in the original text, they are not defined as passive subjects). The annual tax rate is increased from 4% to 5% on the value of the tax base, only in cases of occupation or use of public water domain land; ascending said amount to 100% in case of use of hydraulic public domain. Regarding the tax quota, say that it is that amount that represents the tax, in our case, the result of applying the aforementioned tax rate to the tax base, according to the conditions described in article 40. We must take into account that said tax quota will be determined by each Basin Organization, taking into account various intrinsic conditions of the territory in which said "canon" is applied. Add that as described in section 3, the accrual will occur with the initial granting and the annual maintenance of the concession or authorization and will be payable in the corresponding amount and in the terms indicated in the conditions of said concession. or authorization.

At this point, highlighting for its importance Title VI of the economic-financial regime for the use of the hydraulic public domain, the "discharge control canon" (art. 41 of the Water Law of 46/1999),[26] was modified Article 105 of Law 29/1985 called "discharge fee" and, later, regulated by Article 113 of the Consolidated Text of the Water Law, Royal Legislative Decree 1/2001. Similarly, we must bear in mind Royal Decree 849/198. The wording of article 105 of Law 29/1985 is modified and sections 2, 4 and 7 are incorporated.

When analyzing the above information, we can highlight that we are facing a Canon, which regulates authorized discharges, exclusively, made directly by the use or employment of the hydraulic public domain. Unlike the Water Law 29/1985, in this text it is stipulated that, the fact that gives rise to the tax obligation, will be all those discharges (authorized or not) that are carried out directly or indirectly in the riverbeds, whatever be the nature of these, as well as those carried out in the subsoil or on the ground, rafts or excavations, through evacuation, injection or deposit (in accordance with art. 92 Law 29/1985).

It states that the pollution load of the discharge is expressed in pollution units, that is, the total volume of authorized discharges, with a taxpayer who in the previous text stipulated that they were the ones obliged to bear other economic charges, produced for authorized discharges in the hydraulic public domain; while in this modification of Law 29/1985, it is established that the passive subjects will be all those who have carried out the discharge, produced by the authorized discharges in the hydraulic public domain. The tax rate is determined by the control price of the authorized discharges, which will be determined by the unit price resulting from multiplying the basic price per cubic meter by a coefficient of increase or decrease, which will be established by regulation depending on the nature, characteristics and degree of contamination of the discharge. The tax rate, for its part, is obtained by multiplying the tax base (since there are no reductions), by the type of tax, that is, multiplying the volume of authorized discharge by the unit price of discharge control; noting, finally, that the discharge canon is accrued on December 31 of each year, its tax period being the calendar year itself, except for detailed exceptions.

Finally, it is necessary to highlight Title VI of the economic-financial regime for the use of the public water domain, the "regulation canon and the pricing of water use" (art. 42 of the Water Law of 46/1999)[27], which modifies article 106 of Law 29/1985 and, subsequently, has been regulated by article 113 of the Consolidated Text of the Water Law, Royal Legislative Decree 1/2001. It should be noted that the wording of sections 1, 2 and 5 of article 106 of Law 29/1985 is modified, and sections 6 and 7 are added.

When analyzing the above information, we find ourselves, again, before two taxes, which are the "regulation fee" and, on the other hand, what we can describe as "special contributions", which is the "pricing of water use". The use or enjoyment, direct or indirect, of surface or groundwater regulation works carried out totally or partially, at the expense of the State, and the use and enjoyment of other specific hydraulic works carried out entirely at the State's expense. The expected amount of operating, conservation and administration expenses by the organizer, of the works carried out. The passive subject assumes that all those who benefit directly or indirectly from surface or groundwater regulation works carried out totally or partially, at the expense of the State, and the use and enjoyment of other specific hydraulic works carried out entirely at the State's expense, with a tax rate of 4% of the value of the investments made by the State, duly updated, taking into account the technical amortization of the works and installations and the depreciation of the currency, in the manner determined by regulation, and a tax quota that will be the result of applying the tax rate to the tax base, that is, apply the tax rate to the amount of operating, conservation and administration expenses by the organizer, of the works carried out, all of this taking into account the regulations of the River Basin Hydrographic Plans. Similarly, we must bear in mind that the legislator sets the maximum and minimum limits for this factor, but the rules for its calculation refer to the regulations. The temporary imputation criterion of said canon will be determined by the exact moment in which the use or enjoyment of said hydraulic works occurs.

After the above, we must highlight the important repercussions that the modifications in the economic instruments of the Water Law 46/1999 entail, in the Environment. And it is that, as we well know, the reform of the Water Law is justified by the need to tackle, through new, more effective instruments, the problem of water management in Spain and to deal with it, not only from a quantitative perspective, but also qualitative, to the new demands currently demanded by the administration of water resources. As the Statement of Motives of the Law makes clear, "the practical application of the 1985 Water Law has made it possible to verify both the existence of various practical problems in water management at the national level, which must be resolved with a view to the future, as well as the absence in it of effective instruments to face the new demands in relation to said resource, both in quantity, given that its consumption increases exponentially, as in quality, taking into account the evident need to deepen and improve the protection mechanisms existing in the 1985 Law".

For this reason, "the greater demands imposed by both European regulations and the sensitivity of Spanish society" justify the need for Public Administrations to undertake "the articulation of suitable legal mechanisms that guarantee the good ecological status of the assets that make up the hydraulic public domain, through various instruments, such as, among others, the establishment of a much stricter regulation of discharge authorizations, so that these can truly constitute an effective instrument in the fight against contamination of inland waters, or the regulation of ecological flows as a general restriction on all exploitation systems" From this point of view, the reform intends to incorporate a new perspective of water economy into water management, while promoting a policy of saving consumption and attending special way to the environmental demands of protection of the hydric resource. If the 1985 Water Law was a milestone in our Water Law, among other aspects, due to the incorporation of new approaches that took into account the ecological value of water resources, the Amendments to the Water Law of 46/ 1999 delves into this ecological dimension of inland waters, giving greater importance to its environmental values.

However, this sensitivity is not really evident in the various economic instruments used in the legislation; The concern for the environment is made visible in article 84, which sets as objectives the protection of the hydraulic public domain through: 1) the prevention of the deterioration of the ecological state and contamination of the waters; 2) to establish a water quality control program in the hydrographic basins; 3) prevent the accumulation of toxic compounds in the subsoil; 4) Recover aquatic systems. Similarly, article 90 of the Water Law states that the Basin Organizations, in the concessions and authorizations they grant "will adopt the necessary measures to make the use compatible with the environment and guarantee the ecological flows or environmental demands foreseen in planning".

And, finally, in the aforementioned Water Law, modifications are made in environmental matters regarding discharges and the reuse of treated water; and the role that the different Administrations will have in the regulation and action in said matters. But, economic measures, are not really established in all the established principles, for its implementation. In fact, we would only have to make a distinction between the different environmental protection measures, both prohibitive, preventive -authorizations and concessions-, stimulating -subsidies and exemptions-, dissuasive -inspection and expropriation-, compensatory -compensation- and repressive -sanctions- (but focuses almost exclusively on preventive and repressive), to realize that it is not a complete Law in economic terms and not all the necessary modifications have been made in the text that attacks us (Water Law 46/1999), so that it is.

So that, we can affirm that, only article 41 of the Water Law 46/1999, adjusts to the protection and conservation of the Environment, trying to control and manage the polluting processes of the discharge regime, although we could attribute said protectionist character that non-preventive, to article 40 of the Water Law 46/1999 due to its parafiscal nature, although this does not finish justifying that it is destined to the own conservation of the environment. What is clear is that the quantification of the taxable events included in the economic mechanisms are not large enough to achieve, in the medium term, the practical eradication of the misuse of water resources. Let's stop at the Consolidated Text of the Law of 2001.

CHANGES TO THE ECONOMIC INSTRUMENTS OF THE WATER LAW AT THE BEGINNING OF THE 21ST CENTURY

The modification of the Water Law 46/1999 represented an important change in the legislation of the economic and financial instruments of the water resource in our country, whose purpose emanated from the constitutional interest of "ensuring the rational use of all natural resources, with in order to protect and improve the quality of life and defend and restore the environment, based on the indispensable collective solidarity" (art. 45.2 CE). For this reason, we must point out that in the Royal Legislative Decree 1/2001, of July 20, which approves the revised text of the Water Law, the same instruments are established as in the modification of the Water Law 46/1985, except for some small issues among which it is worth highlighting those included in article 115 (since articles 113 and 114 do not show any modification, except that currency changes from pesetas to euros are included and in article 113.6, article 96 is no longer cited, but rather article 100 of this Law).

It is worth highlighting Title VI of the economic-financial regime for the use of the public hydraulic domain, the "Economic and Administrative Nature of settlements" (art. 115 Consolidated text of the Water Law, Royal Legislative Decree 1/2001)[28], which modifies Article 107 of Law 29/1985. Although it is not an economic instrument, it does regulate the operation and application thereof, collected in the articles of Title VI.

And it is that, in Art. 115, one of the most relevant issues, and one that may have a greater impact on the environment -perhaps not always in a positive way-, is that in the self-assessment process, the speeding up of collections is allowed and establishing facilities to make the payment of said canon jointly by the beneficiaries of the service, as long as they are grouped in a Community of Users or another type of group; this can even enhance the sharing of needs vs. limitations of its public water domain, thus favoring the awareness of all users (even if it is pressure or coercion).

It is essential, however, to point out that the modifications in terms of economic instruments, in the Spanish Water Law, through the approval of the Consolidated Text of the Water Law, Royal Decree 1/2001, have been practically null and the repercussion of these on the environment is relatively limited. Such affirmation is based on the idea that, although it is true that said Consolidated Text of the Water Law includes the pricing and payment for the damages that may be caused to the environment, directly or indirectly affecting the the water resource itself or on the rest of the biotic, abiotic and anthropic factors that make up a landscape or an ecosystem contemplates the prevention and conservation of the environment; at most, emphasis is placed on the possibility of restoring, regenerating or recovering what has already been damaged, although the "how" to do it is not very clear. Likewise, it should be noted that the types of taxes included in this legislation differ quantitatively from the costs themselves and the lack of equity that emanates from the very principle of territoriality to which reference is continually made, when delegating administrative powers in terms of collection vs. penalties, with an excessively low tax rate to be applied, which makes it cheaper to cause damage, pollute or degrade the public water domain itself, than to prevent and preserve its deterioration.

For all these reasons, we must return to the idea that, in water taxation, and more specifically in the Spanish Water Law, the so-called "canons" are being used indiscriminately (as stated by numerous experts in the field) which, in most cases, are public prices that entail a series of very negative vices. This parafiscal levy is paid for the occupation and use of the public asset, paying attention to the market value or the private utility that has been obtained, therefore, since it is a collective asset (both the hydraulic works and the supply of the water resource), it turns out that the contribution made supports global public spending.

Regarding the proposals for action, we can affirm that an interesting trend within this water sector is the reorientation of these classic tax figures, eminently tax collection, and partially representative of the application of the "polluter-pays" principle, towards other figures more preventive measures applied to the use of the environment, taking into account the principle of "who uses the resources, pays". Precisely along these lines, the possibility of changing the traditional discharge fees for a broad fee on the use of the public hydraulic domain may be interesting, in such a way that, instead of insisting on taxing the polluting capacity of residual water, would try to tax, the water consumed that finally becomes a contaminated waste. With this orientation, it is aimed at optimizing water as a scarce resource, introducing the price factor to rationalize management throughout its entire cycle of use; thus, a new canon for the protection of the public domain could be proposed, which would significantly expand the existing one, and replace the current dumping canon which, due to certain circumstances, has proven insufficient, both due to its low collection capacity and its weak operability.

A new legislation on water resources in which the economic-financial regime for water is framed within the Water Framework Directive itself and, therefore, is referenced by the Hydrographic Basins themselves, but with certain modifications, since the intrinsic characteristics of Hydrographic Basins can be used or used to study the general conditions of the Spanish territory and, extrapolating to the whole of Spain, a quantification can be established (a tax base) that is optimal in terms of collection and in environmental terms (protection, prevention and conservation). For this, it is essential to generate not only a new policy, but also to lay its foundations through a set of plans, programs and projects that allow not only to collect but also to evaluate and manage the consumption (supply, purification, extraction,...) of water, in order to carry out optimal, equitable, efficient and, above all, sustainable (economic, ecological and social) comprehensive planning, and in which strategic environmental assessment (today quite forgotten) plays a fundamental role.

With all of the above, the proposal for the unification of the tariff for water resources in Spain would be marked by the following conditions: first, that a new Water Policy be configured at the National level, in which the stipulation of the general canon of the Tax levy on water resources is established by the State itself, and is the same for all regions of Spain, regardless of their climate, hydrography, lithology, edaphology, geomorphology,... Secondly, that the collection powers fall on the Provinces, instead of the Autonomous Communities, since it is easier to levy a more or less small and homogeneous territory (although decentralization can lead to various complications, in this case it could be left in the hands of the Provinces the competence to collect taxation for water consumption, and that this, in turn, be controlled by the Autonomous Communities, to avoid corruption of the system). Finally, the purpose of the collection will not only be the recovery of costs (as stipulated in article 9 of the Water Framework Directive) but it will also be used for the protection of the environment (and, therefore, of ecosystems), the recovery of costs, the efficiency and fairness of tax collection and, something fundamental, avoiding the waste of water in some regions, while others suffer the gloom of scarcity, with which the proceeds will serve to finance the creation of new infrastructures (as well as control processes for obtaining water privately, irregularly and illegally) in the areas of the north of the peninsula and, in turn, generate water transport systems to the southern areas of Spain.

So that, after laying the foundations for the reality of Water Policies in our country, from an economic perspective, through which to configure a new model of sustainable development and, knowing and evaluating the set of elements, means and instruments to use in the management process of water resources, let us dwell on the study of the constitutional fit of hydrological planning in Spain as a support for land use planning.

CONCLUSION

In the present research we have used the word "politics" as an adjective whose purpose is to show that economic phenomena cannot be disintegrated or disassociated from the social environment in which they are inserted, that is, that the Economy, Geography or Law must be considered as open sciences, or as open systems (as Ludwig von Bertalanffy collects in his General System Theory). In fact, after the foregoing, we can conclude that the treatment of Water Policies must be governed by the precautionary principle, without forgetting that this, part of the scientific uncertainty of the effects that certain activities can cause on the environment and human health. In this way, it refers to an approach to risk management according to which, in the event that a policy or action could cause harm to people or the environment and there is no scientific consensus on the matter, the policy or action in question should be abandoned. The measures adopted must not be disproportionate in relation to the level of protection sought or attempt to reach a level of zero risk, which rarely exists. They must be consistent with similar measures already adopted in similar circumstances and must be maintained as long as the scientific data remains incomplete, imprecise or inconclusive, and as long as the risk is considered important enough not to accept that society assume it.

Likewise, it is concluded that, without generating conflict with what has been indicated so far, if we had the intention of considering the environment as a "fundamental right", it should be considered as a management policy. Well, in this way, it would fit into the set of economic policies aimed at establishing and implementing the economic order, that is, the set of rules of economic operation. Although the relationship between the environment and Economic Policy -in addition to its inter and intradisciplinary analysis-, can be framed in specific policies, either as instrumental policies -fiscal, monetary, foreign, etc.- or as sectoral policies (industrial, agrarian, transport, education, tourism, housing,…) In this way, we can affirm that the Economic Policy is linked to the environment, in the search for solutions to negative externalities, as well as to enhance positive externalities. So that, in the relationship between the economic, social and environmental aspects, the Economic Policy is configured as the basis for the global integration of each and every one of these elements; configuring an environmental strategy that inserts active environmental policies. Although, we must keep in mind that said elaboration is not exempt from complications such as intervention priorities (since the so-called "value judgments" that we indicated previously can -and, in fact, do-, mark the decisions of different economic agents) or uncertainties of a scientific and technical nature. For this reason, both in the decision-making process and in action, policies must have the capacity to decree hierarchies between the set of activities, by raising awareness among all economic agents of the risks, damages or impacts of said environmental degradation processes, and, with the scientific argumentation of the expected impact of the environmental pollution process; issues all of which are revealed and highlighted in the Environmental Action Programs of the European Union.

REFERENCES

Buchanan, J. M., & Stubblebine, W. C. (1962). Externality. *Economica*, *29*(116), 371. doi:10.2307/2551386

Consolidated Text of the Water Law, Royal Decree Law 10/2001, of July 20.

Embid Irujo, A. (2002). *Evolution of Water Law and Policy in Spain* (Vol. 1). Civitas.

Embid Irujo, A. (2007). *Dictionary of water law*. Iustel.

Embid Irujo, A. (2016). *Thirty years of the 1985 Water Law*. Editorial Thomson Reuters Aranzadi.

Jané Solá, J. (1974). For an Operational Structure of Economic Policy. *Spanish Journal of Economy*.

Jiménez Herrero, L. (1998). *Trends in the use of economic and fiscal instruments in environmental management. Reflections on the European Union and the Spanish case*. Cedecs Financial and Tax Law.

Madani, K. (2010). Game theory and water resources. *Journal of Hydrology, 381*.

Martin Simon, J. L. (1983). The meaning of externalities. Economy Notebooks, 11(32).

Mataix, C., Ezquerra, I. & Sobrerón, M. (2019). Accelerating transformation. *Foreign Policy Review, 190*.

Ministry of the Environment. (1998). *White Paper on Water in Spain*. Synthesis document.

Sotelo Navalpotro, J. A. (2001). *Environmental Europe*. Oxford University Press.

Sotelo Navalpotro, J. A., & Sotelo Pérez, I. (2018). Water and environment: Constitutional legal framework of water matters in nowadays. Anales de la Facultad de Ciencias Jurídicas y Sociales de la Universidad Nacional de La Plata, 15(48).

Sotelo Pérez, M., & Sotelo Pérez, I. (2014). Planning and management of water in Spain, present day. *Observatorio Medioambiental*, (17), 375–408.

Spanish Water Law, 29/1985, of August 2.

Spanish Water Law 46/1999, of December 13, modifying Law 29/1985, of August 2, on Water.

Summer, L. H. (1991). The case for corrective taxation. *National Tax Journal*, 44.

Villamil Serrano, A., & Maties García, J. (1998). Economic Policy of the Environment (R. Areces, Ed.). Academic Press.

ENDNOTES

1 Treaty on the Functioning of the European Union. Title XX. Environment. Article 191.1. (former article 147 TEC). "Union policy on the environment shall contribute to pursuit of the following objectives; preserving, protecting and improving the quality of the environment; protecting human health; prudent and rational utilisation of natural resources; promoting measures at international level to deal with regional or worldwide environmental problems, and in particular combating climate change".

2 In any case, we must not lose sight of the international objectives of European environmental law, with the fundamental purpose already mentioned of supporting sustainable development in the economic, social and environmental spheres of developing countries, with the fundamental objective of eradicating poverty (Art. 3.5 y Art. 21 TFUE). On the other hand, it can be seen that the objectives that have to do with the protection of health and the use of natural resources are redundant, to the extent that they should normally be considered included in the protection of the environment. "The reason for its inclusion was to avoid a strict interpretation of Community competence in the matter, so that, as the objectives continue to be foreseen, at the substantive level, the European Union has the competence to approve practically any type of measure related to environment" (Lefebvre, F, 2018).

3 Directive 2004/35/CE of the European Parliament and of the Council of 21 April 2004 on environmental liability with regard to the prevention and remedying of environmental damage, establishes the regulatory bases that are governed by the principle of "the polluter pays", through which, a framework of environmental responsibility is created that allows the preservation and repair of damage caused by a natural or legal person, public or private, that carries out a professional activity.

4 The principle of prevention, also called "principle of not causing transboundary environmental damage, constitutes one of the original rules of International Environmental Law, and lays the foundations or foundations for the general obligations of all States to safeguard the protection of the environment of their jurisdictions and of the areas that go beyond their areas of control, whatever the activities carried out in their territory.

5 Treaty on the Functioning of the European Union. Title XX. Environment. Article 191.2. (former article 147 TEC). "Union policy on the environment shall aim at a high level of protection taking into account the diversity of situations in the various regions of the Union. It shall be based on the precautionary principle and on the principles that preventive action should be taken, that environmental damage should as a priority be rectified at source and that the polluter should pay. In this context, harmonisation measures answering environmental protection requirements shall include, where appropriate, a safeguard clause allowing Member States to take provisional measures, for non-economic environmental reasons, subject to a procedure of inspection by the Union". Similarly, the principle of "who pollutes, pays" is included in Law 26/2007, of October 23, on environmental responsibility, which "regulates the responsibility of operators to prevent, prevent and repair environmental damage, in accordance with Article 45 of the Constitution and with the principles of prevention and the «polluter pays»"(Chapter I. Art. 1. L26/2007).

6 Market failures are those situations that produce an inefficient allocation of resources by the market. Although the concept has its basis in the philosophical studies of Henry Sidgwick during the nineteenth century, it was not until the mid-fifties of the last century that it began to be used recurrently in economic studies. The main causes that give rise to these inefficient allocations are, together with externalities, the existence of imperfect competition -such as monopoly, oligopoly, monopsony, oligopsony-, the existence of inappropriate market structures, an inadequate calculation of costs and benefits, existence of incomplete and asymmetric information, public goods or price discrimination.

7 It should be noted that most of the goods of an environmental nature -and, more especially if we mention water-, have gone from being "pure public" or "free" and being available in an unlimited manner, to being incorporated as "economic" goods. ", thus becoming scarce and limited goods

(in the specific case of water resources, essential for human life, despite speaking of a good that is available in three quarters of the earth's sphere -and is used in the production of practically all goods and services, in addition to its direct consumption-, highlights its limitation in its liquid state, such as fresh, drinking water, not polluted or degraded,..., so its availability in nature does not meet the necessary conditions for its direct consumption), so the environment is economically computable. Although, previously, natural assets, considered resources without economic value, that is, without market price stipulation, and used as publicly owned assets (lacking property right delimitation) were -and, in part, continue to be -, overexploited, wasted and dilapidated, giving rise to what Garrett Herdin came to call "the tragedy of the commons".

8 All this is erected and sustained "on pillars that range from configuring it in market terms, with adjustment variables such as prices and real wages, to the global configuration of macroeconomic relations based on the general microeconomic equilibrium, going through the use of the hypothesis of the neutrality of money and the idea of interrelation between aggregate demand and supply. The relative price system enables the simultaneous equilibrium of all markets, since the adjustment is timeless, the agents (consumers and businessmen) take into account an optimization program, there is a regime of pure and perfect competition, the price information it is also perfect, in addition to free, and there is instant and total price flexibility; reaching the Paretian optimum almost "spontaneously".

9 In the real world, what prevails are situations of monopoly or oligopoly, to whose disturbing effects are added the disturbances that hinder the correct development of the adjustment mechanisms that allow obtaining equilibrium.

10 A tenor de los estudios realizados por el ínclito economista David W. Pearce, la reducción continua y contante de la capacidad de carga de los ecosistemas ha de tenerse en consideración, a la hora de desarrollar o establecer una actividad económica, que es necesario conocer el zenit a partir del cual los costes externos dan paso a la degradación del medio ambiente. Ya que, tras un sólido análisis establece que los niveles de producción que sea seguro, en términos ecológicos, no corresponde con los niveles de producción económica que conduzca a un óptimo –en términos paretianos-, alcanzando lo que se denomina un "desfase" ecológico y, por ende, un proceso de inestabilidad ambiental.

11 This is the point at which the optimal investment for environmental damage control is located; above this, the costs of control and management are excessively high compared to the damage avoided and, below this, the damage is higher than the costs paid to avoid it.

12 The neo-Keynesian school recovers the scientific discourse of the most significant aspects of macroeconomics -study of economic cycles, market regulation, fiscal and monetary policies, wages or growth-, from perspectives different from those established by J.M. Keynes in his General Theory, although maintaining his thought and essential contribution on issues such as full employment and growth; and, introducing the precepts of neoclassicism, more specifically, the studies of Arthur Pigou, Alfred Marshall and León Walras. Among the set of precepts developed, it is worth mentioning the IS-LM model of John Hicks, the analysis of "externalities" or the studies related to the "Economics of Well-being". Among the main representatives of this current of economic thought, Paul Samuelson stands out –Nobel Prize winner in economics, who established the term neo-Keynesian, to denominate the neoclassical economic visions with the Keynesian ones-, John Hicks, Franco Modigliani, Robert Solow, Robert Mundell, James Tobin, Paul Krugman or Joseph Stiglitz.

13 Economic theory that states that all economic agents aim to achieve maximum levels of profit or utility, according to the resources available to them.

14 In principle, all economic activity produces a cost for society -"social cost"-, but, in general, part of that cost is compensated by the economic agent that originates it, through the private costs of the activity (energy, labor, raw materials,…); although, social costs linked to the activity are also generated that, in theory, "punish" society and are not compensated by the cause of the environmental impact (such as environmental degradation). In this way, when the social cost is higher than the compensated social cost, that is, the private cost, we find ourselves with a negative externality, imposed exhaustively and without market transaction -in an extra-commercial way-, to society as a whole.

15 In addition, these discrepancies are not reflected in their price, an issue that cannot be taken into consideration when dealing with "public goods and more specifically of the so-called "pure public goods" (term coined by James Buchanan in the sixties of the last century).

16 The purpose of public authorities is to control or eliminate negative externalities and enhance positive externalities. In this way, it will use instruments to enhance or penalize said situations through the implementation of a tax system -based on the so-called "Pigouvian taxes"-, through regulatory instruments, price and market control, promoting subsidies and grants, etc.

17 The Nobel Prize in Economics, Ronal Coase, established what has come to be called "the Coase Theorem" -consolidating himself as one of the fundamental pillars of economic analysis of Law-, through which it establishes that the intervention of the public powers is not necessary, as long as certain precepts are fulfilled that allow the regulation of the external effects produced by one or several economic agents on another and others, when carrying out their economic activity. In this way, it stipulates that as long as property rights are clearly defined and transaction costs are eliminated -or at least minimized-, as well as equality between the parties and clearly defined agreements-, it will be possible to achieve an optimal reallocation of rights towards those who value them more, and, therefore, establishes that with a clear definition or delimitation of property rights, both the origin of the damages and the one that originates them can be identified, Thus, the intervention of the public authorities is unnecessary. All this becomes especially complex when we are referring to the so-called "public goods", since these are established as non-rival, non-exclusive, non-divisible goods with zero cost, that is, free.

18 "Public good is one for which the same unit can be consumed by more than one individual (non-rival consumption).

19 Special mention should be made in this regard, what was stated by Professor Antonio Embid Irujo (2007), who affirms the following precepts; namely: "I believe that in few countries in the world the debate on the public or private ownership of certain types of inland waters will have reached the virulence –and, at the same time, the depth and even sophistication in reasoning- of ours. And this based, however, on legal premises that formally configure the public property of water as the general rule and as an exception the possibility of private property, at least in an initial consideration and without carrying out the logical work of deepening and interpretation. on the corresponding legal framework. Public ownership of waters in Spanish law, in principle, which would coincide, also in principle, with the general trend in very different legislations to walk along the path of affirming public ownership of waters (Embid Irujo, 2002) or, at least, in the recognition -and its consequent legal consequences- of the difficulty of having private property over them with the usual characteristics of civil property and this due to the particularities of the medium, the water, over

which it would be exercised. which leads to, in any case, speaking of water as the most relevant of the special properties. This seems evident, as is the fact that these difficulties in the exercise of private property are, without a doubt, very important, which cooperates, as a reaction, in achieving the actual publication of a legal relationship that, initially, has tended to be private in relation, at least, to some kinds of water (not to that of rivers or continuous surface currents, for example, where the characteristic of public or common property and the use of all has always been present). On the other hand, and apart from the physical characteristics of the environment, it is also necessary to highlight as a determining cause of the result that I notice that of finding ourselves before a natural resource, essential for the development of human life, the economy, the ecosystems, the that in reality and above all causes an inevitable intervention of the Public Powers either to achieve its preservation or for its more rational use with the logical result of a real, material publication of the set of waters as a note some doctrine built on the basis of a very extensive study of comparative law even when the expression "public property" is not always formally used as a predicate of the "water" genre.

20 The main instruments used in the protection and conservation of natural resources throughout the last decades have been, mainly, the normative and economic instruments, applying, at the present time, said claims or realities to water resources, consolidating or, at least, trying to consolidate Water Policies that allow controlling, regulating and guiding the economic activities carried out, both within the market and outside it, by the different economic agents, with respect to their consumption, preferences or water needs.

21 In principle, although the theory of "internalizing" the damage must be neutral, it does not take into account the social conflicts and the impact on the natural environment that causes environmental degradation, and the so-called "ecological crises" that originate (which suppose a enormous structural transformation of the environment). Although, from the neoclassical analysis, if the degradation of the environment is considered as a "negative externality" since it exerts a pernicious impact that occurs outside the "market" and that originates social costs that are not economically compensated and, These social costs are higher than the private costs, giving rise to great conflicts between social and private interests.

22 As stated in this text, "1. The occupation or use that requires authorization or concession of the assets of the hydraulic public domain, referred to in sections b) and c) of article 2 of this Law, will be taxed with a canon destined to the protection and improvement of said domain, whose application will be made public by the Basin Agency. The water concessionaires will be exempt from the payment of the fee for the occupation or use of the public domain lands necessary to carry out the concession. 2. The taxable base of this levy will be the value of the good used, taking into account the performance it reports. The annual tax rate will be 4% of the value of the tax base. 3. This canon will be managed and collected, on behalf of the State, by the Basin Organizations, who will report to the Ministry of Economy and Finance periodically in the manner in which it determines." (Article 104 of Law 29/1985).

23 As stated in this text, "1. Authorized discharges, in accordance with the provisions of articles 92 and following of this Law, will be taxed with a canon destined for the protection and improvement of the receiving environment of each hydrographic basin. 2. The amount of this levy will be the result of multiplying the pollution load of the discharge, expressed in pollution units, by the value assigned to the unit. A unit of contamination is understood as a conventional pattern of measurement, which will be established by regulation, referring to the contaminant load produced by the

discharge-type of domestic water, corresponding to 1.000 inhabitants and a period of one year. Likewise, through regulations, equivalence scales will be established for wastewater discharges of another nature. The value of the pollution unit, which may be different for the different rivers and river sections, will be determined and reviewed, where appropriate, in accordance with the provisions of the Hydrological Plans regarding the quality of continental waters, so that the financing of the works necessary to comply with said forecasts be covered. 3. This canon will bc received by the Basin Organizations and will be allocated to actions to protect the quality of the waters that have been foreseen in the Basin Hydrological Plans, for which purpose it will be made available to the competent bodies. 4. When the passive subject of the dumping canon is obliged to bear other economic charges, already established or that may be established by the Autonomous Communities or by the Local Corporations, in the exercise of their powers, to finance public wastewater treatment plans or programs, the Water Council will determine annually the deductions that must be made in the amount of the discharge canon" (Article 105 of Law 29/1985).

24 As stated in this text, "1.Those benefited by the surface or groundwater regulation works carried out totally or partially, at the expense of the State, will pay a canon destined to compensate the contribution of the State and meet the expenses of exploitation and conservation of such works. 2. The benefits for other specific hydraulic works carried out entirely at the expense of the State, including those for correcting the deterioration of the public hydraulic domain, derived from its use, will satisfy a levy for the availability or use of the water destined to compensate the investment costs and attend to Expenses for exploitation and conservation of such works. 3. The amount of each of the levies will be set for each budget year, adding the following amounts: • The total expected operating and maintenance costs of the works carried out. •The administrative expenses of the managing body attributable to said works. • 4 percent of the value of the investments made by the State, duly updated, taking into account the technical amortization of the works and installations and the depreciation of the currency, in the manner determined by regulation. 4. The individual distribution of said global amount, among all those benefited by the works, will be carried out in accordance with criteria of rationalization of the use of water, fairness in the distribution of obligations and self-financing of the service, in the manner determined by regulation. 5. These levies will be managed and collected, on behalf of the State, by the Basin Organizations, who will report to the Ministry of Economy and Finance periodically in the manner in which it determines" (Article 106 of Law 29/1985).

25 As stated in this text, "1. The occupation, use or exploitation of public hydraulic domain assets included in sections b) and c) of article 2 of this Law, which require a concession or administrative authorization, will accrue in favor of the competent Basin Agency a rate called utilization fee. of assets of the hydraulic public domain, aimed at the protection and improvement of said domain. The water concessionaires will be exempt from the payment of the canon for the occupation or use of the public domain lands necessary to carry out the concession. 2. The accrual of the rate will occur with the initial granting and the annual maintenance of the concession or authorization and will be payable in the corresponding amount and in the terms indicated in the conditions of said concession or authorization. 3. Concessionaires or authorized persons or, as the case may be, those who are subrogated in their place will be passive subjects of the canon. 4. The tax base of the levy will be determined by the Basin Agency according to the following assumptions: a) In the case of occupation of land in the hydraulic public domain, by the value of the occupied land, taking as a reference the market value of the adjoining land. b) In the case of use of the hydraulic

public domain, for the value of said use or of the benefit obtained with it. c) In the case of the use of public hydraulic domain assets, for the value of the materials consumed or the utility reported by said use. 5. The annual tax rate will be 5% in the cases provided for in letters a) and b) of the previous section, and 100% in the case of letter c), which will be applied to the value of the resulting tax base. in each case. In the event of inter-community basins, this fee will be collected by the Basin Agency or by the State Tax Administration, by virtue of an agreement with the latter. In this second case, the State Tax Administration Agency will receive from the Basin Agency the pertinent data and censuses that facilitate its management, and will periodically inform it in the manner determined by regulation. The canon collected will be made available to the corresponding Basin Agency" (Art. 40 de la Ley 46/1999).

26 As stated in this text, "1. Discharges into the public hydraulic domain will be taxed at a rate intended for the study, control, protection and improvement of the receiving environment of each hydrographic basin, which will be called the discharge control canon. 2. Those who carry out the discharge will be passive subjects of the discharge control canon. 3. The amount of the discharge control fee will be the product of the authorized discharge volume times the discharge control unit price. This unit price will be calculated by multiplying the basic price per cubic meter by a coefficient of increase or decrease, which will be established by regulation based on the nature, characteristics and degree of contamination of the discharge, as well as by the highest environmental quality of the physical environment in which it pours. The basic price per cubic meter is set at 2 pesetas/cubic meter for urban wastewater and 5 pesetas/cubic meter for industrial wastewater. These basic prices may be revised periodically in the General State Budget Laws. The increase coefficient of the basic price may not exceed 4. The discharge control canon will accrue on December 31, the tax period coinciding with a calendar year, except for the year in which the discharge authorization occurs or its cessation, in which case, the canon will be calculated proportionally to the number of days of validity of the authorization in relation to the total of the year. During the first quarter of each calendar year, the canon corresponding to the previous year must be settled. 5. In the event of intercommunity basins, this fee will be collected by the Basin Agency or by the State Tax Administration, by virtue of an agreement with the latter. In this second case, the State Tax Administration Agency will receive from the Basin Agency the pertinent data and censuses that facilitate its management, and will periodically inform it in the manner determined by regulation. The fee collected will be made available to the corresponding Basin Agency. 6. When the existence of a discharge is verified, whose person in charge lacks the administrative authorization referred to in article 92, regardless of the corresponding sanction, the Basin Agency will liquidate the discharge control fee for the non-prescribed exercises, calculating its amount by indirect estimation procedures in accordance with what is established by regulation. The discharge control canon will be independent of the canons or rates that the Autonomous Communities or local Corporations may establish to finance the sanitation and purification works" (Article 41 of Law 46/1999).

27 As stated in the regulations, "1. Those who benefit from the surface or groundwater regulation works, wholly or partially financed by the State, will pay a regulation fee to offset the investment costs borne by the State Administration and cover the exploitation and conservation expenses of such works. 2. Those who benefit from other specific hydraulic works financed totally or partially by the State, including those to correct the deterioration of the public hydraulic domain, derived from its use, will pay for the availability or use of water a levy called the water use rate, intended to offset the investment costs borne by the State Administration and to meet the operating and conservation

costs of such works. 3. The amount of each of the levies will be set for each budget year, adding the following amounts: a) The total expected operating and maintenance costs of the works carried out. b) The administrative expenses of the managing body attributable to said works. c) 4 percent of the value of the investments made by the State, duly updated, taking into account the technical amortization of the works and installations and the depreciation of the currency, in the manner determined by regulation. 4. The individual distribution of said global amount, among all those benefited by the works, will be carried out in accordance with criteria of rationalization of the use of water, fairness in the distribution of obligations and self-financing of the service, in the manner determined by regulation. 5. In the case of intercommunity basins, the levies provided for in this article will be managed and collected by the Basin Agency or by the State Tax Administration, by virtue of an agreement with the latter. In this second case, the State Tax Administration Agency will receive from the Basin Agency the pertinent data and censuses that facilitate its management, and will periodically inform it in the manner determined by regulation. The fee collected will be made available to the corresponding Basin Agency. 6. The liquidating body for the fees and levies will introduce a corrective factor for the amount to be paid, depending on whether the beneficiary of the hydraulic work consumes in amounts higher or lower than the reference allocations established in the basin hydrological plans or, as the case may be, in the regulations that regulate the respective sectoral planning, especially in terms of irrigation or other agricultural uses. This corrective factor will consist of a coefficient to be applied to the settlement, which may not be greater than 2 nor less than 0.5, in accordance with the rules determined by regulation. The Basin Agency will approve and issue the settlements regulated in this article in the year to which they correspond" (Article 42 of Law 46/1999).

28 As stated in the regulations, "1. Regulations may establish the self-assessment of the fees or levies mentioned in the previous articles. 2. The acts of approval and liquidation of these fees or levies will have an economic-administrative nature. Without prejudice to the provisions of the regulations governing the applicable procedures, the challenge of the acts will not suspend their effectiveness, the payment of the debit being required by the administrative procedure of urgency. Failure to pay may lead to the suspension or loss of the right to use or take advantage of the hydraulic public domain. 3. The payment of the levies foreseen in this Law, when those obliged to do so are grouped in a community of users or representative organization of the same, may be made through such communities or entities, which are empowered for this purpose to carry out carry out the corresponding collection, in the terms established by regulation" (Art. 115 of the Consolidated Text of the Water Law 1/2001).

Chapter 7
The Need and Development of Dams in India as Well as in the World for a Better Future

Aditya Singh
https://orcid.org/0000-0001-9347-5627
Lovely Professional University, India

ABSTRACT

This chapter explains the concept of dams. It mainly focuses on the dams in India. The chapter discusses the purpose as well as history of dams. The author collected data from various sources in order to perform graphical analysis to support the study. A comprehensive review of various scientific and research papers is done by the author to find the gaps in the research in the recent years. Some new dam projects are also briefly mentioned in the chapter. Plus, the author worked in the direction to find whether the current number of dams in the world, including India, is enough or not. Then, would building more dams replace the use of non-renewable sources of energy completely by itself or not? These are the matters that are mentioned in the chapter.

INTRODUCTION

A dam is any structure which was built across a river, stream or estuary, for water retaining purpose. They are normally built to supply the retained water for different purposes, in order to be used by people for their day to day activities, for industrial activities, and to irrigate semi-arid and arid lands. To generate hydroelectric power, dams are required to increase water's quantity. Similarly, dams help in increasing the depth of a river water in a given place in order to enhance navigation on waterbody, as well as permit ships along with barges to travel smoothly, and also in cases of heavy snowmelt or large storms, it should have the ability to decrease the peak discharge of flood water. They also are able to give a lake, which can be used for activities which are amusing ones, like fishing, swimming as well as boating. There are numerous dams which are constructed not for only one purpose but for two or more purposes, like for instance, the water present in one particular reservoir can be utilized for to assist in the irrigation system

DOI: 10.4018/978-1-6684-6449-6.ch007

of a given area, also to generate hydroelectric power to be used in nearby areas and also for fishing by people. Such water controlling structures are frequently labelled as multipurpose dams. Supporting works which are able assist dams in order to allow them to function in a proper way, consists of movable gates, spillways as well as valves, which control the discharge of extra water downstream from the dams. The dams also consist of intake structures which can supply water to a power station or to tunnels, canals, or pipelines intended to transport watcr storcd by the dam to faraway places. Further supplementary works are locks to allow the passage of ships to go through the dam site or around it, systems used to flush out or clear out the silt which mounts up in the reservoir, as well as fish ladders along with other devices to help fishes in looking for to swim past a dam or around it. Dams can also act as main structures in multipurpose schemes which are intended to save water resources in some particular regions. In developing countries, multipurpose dams have exceptional significance, as a single dam is able to bring numerous benefits like industrial growth, the development of agriculture in that region as well as hydropower generation to be used for other activities. Nevertheless there is an important concern related to the environment has arisen because of dams, as the dams affect the riparian ecosystems as well as migrating fishes in a negative way. Additionally, massive land areas which were people's home initially and covered with vegetation can be submerged because of the creation of large reservoirs. This has raised people's opposition to the dam projects in large groups in many areas, leading to protests in different forms, as many people questioned the concerned authorities about the benefits offered by such projects are adequate enough after such great losses. According to engineering terms, dams can be classified into different categories like based on building materials as well as based on structural types of dams. The type of dam which will be build is generally decided based on the construction materials available in the place or nearby, the valley's foundation conditions, experiences of the concerned engineers, promoters as well as financers who are accountable for the given project, and site accessibility to the transport facilities or to the transport network. According to the dam engineering of the current times, rockfill, earthfill and concrete are the materials among which the concerned authorities of the project has to make a decision to build the dam in a given place. However in the past, jointed masonry was the material responsible for the making of various dams, later on it became outdated and it was replaced by concrete. Huge gravity dams, buttress dams as well as thin arch dams are being built using concrete from several decades in the modern era. With roller compacted concrete's development, high quality concrete was indorsed in sufficient quantities in order to be placed with the equipment type which was initially made to combine, distribute as well as move earthfill. Typically rockfill as well as earthfill dams are normally assembled together in a group which is named as embankment dams, as they create enormous rock as well as earth duncs, which can be gathered into daunting embankments which are manmade structures.

Types of Dams

Dams can be divided into some major categories, which are as follows:

- On the basis of structure of dam
- On the basis of use of dam
- On the basis of size of dam
- On the basis of material used in dam
- Others

On The Basis of Structure of Dam

On the basis of the structure of dam, dams can be further classified into subcategories, which are as follows:

- Embankment dams
- Arch dams
- Barrages
- Gravity dams
- Rockfill embankment dams
- Arch-gravity dams
- Fixed-crest dams

Embankment Dams

These types of dams are built with the help of compacted earth, with two major types of embankment dams, which are earth-fill embankment dams as well as rock-fill embankment dams. The embankment dams are having a structure which makes them rely on their own weight in order to hold back the water force.

Arch Dams

These types of dams acquire stability by help of arch as well as gravity action mishmashes based on the requirement of the situation. They normally need solid dependable supports at the abutments like the side wall of a canyon or buttress, which is essential for their existence. Places having narrow canyon along with steep side walls which are made of sound rocks, are generally preferable for Arch dams to be built. There are two major types of arch dams, which are constant radius dams as well as constant angle dams, which are used according to the situation and need.

Barrages

Barrages are special types of dams, which have a line of large gates which are being used for opening as well as closing the amount of water that needs to be released or passed through it. Gates are generally located between the flanking piers which are accountable to assist the load of water, it helps in controlling as well as stabilizing the flow of water which is provided for the irrigation purposes. There are another kind of barrages which are called Tidal barrages, as they are constructed at the mouths of lagoons or rivers, in order to stop tidal incursions or to use the tidal flow for the purpose of tidal power.

Gravity Dams

These types of dams uses gravitational force to hold themselves in place against water's intense push, as the huge mass of the dam is being pulled in the downward direction because of the Earth's gravity, which makes the whole dam stable even against a great push. When there is an availability of solid foundation, then gravity dam is one of the best choice for that place. They are also categorized as hollow or solid, and requires masonry or concrete to be used for their construction.

Rockfill Embankment Dams

These types of dams are embankments made of compact free-draining granular soil using impervious zone. Since, the earth which is utilized in its construction, frequently consists of high percent of large size particles, that's why the name rock-fill is used. The impervious zone can be on the upstream face or inside embankment, in the latter case, it is known as core. For earthquake regions, rockfill embankment dam is the best choice, because of its high resistance against earthquakes. There are two categories, which are known as Concrete fill dams and Earth fill dams which won't be covered in this subsection of the chapter.

Arch-Gravity Dams

These types of dams are a hybrid of gravity dams and arch dams, because both of them are combined to make it. They are constructed in areas where huge amount of water flow is present and the materials available are not sufficient to construct a pure gravity dam. Resources are easily saved in their construction and they are a good choice for earthquake regions.

Fixed-Crest Dams

These types of dams are generally a barrier made of concrete across a river. They are designed and constructed in order to maintain a given depth in the waterbody of a river, to allow smooth navigation. The people on small boats are having a risk while travelling over them, as it is hard for them to escape from any induced current and the small size boat is easily affected by such induced current in the waterbody.

On The Basis of Use of The Dam, it is subdivided into different parts, which are as follows:

- Weir
- Tailings dam
- Saddle dam
- Diversionary dam
- Check dam
- Underground dam
- Dry dam

Weir

Weir is a small dam which is also known as overflow dam and it is normally build in a channel of a river in order to make an impoundment lake, which if further utilized for the purpose of water abstraction as well as to measure the flow of water or for retardation purpose.

Tailings Dam

This type of dam is generally an earthfill embankment dam, which is utilized to store tailings that are being produced in the course of mining operation being performed after the unprofitable portion is being detached from the profitable portion of the ore. Tailings dams are more economical in comparison to traditional water retention dams, to serve this purpose. Majorly, there are 3 raised tailings dam designs, which are centreline, downstream as well as upstream, which are named based on the crest's movement at the time of raising.

Saddle Dam

This type of dam acts as a secondary or supporting dam, which is built restrain the water of a reservoir made by a main dam, due to two reasons, either to increase the efficiency of that reservoir by limiting its boundary or to allow higher raise of water as well as storage of water.

Diversionary Dam

This type of a dam is used in order to divert a part of the river's flow or to diver the whole river's flow from the natural course of the river. This diverted water of the river is sent for the production of hydro-electric power and/or into a tunnel or canal for the purpose of irrigation in that region.

Check Dam

This type of dam is used to decrease the flow velocity as well as to handle the erosion of the soil. This type of dam is also effective to slow down the flow of river water during a flood upto a certain extent. By controlling the soil erosion in that area, the nearby doesn't lose much of it fertility as compared in the normal conditions.

Underground Dam

This type of dam is constructed to trap groundwater as well as help in storing most or all the groundwater below the surface for prolonged use in a local area. Even to prevent the saltwater to mix with freshwater aquifer, underground dam is required to serve this purpose. Generally, the areas where water resources are scarce as well as need to be stored like on islands or in deserts, this type of dam is used. They are mainly of two types, which are sand storage as well as subsurface dams.

Dry Dam

This type of dam is called as a flood retarding structure and its purpose of construction is to control or stop the floods in a given area. Generally, it won't stop the flow of water as well as it will permit the water to flow smoothly, but in the time of intense flow of water, it will control or stop it, to avoid the situation of flood in downstream.

On the Basis of the Size of the Dam

It can be further subdivided into different types which are as follows:

- Large dams
- Small dams

Large Dams

Large dams are those which have a height of fifteen metres or more counting from their lowest foundation to their crest, or if their height is in between five metres and fifteen metres as well as containing more than three million cubic metres capacity to hold water, according to the definition of the (ICOLD) International Commission on Largc Dams. Additionally, dams which are having a height of more than one hundred fifty metres, are called as Major Dams.

Small Dams

Small dams are those dams which have numerous uses, which includes protection from the floods, hydropower production, storage of water, but also catching runoff water for future utilization which is helpful for farms especially during the time of dry season, and so on. Small dams produce many benefits and they don't even need to displace the local people like the large dams.

On the Basis of Materials Used in Dam

There are two major categories which are discussed below:

- Timber dams
- Steel dams

Timber Dams

These types of dams are made of timber named material, they were really popular in the initial period of the industrial revolution as well as in the frontier areas, because using timber as a material to construct dams was easier and the construction time to build the dams was also considerably less. Now in the current times, timber dams are hardly constructed as their height is limited as well as they have lifespan generally short. They are only built when timber as a material is found in huge quantities in and around the given area as well as other materials are difficult or costly to transport to the given area along with the aim to not focus on the longevity of the dam. They also have two major variants like Timber Plank Dams as well as Timber Crib Dams.

Steel Dams

These types of dams used steel as a main material to construct the dams, initially around the beginning of the twentieth century, they just started as an experiment, where load bearing beams as well as steel planting at an angle were made of steel to be used in the construction of dams. The experiment was a failure, as steel dams couldn't meet the expectations of being permanent structures along with being sturdier than the above mentioned Timber Crib Dams in addition to remain cheaper than concrete, masonry or earthworks, which are used to build dams.

Other major types:

- Natural dams

- Cofferdams
- Beaver dams

Natural Dams

Natural dams are those dams which were made by the natural geological forces. For instance, the flow of lava through a given area can naturally create Lava dams, here generally the outlet of a lake or the path of a stream is cut-off to form an impoundment of waterbody naturally. Similarly, glacial activity as well as natural disasters including landslides in addition to earthquakes can form natural dams. However, at times their sudden appearance can pose danger to the existing infrastructure as well as human settlement in that given area. Additionally, if these types of natural dams collapse suddenly at any time, then they will easily destroy the surrounding infrastructure as well as the settlements of the humans in that region.

Cofferdams

These types of dams are normally temporary barriers, which are built to eliminate water from a given area, which is generally submerged. They are normally made of steel sheet piling, concrete or wood. Additionally they are built on the foundation of bridges, permanent dams and so on; after the project is completed they are generally removed or demolished, except in the case of the area needing constant maintenance which is a rare case.

Beaver Dams

Beaver dams are built by beavers, consisting of sticks as well as mud, enough to flood a specific area which is habitable in nature. This allows beavers to navigate near or below the surface as well as to continue to be hidden or in order to protect them from their predators. Plus, during winter season, this permits the beavers to obtain food for their survival.

Objectives

Some major objectives of this book chapter are as follows:

- The book chapter explains the concept and types of dams in brief.
- The chapter explains the history and evolution of the dams to reach the current times.
- This book chapter explains the need to develop dams not only in India but also in the world in the future.
- The chapter will help the readers to understand when, where and what type of dam will be required in any given region in the future.
- This book chapter also focussed on discussing the challenges of some new dam project proposals in brief.

Motivation

This book chapter briefly explains the concept of dams in general, as well as it mentions the various types of classification of dams that can be done on the basis of different parameters. The author also performed a thorough review of literatures related to the recent published papers of researchers from different parts of the world in regards to dams. It briefly explains the working of a dam and gives the idea to the readers about the types of dams required in a given place or region depending upon the requirements of that given area. In the recent years, in the country India, a number of dam projects were proposed and there are many dam projects which are still being constructed. However, there were also some proposed dam projects which are being cancelled after the government of India, saw the protest of people against the development of those projects. Then, there are dam projects, which were completed in the recent years. India is already a leading country in the world in regards to the number of dams, after China and USA. At present the current Central Government of India, is in the favour of building more dams in the country wherever necessary, but it is necessary to understand the challenges and disadvantages of dams before constructing them. This book chapter had covered all these aspects which should be considered before the construction of dams, and it also explained why there is need to develop dams not only in India but in the rest of the world wherever necessary for a better future.

What Is the Importance of Dam and How Can It Promote the Use of Renewable Sources of Energy and Sustainability?

Dam are constructed for different purposes and there is a wide variety of dams which are built based on the requirements, location and availability of resources. After construction, dams are able to support irrigational facilities on a large scale, it also helps in transporting water to faraway and remote places. Plus, the recreational activities as well as inland navigation are also improved and so on. However, it also replaces the non-renewable sources of energy with hydroelectric power which is a renewable source of energy as well as clean and safe one. By this way, it promotes sustainability in a way upto certain extent. In comparison other sources of renewable sources of energy, dams after construction doesn't need much technical maintenance on a frequent basis.

ORIGIN, HISTORY, AND EVOLUTION

According to the records, the oldest dam in the world is said to be a masonry as well as earthen embankment present at Jawa, which is located in the Black Desert of the current Jordan. This dam was built around in between 4000 BCE to 3001 BCE, which can be considered around 3500 BCE, in order to restrict the water of a small stream as well as to help in the irrigation of the land present in the downstream. Based on evidence found, there was another similar type of dam which was constructed around 2700 BCE at Sadd el-Kafara, which is around 30 Km south of Cairo present in Egypt. Then there is a rockfill embankment in Syria, which is considered to be the oldest dam which is still in use and it was constructed around 1300 BCE, for the purpose of irrigation. Then, between 700 BCE and 250 BCE, according to the records, Assyrians, Babylonians as well as Persians also constructed various dams. Of course, remains of the old Ma'rib dam can be found in Yemen and then based on records, numerous dams were constructed in India, Sri Lanka as well as in China around the same period. Even though the Romans were considered as skilled civil engineers, they weren't impressive in regards to the evolution of dams specifically in terms of the height of the dams as well as the number of dams constructed. Byzantine engineers played a major role in the evolution of dams, after they construction of a dam in around 550 CE, which is at present located around the Turkish-Syrian border. In Sri Lanka after the fifth century Before Common Era, constructed numerous earthen dams of moderate height to support widespread irrigational facilities. Well, many old dams are still used in the country. Around two hundred forty Before Common Era, a massive stone structure was constructed across the Jing River present in China. Diamonike Dam is an old dam which was constructed in one thousand one hundred twenty eight Common Era in Japan. Then, several dams were also constructed in India as well as in Pakistan. The World's 1st large scale thin arch dams can be seen in Iran (earlier known as Persia) because of the existence of Kebar as well as Kurit dams. In Italy the dam construction restarted during the fifteenth century and in the sixteenth century, as well as in Spain on a large scale. Tibi Dam in Spain was major curved gravity structure for almost 300 years throughout the Western Europe, but it was later on surpassed by France's Gouffre d'Entfer Dam. During the seventeenth century, high Elche Dam was constructed in Spain, which was an advanced thin arch masonry structure. In case of British islands as well as in the northern Europe, the dam were constructed with only modest heights until the industrial revolution. The design as well as construction of the dams were on the basis of empirical knowledge as well as experience until the mid-nineteenth century. Many famous scientists like Galileo, Newton, Leibniz, Hooke, Bernoulli, Euler, Coulomb as well as Navier made major contributions in the advancements of the dams. Civil Engineering Professor Rankine of Scotland's University of Glasgow in the 1850s effectively proved the way applied science can be of significant help to the field engineers. Similarly, Sazilly as well as Zola made a major contribution in the evolution and the making of modern dams. With technological advancements in the 20th century, USA and Western Europe constructed a number of major modernized dams. Similarly, China, India and other countries also constructed several major modernized dams later on.

LITERATURE REVIEW

Coughenour et al. (2022) studied in gumshoe of the dam about the Little Conemaugh river as well as its south divergence to understand the floods happening in the past as well as that may happen in the future. They considered the disastrous flooding of Johnstown and the surrounding areas during the late 19th century, first and second half of the 20th century, for their study. They used historical stream gage data as well as data from the novel gage on the south divergence of the river, along with the Nexrad rainfall data in order to measure the watershed response in addition to deliver a new analysis of flood hydrography to perform their study.

Llamosas and Sovacool (2020) studied the transboundary dams and they worked on a methodical review of the governance dynamics, benefits as well as the drivers of them, in order to understand the hydropower's future. To conduct their study, they considered one thousand two hundred sixty four peer-reviewed papers which were published in between the years 2009 – 2019, in regards to the above mentioned dams. They found out that the majority of the papers were only aimed at problems in regards to the allocation as well as the management of the water, while only smaller number of the papers were aimed at temporal as well as spatial variation, scope of hydropower benefits, in addition to equity as well as justice dimensions.

Grubert (2020) worked on enabling the analysis of hydroelectricity by connecting national record of dams as well as energy info management data, in order to understand the future of energy along with traditional hydroelectricity. She found that reservoir based hydroelectricity as well as traditional dams are cheap, readily available zero carbon resources which could enable periodic storage, incorporation of highly irregular renewable electricity and so on.

Ding et al. (2022) studied response surface methodology in order to optimize the cemented sand as well as gravel dam profile. They selected the profile form of the above mentioned dam as their object of research in their paper, in addition to they performed the weight optimization analysis using the above mentioned method on the 3 factors of dam which are slope ratio, height as well as cemented material content.

Vasques et al. (2022) performed their study in the semi-arid regions of Brazil and they worked on an underground dam, with the goal to use ground penetrating radar model in order to map as well as identify the regolith's depth, authenticate the position, size as well as compute the volume of the water reservoir of a forthcoming underground dam.

Heidarzadeh and Feizi (2022) studied the Toddbrook dam present in England as well as about its concrete spillway's failure at the time of the Aug flood in the second last year of the second decade of the 21st century, and they prepared a cascading risk model for it. They performed a combination of desk studies, field surveys as well as numerical modelling in order to analyse the incident as well prepare their above mentioned model.

Peluso et al. (2022) studied the hydroelectric dams and they found out the negative effects of climate change on the neotropical fishery resources can be worsened by them. They performed their study in the country Brazil and specifically on the upper Paraguay River basin which also includes mega wetlands of the world like Pantanal. In this area the biodiversity is rich in nature and migratory fishes are essential in the ecosystem which is responsible for the ecosystem services like fisheries. They stated that the route of migratory fishes will be blocked by the presence of artificial dams, which will have drastic change in the future.

Rivera-Trejo et al. (2022) worked on forecasting more or less the penalties of constructing a dam which is a planned one particularly, and also present in a river catchment of pristine sand gravel in nature. They conducted their study in Rio San Pedro, which is located in the Northwest Mexico. They focused on the river characterization previous to the interference done due to dam construction within its channel.

Flaminio et al. (2021) worked on whether dam is required or not in the Anthropocene era. They stated that various studies in the past showed disadvantages and unsustainable approach of dams, but in the recent decades the demand of clean energy which dams provide by giving hydropower energy has supported the need of dams in the current times. To clear this long going debate, they inspected numerous articles from the year 1945 to the year 2019, which were based on dams and available in the public domain. They found that over the years there was always an addition of new points against and in support of dams by the concerned parties.

Yang et al. (2022) studied the 2 consecutive landslide dams in the year 2018 and they performed a risk assessment which was a quantitative one on them. Their study area was Jinsha River present in China. They used empirical formulations with a human risk assessment model in order to calculate the economic as well as human losses because of the above mentioned dams. Their results showed the breaching floods increased after the 2nd landslide dam even more but because of the adequate time lag given by the long distance between the dam site and residential area, caused only minor loss of human lives.

Prakash et al. (2022) worked on novel progress of the health monitoring technology of concrete dams. They reviewed numerous models which were made in the recent years for the purpose of dam health monitoring and they categorized them into three different categories like data driven, numerical as well as the hybrid ones. They found that the numerical models are the most accurate ones compared to other models and it can be used in the case of monitoring data is unavailable; whereas if long term data has been gathered for a dam, then data driven models is a suitable choice. However if the mechanism of the deterioration of the dam is partly known as well as inadequate monitoring data can be found, then the hybrid models is the best choice.

Hancock and Coulthard (2022) studied the tailings dams and they evaluated the valley fill design's longstanding erosional stability. They used CAESAR – Lisflood model, which is a computer based landscape one in order to measure the failure of an imaginary tailings dam due to erosion, to complete their study. They found that infill as well as erosion will carry on to decrease a structure's integrity as the time passes. They concluded that continuous maintenance as well as monitoring are required in the case of tailings dams.

Bilali et al. (2022) studied dam break flood and they worked on the way to simulate the loss of life in the city areas due to it, using a real probabilistic method. They used HEC – LifeSim, MCS which is known as Monte Carlo Simulation as well as HEC – RAS 2D models, in order to propose an integrated method to simulate the dam break with the case of risk of flood by using probability instead of performing an analysis of some scenarios in regards to dam break.

Zeng et al. (2022) studied UAV remote sensing and they calculated sediment deposition volume present behindhand of check dams centred on it. They considered area which suffers with the highest soil erosion in the world for their study, then they proposed a new method by merging together the simulate submerging analysis with unmanned aerial vehicle photogrammetry, in order to calculate approximately the sediment silted due to check dams especially in the case of a topography which is complex in nature.

Lee et al. (2022) studied TSIKO as well as WQIEUT and after the Yeongju Multipurpose Dam's construction was finished they applied them in order to perform an assessment which was a wide-ranging water quality one. Their work area of Naeseong Stream in South Korea and they were able to assess the alterations in the quality of the river water in addition to the aquatic environ because of the above mentioned dam's construction which lasted for more than 8 years.

Villablanca et al. (2022) studied the river named Chilean and they worked on large dam's hydrological effects on it. They took data from official gauging stations in addition to they performed an analysis of flow records which was a multi temporal scale one, in order to typify the changed and unchanged river regimes to find out the persistence as well as magnitude of the hydrological changed downstream which happened because of the presence of large dams in the country Chile.

Freitas et al. (2022) studied the small scale localized dams which have the ability to generate large scale regional impacts in the river named Amazon. They stated that overfishing, deforestation, large hydroelectric dams as well as climate change have been identified as major problems in the Amazon basin. Nevertheless numerous small scale happenings or events have the ability to exercise major cumulative – environ impacts. They then performed a spatial analysis on numerous small scale dams which are used for aquaculture as well as hydropower and they found that such dams are higher in density in the given areas which is significantly affecting the aquatic environ.

Beiranvand and Rajaee (2022) studied dam's pore water pressure as well as seepage and they used single as well hybrid models by applying AI in order to forecast them. They stated that neural network models are over twenty seven percent popular in predicting pore water pressure as well as leakage in dams, whereas machine learning models are the most popular ones with over thirty seven percent in hybrid models are the third most popular with slightly over twenty one percent in comparison to other techniques.

Bu et al. (2022) studied the high concrete face rockfill dams and they performed a risk analysis on it which was a semi quantitative one, particularly under the condition of a seismic action. They stated that no such earlier study was performed like theirs and they concluded that their study results will become the technical support in addition to the theoretical basis for the seismic risk assessments, seismic designs as well as risk decision making for high concrete face rockfill dams in the future.

Xiao et al. (2022) performed the study on failure of Sardoba dam in the year 2020 present in Uzbekistan. They examined its failure on the basis of 3 sets of Earth observation data, which comprises of multi geometry Sentinel -1 SAR data, Satellite altimetry products like ICESat – 2 data, as well as optical images taken from GPM in addition to Sentinel – 2 satellites products, to accomplish the aim of their study. They stated that they worked on analysing the conceivable reasons of its collapse by considering the terms of both factors which includes human as well as physical.

RESEARCH GAPS

After reviewing a number of scientific and research papers published in journals in the recent years, the author found that there was an absence of research done on the recent ongoing dam projects in not only India but in other countries also. There was also absence of published scientific papers of researchers regarding the evaluation of the new proposed dam projects in India as well as in the rest of the world.

MAIN FOCUS OF THE CHAPTER

This chapter focuses on explaining the concept of dams and the requirement to construct as well as develop dams in the country, India and also in the whole world. It highlights the various types of dams in brief and the chapter explains the origin, history and evolution of dams over thousands of years, to give a rough idea to the readers. This book chapters also considers a number of research papers published in recent years to find out the appropriate gaps in the research. It also mentions stages of a dam construction briefly and it discusses some major advantages as well as disadvantages along with the challenges during the construction of dams and reservoirs. The book chapter discusses some new dam proposals in India and some graphical analysis to show the requirements of dams or whether they are required to develop further.

ISSUES, CONTROVERSIES, PROBLEMS

The construction of dams comes up with not only advantages but with various issues, problems and controversies. This fact can't be ignored that the local people are forced to leave their homes as well as their traditional source of income and lifestyles are drastically affected in a negative way most of the time. The compensation is either mostly delayed or less than the promised amount, which is ethically and economically wrong. Often due to the dam projects, a lot of politics, mass protest, and involvement of different agencies for different purposes creates a huge uproar and diverts the attention of the people from the original purpose of the project or the appropriate actions to be taken. Additionally, habitat loss, diversity of flora and fauna is really lost after the construction of dam projects, particularly the indigenous species of the place. The author briefly points out these things in the chapter, to make the readers think whether any new dam project is required or serving its supposed purpose or not.

PROPOSALS OF SOME NEW DAMS IN INDIA

In this section some new dams which were proposed in the country India in the recent years are discussed briefly. In India, the State Governments have the power to plan, execute, fund as well as maintain any water resource project, depending upon the primacies and the resources owned by them. The Central Government or Government of India has a limited role in this regard, which is also mostly to provide technical support to the State Government whenever required during the project. However, the Central Government can partly finance such projects of the State Governments under the already established schemes which are being executed by the Ministry of Jal Shakti. Nevertheless in case of a medium in addition to major irrigation project which is to be planned on Interstate River systems, then Central Water Commission is supposed to do appraisal of the project for its feasibility especially the techno economic one.

- **Par-Tapi Narmada Link Project:** It was proposed a major or interlinking project, which was involving the states like Gujrat and Maharashtra. However, recently the Gujrat state government has backed out after the continuous opposition from the public.
- **Eastern Rajasthan Canal Project:** It is a major project which was proposed and only involves the state of Rajasthan.
- **Attapady Irrigation Project:** It is a medium category project which was proposed in the state of Kerala.
- **KalasaNala Diversion Scheme:** It is a drinking water project which was proposed in the state of Karnataka.
- **Phina Singh Medium Irrigation Project:** It is a multipurpose project which was proposed in the state of Himachal Pradesh.

There are 16 other such dam projects which were proposed earlier and are under consideration by Central Water Commission. In the state of Rajasthan, there were total 3 dam projects were proposed, similarly in Himachal Pradesh and Karnataka. However, in Odisha 4 dam projects were proposed; whereas in states like Madhya Pradesh, Kerala, Bihar, Assam, Gujrat/Maharashtra, only 1 dam project was proposed in each of them.

METHODOLOGY

In this section the major defined issues which were observed during the course of some of the new dam proposals were faced, which are as follows:

- The protests in the year 2022 by tribal in Gujrat made a political pressure on the state's Chief Minister at that time, to cancel the Par-Tapi-Narmada dam project. The future elections were expected to be affected greatly otherwise.
- Obstruction of the natural ecology of the mentioned rivers were expected to happen due to the project, leading to huge environmental costs.
- The project cost of Phina Singh Medium Irrigation Project increased more than thrice the initial cost by the beginning of the year 2023.
- Ecological reasons as well as huge project costs were one of the major reasons of protest in the case of Attapaday Irrigation Project in Kerala state.
- At present the DPR of the earlier project is waiting for the approval of Central Water Commissions.
- In the year 2022, the Central Government told the Rajasthan State Government to stop the Eastern Rajasthan Canal Project, until and unless the State Government take the consent of the neighbouring states for the project.

RESULTS AND DISCUSSION

In this section, the author had collected data from different sources for the purpose graphical analysis and to further support the study in the chapter.

Table 1. Comparison of existing dams and dams under construction in India

Status of Dams in India	Total Number
Completed and Operational	5334
Under Construction	411

Source: National Register of Large Dams (2019)

Table 2. Comparison of Indian states with the most existing dams in the country

Indian State With Most Constructed Dams	Number of Dams
Maharashtra	2117
Madhya Pradesh	899
Gujrat	620
Chhattisgarh	249
Karnataka	230

Source: National Register of Large Dams (2019)

Table 3. Comparison of the Indian states with the most dams which are under construction at present in the country

Indian States With the Most Dams Under Construction	Total Number of Dams
Maharashtra	277
Jharkhand	24
Andhra Pradesh	17
Telangana	16

Source: National Register of Large Dams (2019)

Table 4. Showing top countries with the highest dam capacity (km^3) by the year 2019 in the world

Top Countries With the Highest Dam Capacity	Total Dam Capacity km3 by 2019
Canada	841.51
China	830.46
Russian Federation	801.54
USA	735.87
Brazil	700.44
India	247.46

Source: Knoema

Table 5. Showing top countries with the highest dam capacity (km^3) by the year 1970 in the world

Top Countries With the Highest Dam Capacity	Total Dam Capacity km3 by 1970
USA	622.8
Canada	511.3
Brazil	426.4
Egypt	168.2
Ghana	148.1
Iraq	127.1
India	105

Source: Knoema

Table 6. Comparison of two major countries with high dam capacity (km^3) from 1970 to 2019 in regards to total dam capacity (km^3)

Two Major Countries With High Dam Capacity	Total Dam Capacity km3 by 1970	Total Dam Capacity km3 by 2005	Total Dam Capacity km3 by 2012	Total Dam Capacity km3 by 2019
India	105	224	237.96	247.46
China	0	562.4	674.54	830.46

Source: Knoema

Table 7. Showing comparison of some major dams in India

Some Major Dams in India	Height (m)	Length (m)	Reservoir Capacity (Million m^3)	Capacity Installed (MW)
Tehri Dam (Uttarakhand)	260.5	575	2590	1000
Bhakra Nangal Dam (HP and Punjab)	226	520	9253	1325
Hirakud Dam (Orissa)	61	4800	5896	347.5
Nagarjuna Sagar Dam (Telangana)	124	4863	11560	816
Sardar Sarovar Dam (Gujrat)	163	1210	9498	1450

Source: Study IQ

Table 8. Showing comparison of the world's largest dams by reservoir size

Some Major Dams in the World by Reservoir Size	Height (m)	Length (m)	Reservoir Capacity (Billion m^3)	Capacity Installed (MW)
Kariba Dam (Zambia and Zimbabwe)	128	579	185	1320
Bratsk Dam (Russia)	125	1500	169	4500
Akosombo Dam (Ghana)	114	660	148	1038
Daniel Johnson Dam (Canada)	214	1314	140	1596
Guri Dam (Venezuela)	162	7500	135	10200

Source: We Build Value, Volta River Authority, Wayback Machine

Table 9. Showing causes of dam failure in USA in (%)

Major Causes of Dam Failure in USA	In (%)
Overtopping	34
Piping	20
Foundation Defects	30
Others	16

Source: Association of State Dam Safety Officials

Table 10. Showing the increase in the increase in the hydropower installed capacity of the world from 2014 to 2021

Global Hydropower Installed Capacity	In (GW)
Year 2014	1036
Year 2021	1360

Source: Statista

Figure 1. Graph showing comparison of existing dams and dams under construction in India
Source: National Register of Large Dams (2019)

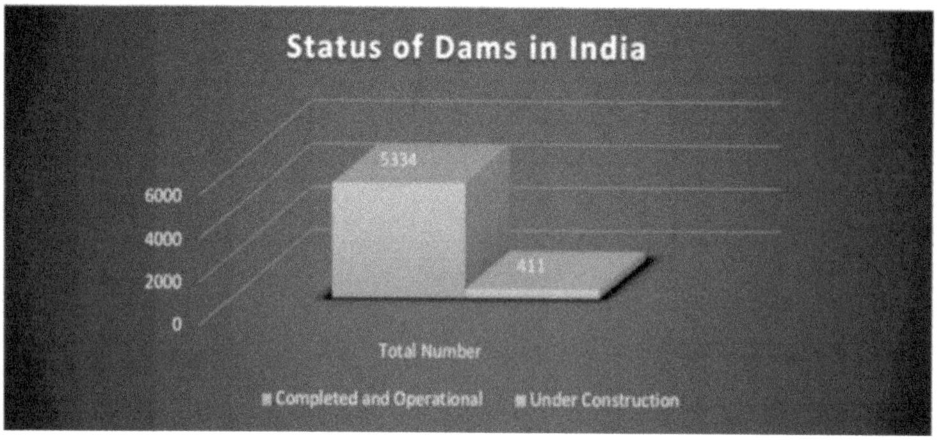

Figure 2. Graph showing comparison of Indian states with the most existing dams in the country
Source: National Register of Large Dams (2019)

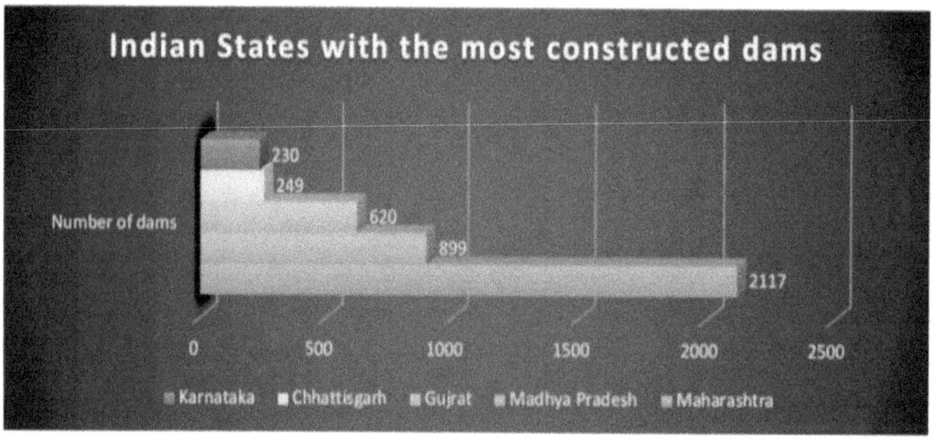

Figure 3. Comparison of the Indian states with the most dams which are under construction at present in the country
Source: National Register of Large Dams (2019)

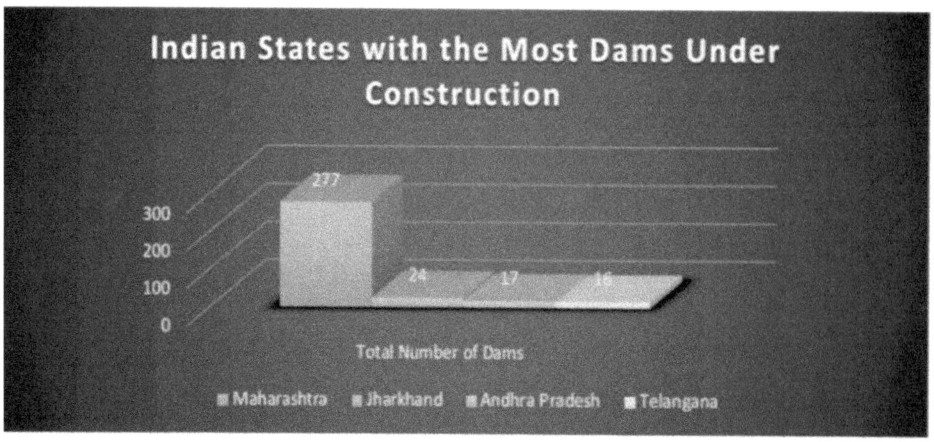

Figure 4. Showing top countries with the highest dam capacity (km³) by the year 2019 in the world
Source: Knoema

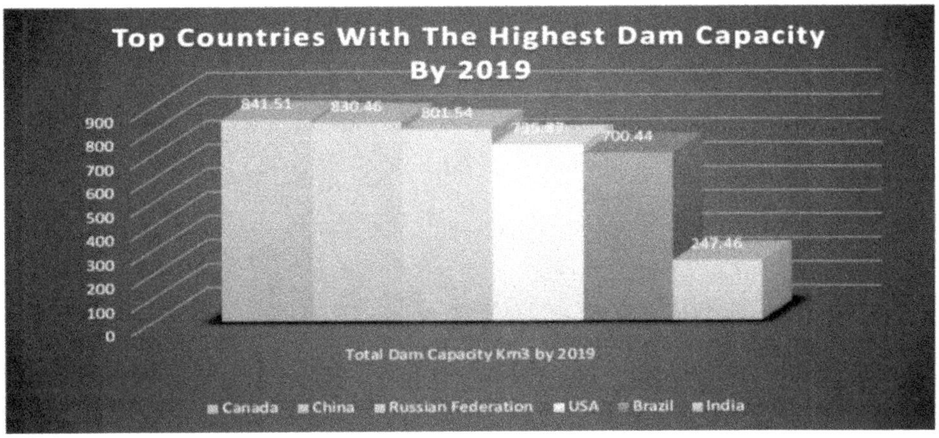

Figure 5. Showing top countries with the highest dam capacity (km³) by the year 1970 in the world
Source: Knoema

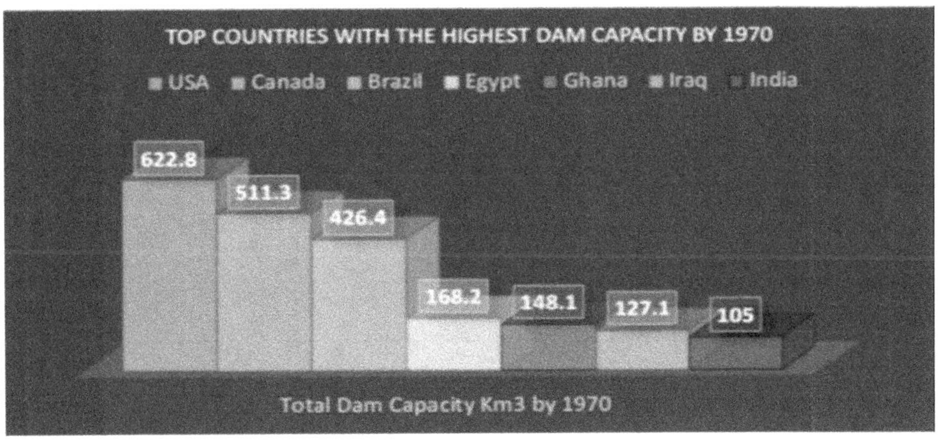

Figure 6. Comparison of two major countries with high dam capacity (km³) from 1970 to 2019 in regards to total dam capacity (km³)
Source: Knoema

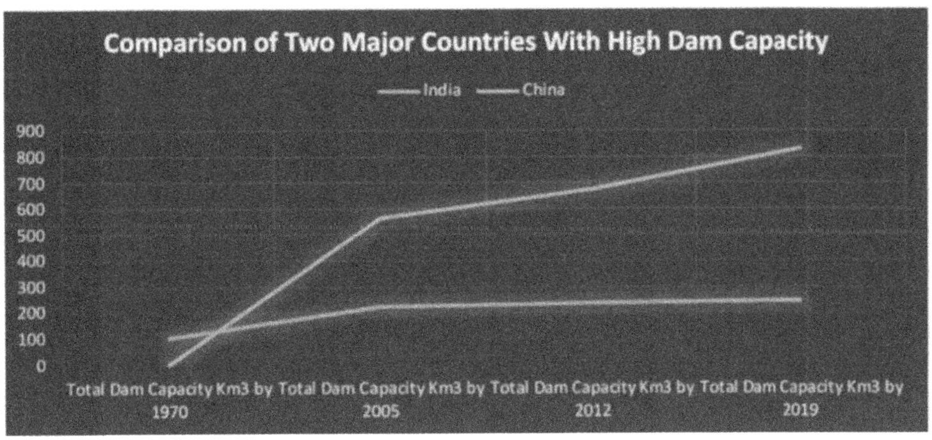

Figure 7. Showing comparison of some major dams in India
Source: Study IQ

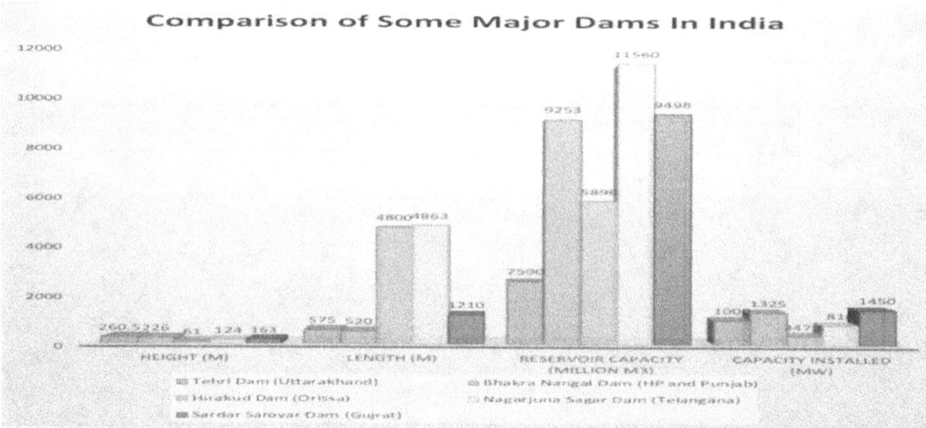

Figure 8. Showing comparison of the world's largest dams by reservoir size
Source: We Build Value, Volta River Authority, Wayback Machine

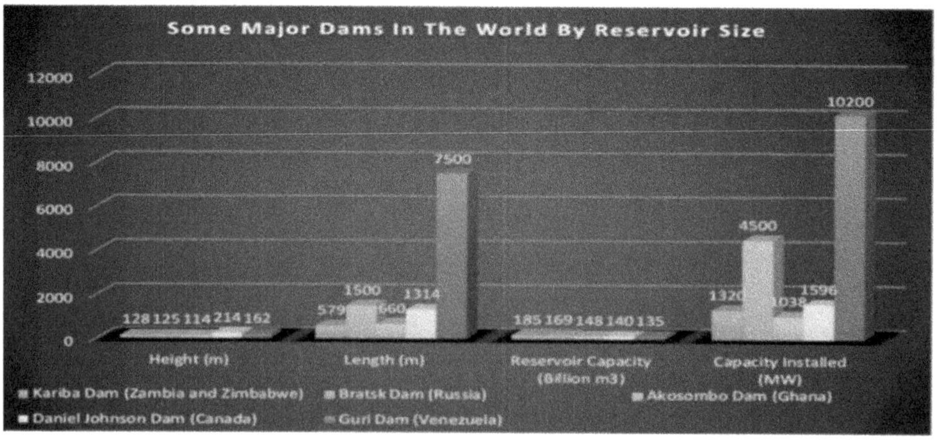

Figure 9. Showing causes of dam failure in USA in (%)
Source: Association of State Dam Safety Officials

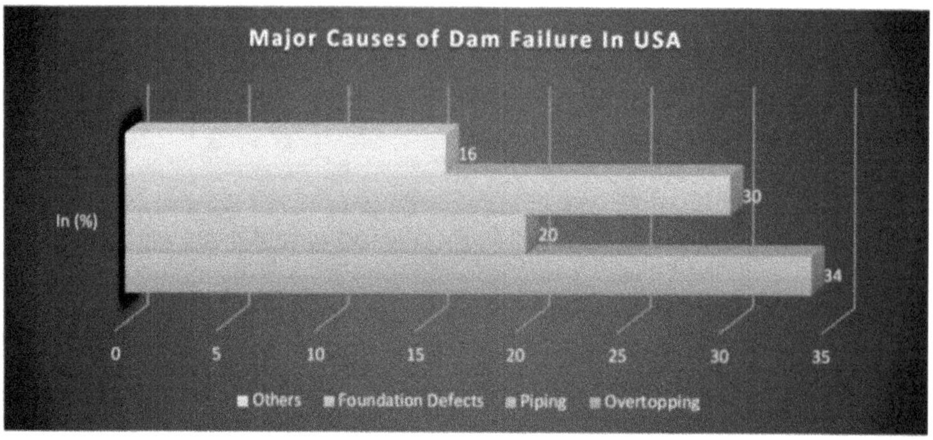

Figure 10. Showing the increase in the increase in the hydropower installed capacity of the world from 2014 to 2021
Source: Statista

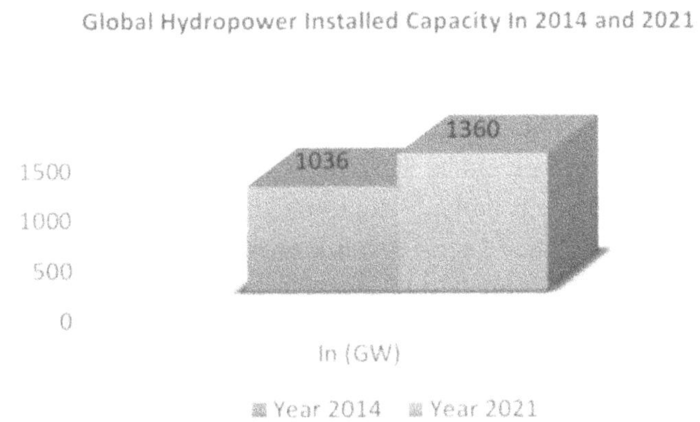

Major Advantages of Constructing Dams and Reservoirs

In this subsection, some major advantages of dams are discussed which are as follows:

- According to Figure 1, there are over 5000 dams which are already completed and operational in India, whereas over 400 dams are still in under construction. It is clear that India is the top 3 country in the world with the maximum number of dams and still more dams are required for different purposes which is why India is still constructing more than 400 dams apart from the existing ones.
- According to Figure 2, Maharashtra is the leading state in India, with most number of dams and it is also one of the most developed states of India. Madhya Pradesh and Gujrat are also the top 3 states of India with most number of dams in the country. These are also the neighbouring states which are also helping each other's development, through dams. The development in these states were rapid after presence of numerous dams especially the large scale and medium scale ones.
- Based on Figure 3, Maharashtra is again the leading state in India, in most number of dams which are under construction with a big margin compared to other states. Maharashtra's development can be associated with the development of dams and it is clear that the existing significant number of dams are not sufficient for the state, due to which 277 dams are still under construction in the state of varying sizes for various purposes. Then, Telangana is also one of the top states with most dams which are under construction, this new state is also developing at a rapid rate.
- According to Figure 4, Canada is the leading country in the world with the highest dam capacity (Km^3) by year 2019, additionally USA is also in the same list in the top 4 countries of the world. These two countries are major developed countries in the world and their development can also be associated with the development of such large size dams. In the same list, China, Russia, Brazil and India can also be seen in the top 6 countries in the world and these countries are the major developing countries, whose development can also be linked with the existence of major dams.

- Based on Figure 5, in year 1970, USA was the leading country with the highest dam capacity and it was also a powerful developed country at that time. It is safe to say that USA's development even more than 5 decades ago relied considerably on huge capacity of dams.
- Based on Figure 6, China and India had a significant increase of high dam capacity from the year 1970 to 2019 and these two countries also saw major development in that period of time. It is safe to say that the development of dams played a major role in the significant development of these two countries. Plus, China which focused even more on the construction and development of dams than India, also saw higher development of the country than India.
- According to Figure 7, some major dams of India were compared, Tehri Dam, Bhakra Nangal Dam, Hirakud Dam, Nagarjuna Sagar Dam and Sardar Sarovar Dam, were shown. These dams not only played a big role in the development of the areas where they were situated but also the whole state and also affected the development of the neighbouring states as well as the whole country's development upto a certain extent.
- In Figure 8, world's largest dams according to reservoir size were shown and it is safe to say that Kariba Dam, Bratsk Dam, Akosombo Dam, Daniel Johnson Dam and Guri Dam, clearly supported not only the development of the countries where they were constructed but they also transformed the livelihood of the people significantly.
- In Figure 10, global hydropower installed capacity was compared from 2014 to 2021 and there was a significant positive trend in the mentioned seven years' timeframe. This also shows that in the recent years worldwide many countries are still focusing on the construction of dams for the country's development.

Major Disadvantages and Challenges in the Construction of Dams

In this section some major disadvantages and challenges of dams are discussed which are as follows:

- According to Figure 1, since India is already having over 5000 dams of different sizes for various purposes so not a significant percent in increase in the number of dams can be expected in the country in the future. Plus, it can also be said that a considerable population of natives and their livelihood were affected because of many of the existing dams as well as the under construction ones, especially the tribal.
- From Figure 2 and Figure 3, it is clear that the natives or tribal in particular were displaced and lifestyles were dramatically affected in the states like Maharashtra, Madhya Pradesh, Gujrat, Chhattisgarh, Karnataka, Jharkhand, Andhra Pradesh and Telangana. Flora and fauna of those places went drastic changes and villages were being completely displaced, which also disturbed the native humans as they had to adapt drastically to new lifestyles and way of livelihood as repercussions.
- According to Figure 9, Overtopping is the greatest cause for failure of dams in USA, followed by foundation defects and then piping. These causes are supposed to be taken care of carefully in order to reduce the possibilities of dam failures in USA.

LIMITATIONS

In this section, the author has discussed the limitations of the book chapter. The chapter focussed on explaining the different types of dams and categorised them on different basis. However, the chapter didn't cover those mentioned types of dams in detail. It also mentioned the challenges faced in some of the new dam projects in short, but not in much detail. Parts of some new proposed dams were briefly discussed in the chapter, but it didn't give any detailed elaboration about them. Mostly the chapter focused on the country India in particular and then rest of the world in brief.

SOLUTIONS AND RECOMMENDATIONS

The chapter focussed on highlighting the various benefits of the dams in India and in the world, but the author also discussed several problems, challenges and issues of constructing a dam, which can't be ignored. It is required that all the problems and issues caused by a dam project needs to be solved, which can significantly reduce the controversies. Then, the advantages of the dam project should overwhelm all the possible disadvantages or problems caused by the project. Thorough feasibility analysis by the experts and the concerned authority is a must before starting any project and it is essential to think that whether any specific dam project is needed or not in the first place.

FUTURE RESEARCH DIRECTIONS

Dams have been an area of research for a long time in the history and due to which it evolved tremendously over the thousands of years, especially during industrial revolution and afterwards. With more technological advancements in the future, it is possible to further evolve the dams and improve its designs further as well as it may be possible that new materials can be used and replace the current ones in the construction of dams. Plus, more research is required in order to evaluate and observe the problems in the rising number of dams all over the world in the recent years.

CONCLUSION

This book chapter explained the general concept of dams and categorized them in different types based on different parameters. The author briefly mentioned the importance of dams as well as the way they promote towards sustainability and use of renewable source of energy in the chapter. The book chapter discussed the history and evolution of dams over thousands of years briefly. The author considered numerous scientific and research papers published in the recent years about dams, in order to find out the gaps in the research. Some major challenges faced in the novel proposed dams in India were discussed in brief in the chapter. Some new dam proposals in the country India are also talked about in this chapter briefly. Then, graphical analysis was conducted by the author by collecting various data from different sources, in order to support the study. Major benefits and disadvantages as well as challenges in the construction of a dam are discussed here in this chapter. The book chapter not only tells the readers about the requirements of dams or their development but also points out the problems or issues which needs to be solved before the start of a dam project in an effective manner. Dams are useful and important until and unless, they are serving their supposed purpose and their benefits overwhelms their drawbacks. However, it is still required to conduct a thorough feasibility analysis of a dam project and decide whether that specific dam project is really needed in the first place. If the dams are built in a proper way and they are really serving their purpose, then dams play an important role in the socioeconomic development of not that particular region but also the whole country.

ACKNOWLEDGMENT

This research received no specific grant from any funding agency in the public, commercial, or not-for-profit sectors.

REFERENCES

Association of State Dam Safety Officials. (n.d.a). https://damsafety.org/media/statistics

Association of State Dam Safety Officials. (n.d.b). https://damsafety.org/dam-failures

Beiranvand, B., & Rajaee, T. (2022). Application of artificial intelligence-based single and hybrid models in predicting seepage and pore water pressure of dams: A state-of-the-art review. *Advances in Engineering Software*, *173*, 103268. doi:10.1016/j.advengsoft.2022.103268

Bhakra Beas Management Board. (n.d.). https://bbmb.gov.in/benefits-of-large-dam-.htm

Bilali, T. (2022). A practical probabilistic approach for simulating life loss in an urban area associated with a dam-break flood. *International Journal of Disaster Risk Reduction*, *76*, 103011. doi:10.1016/j.ijdrr.2022.103011

Britannica. (n.d.). https://www.britannica.com/topic/Grande-Dixence-Dam

Bu, L., Li, Y., Tang, W., Chen, X., & Li, B. (2022). Semi-quantitative risk analysis of high concrete face rockfill dams under seismic action. *Structures*, *45*, 685–694. doi:10.1016/j.istruc.2022.09.013

Byjus. (n.d.). https://byjus.com/bank-exam/dams-of-india/

CivilA. (n.d.). https://www.aboutcivil.org/planning-design-dams

Coughenour, C., Coleman, N. M., & Taylor, A. L. (2022). In the shadow of the dam – Hydrology of the Little Conemaugh river and its South Fork, with insights about past and future flooding. *Heliyon*, *8*(9), e10679. doi:10.1016/j.heliyon.2022.e10679 PMID:36177242

Data Flair. (n.d.). https://www.google.com/amp/s/data-flair.training/blogs/dams-in-india/amp/

DiaryO. (n.d.). https://orissadiary.com/india-has-5334-completed-and-operational-large-dams-while-411-large-dams-are-under-construction-says-bishweswar-tudu/amp/

Food and Agriculture Organization of the United Nations. (n.d.). https://www.fao.org/3/v5930e/v5930e01.htm

Freitas, M., Pereira, S.-S., & Hurd, K. (2022). Death by a thousand cuts: Small local dams can produce large regional impacts in the Brazilian Legal Amazon. *Environmental Science & Policy*, *136*, 447–452. doi:10.1016/j.envsci.2022.07.013

Grubert. (2020). Conventional hydroelectricity and the future of energy: Linking national inventory of dams and energy information administration data to facilitate analysis of hydroelectricity. *The Electricity Journal, 33*(1).

Hancock, G. R., & Coulthard, T. J. (2022). Tailings dams: Assessing the long-term erosional stability of valley fill designs. *The Science of the Total Environment*, *849*, 157692. doi:10.1016/j.scitotenv.2022.157692 PMID:35908711

Heidarzadeh, M., & Feizi, S. (2022). A cascading risk model for the failure of the concrete spillway of the Toddbrook dam, England during the August 2019 flooding. *International Journal of Disaster Risk Reduction*, *80*, 103214. doi:10.1016/j.ijdrr.2022.103214

International Rivers. (n.d.). https://archive.internationalrivers.org/damming-statistics

Irrigation Department Government of Kerala. (n.d.). https://irrigation-kerala.org/attappady-irrigation-project

Knoema. (n.d.a). https://knoema.com/atlas/topics/Water/Dam-Capacity/Total-dam-capacity

Knoema. (n.d.b). https://www.google.com/amp/knoema.com/atlas/India/topics/Water/Dam-Capacity/Total-dam-capacity%3fmode=amp

Knoema. (n.d.c). https://www.google.com/amp/knoema.com/atlas/topics/Water/Dam-Capacity/Total-dam-capacity%3fmode=amp

Lee, P., Park, J. R., Joo, J. C., & Ahn, C. H. (2022). Application of WQIEUT and TSIKO for comprehensive water quality assessment immediately after the construction of the Yeongju Multipurpose Dam in the Naeseong Stream Basin, Republic of Korea. *The Science of the Total Environment*, *819*, 152997. doi:10.1016/j.scitotenv.2022.152997 PMID:35031363

LinkT. (n.d.). https://www.touristlink.com/canada/daniel-johnson-dam/overview.html

Llamosas, C., & Sovacool, B. K. (2021). The future of hydropower? A systematic review of the drivers, benefits and governance dynamics of transboundary dams. *Renewable & Sustainable Energy Reviews*, *137*, 110495. doi:10.1016/j.rser.2020.110495

MachineW. (n.d.). https://web.archive.org/web/20100817021129/http://canada.suite101.fr/article.cfm/manic-5--colossal-temoin-du-genie-quebecois-en-hydroelectricite

Peluso, M., Penha, B., Cassemiro, S., & Fantin-Cruz, K. (2022). Climate change negative effects on the Neotropical fishery resources may be exacerbated by hydroelectric dams. *The Science of the Total Environment*, *828*, 154485. doi:10.1016/j.scitotenv.2022.154485 PMID:35283123

PortalI. W. (n.d.). https://www.indiawaterportal.org/topics/large-dams-barrages-reservoirs-and-canals

Prakash, D., Dugalam, R., Barbosh, M., & Sadhu, A. (2022). Recent advancement of concrete dam health monitoring technology: A systematic literature review. *Structures*, *44*, 766–784. doi:10.1016/j.istruc.2022.08.021

Press Information Bureau. (n.d.). https://pib.gov.in/Pressreleaseshare.aspx?PRID=1782281

Rivera-Trejo, Radecki-Pawlik, & Filipczyk, Priego-Hernandez, & Soto-Cortes. (2022). On predicting of some consequences of planned dam construction in a pristine sand-gravel river large catchment: Rio San Pedro, Nayarit, Northwest Mexico. *Ecological Engineering, 184.*

Statista. (n.d.). https://www.statista.com/statistics/1179170/global-hydropower-capacity/

StudyI. Q. (n.d.). https://www.google.com/amp/s/www.studyiq.com/articles/important-dams-in-india/mp/

The Hindu. (n.d.a). https://www.thehindu.com/news/national/other-states/eastern-rajasthan-canal-project-takes-political-twist-after-centres-directive-to-stop-work/article65600112.ece

The Hindu. (n.d.b). https://www.thehindu.com/news/national/kerala/Setback-to-Attappady-irrigation-project/article14545757.ece

TodayC. (n.d.). https://civiltoday.com/water-resource-engineering/dams/350-advantages-and-disadvantages-of-dams

ValueW. B. (n.d.). https://www.webuildvalue.com/en/infrastructure-news/biggest-dams-in-the-world%20.html

Vasques, R., & Huber, T. (2022). Ground penetrating radar (GPR) models of the regolith and water reservoir of an underground dam in the Brazilian semiarid region. *Journal of Applied Geophysics*, *206*, 104797. doi:10.1016/j.jappgeo.2022.104797

Villablanca, B., Batalla, R. J., Piqué, G., & Iroumé, A. (2022). Hydrological effects of large dams in Chilean rivers. *Journal of Hydrology. Regional Studies*, *41*, 101060. doi:10.1016/j.ejrh.2022.101060

Volta River Authority. (n.d.). https://www.vra.com/our_mandate/akosombo_hydro_plant.php

Wikipedia. (n.d.). https://en.m.wikipedia.org/wiki/Dam

Xiao, J., Jiang, M., Li, Z., & He, X. (2022). New insights into the 2020 Sardoba dam failure in Uzbekistan from Earth observation. *International Journal of Applied Earth Observation and Geoinformation*, *107*, 102705. doi:10.1016/j.jag.2022.102705

Yang, S., Peng, Z., & Soares-Frazao, Z. (2022). Quantitative risk assessment of two successive landslide dams in 2018 in the Jinsha River, China. *Engineering Geology*, *304*, 106676. doi:10.1016/j.enggeo.2022.106676

Zeng, Meng, Zhang, Dai, Fang, & Shi. (2022). Estimation of the volume of sediment deposited behind check dams based on UAV remote sensing. *Journal of Hydrology, 612*(B).

Zigya. (n.d.). https://www.google.com/amp/s/www.zigya.com/blog/7-highest-dams-india-interesting-fact/amp/

Chapter 8

Analytical Hierarchy Process — Based Spatial Data Analysis for Landslide Susceptibility Mapping

Azmat Raja
National Institute of Technology, Patna, India

Gopikrishnan T.
https://orcid.org/0000-0003-4861-0038
National Institute of Technology, Patna, India

ABSTRACT

This chapter aims to develop landslide susceptibility maps for the Sikkim state in India by combining the analytical hierarchy process, geographic information systems, and remote sensing. The delineation of the landslide susceptibility maps has taken into consideration a variety of data such as density of lineament, slope, lithology, aspect, land cover and land use, road buffer, rainfall, and drainage density. Using both Landsat 8 and ground data in a GIS framework, spatial distribution of maps and map layers of required themes were produced. The appropriate weights based on the Saaty's scale were given to these thematic layers in accordance with their respective significance in the occurrence of landslides in the study area. According to the study area's demarcated landslide susceptibility map, the risk levels were very low (12.52%), low (21.12%), moderate (8.05%), high (31.13%), and very high (27.18%). The accuracy of the study region is computed using the AUC curve using the AHP model landslide map and inventory map, which shows good result with 70% accuracy.

DOI: 10.4018/978-1-6684-6449-6.ch008

INTRODUCTION

Natural hazards comprises of Landslides that can occur on both land and water; they are common throughout the world and are primarily concentrated in hilly areas. Landslides are caused by a variety of phenomena, including earthquakes, rainfall intensity, snow melting rate, and human involvement. (Hutchinson, 1988) defines a landslide as the relatively quick mass wasting process that causes the movement of large masses of rock, rubble, or earth downward when one or more external stimuli are present. The landslide can be described as "the mass movement of soil and rock down a slope" (Couture, 2011). Regarding the kind of debris that slides down a slope, this definition of a landslide is more inclusive. Landslide susceptibility is defined as a quantitative or qualitative evaluation of the classification, volume (or area), and spatial distribution of landslides that exist or may occur in a given area. Landslides usually result from a combo of geo - environmental phenomena, as explained by (Varnes, 1984) and (Hutchinson et al., 1985). The Himalayas provide shelter for millions of people from various nations. About 13% of India's territory is exposed to various levels of landslide risk (Geological survey of India, 2009), which also includes the north east state Sikkim. The North East Himalayas region of seven sister states accounts for 0.18 million square kilometres of the total area. Landslides are a common natural phenomenon in India, particularly during the monsoon season. Landslides are sensitive to rock quality, slope geometry, topographical and natural conditions, and disturbances. The landslides can be classified as fall (separation of soil), flow (soil flow), creep (down slope movement of soil), debris flow (quick development of free soil), lahar (volcanic eruption cause ice melting), mudflow, and slides. Rotational slides and translational landslides are the two main types of landslides.

Sikkim is vulnerable to natural disasters due to heavy rainfall, weak geology, and earthquakes. Due to its land and climatic conditions, Sikkim is prone to landslides. It is effected by land cover and land use by humans. There is loss of infrastructure and invaluable human life. Due to the complex geological setup and extensive rainfall in the terrain, the severity of damages caused by there are a lot of abrupt mass movement in several areas of the Sikkim. Every year, thousands of people are killed in landslides in Sikkim Himalaya (Bhasin et al., 2002). In the year 1968, landslides in Sikkim resulted in the deaths of almost 36,000 people (Kaur et al., 2019). Rainfall, seismic activity, and anthropogenic activities are the main controlling factors for slope instability (V. Kumar et al., 2018; Sonker et al., 2021; Taalab et al., 2018; van Westen et al., 2008; Yu & Huang, 2006). Around the world, both directly and indirectly, landslides have a huge effect. In response, this has led to a rise in injuries and fatalities, as well as negative effects on the economy and ecology, in the nations where landslides are most frequent.

Many landslide susceptibility maps have been developed in recent years the use of statistical approaches based on GIS such as Weights of Evidence (WoE) and Frequency Ratio (FR) model. The fact that the findings of these models demonstrated excellent results with great consistency, that they are relatively easy to use (Akgün & Bulut, 2007; Choi et al., 2012; Dahal et al., 2008; Fayez et al., 2018; R. Kumar & Anbalagan, 2016; Lee & Pradhan, 2007; Mersha & Meten, 2020; Park et al., 2013; Pradhan et al., 2010; Sonker et al., 2021; Vakhshoori & Zare, 2016; Yilmaz, 2009).

There are numerous techniques for determining landslide-prone areas. Several publications have made an effort to categorise the LHZ approach (Aleotti & Chowdhury, 1999; Guzzetti et al., 1999, 2005; Kanungo et al., 2009). The three main kinds of landslide susceptibility methodologies are quantitative, quasi-quantitative, and qualitative. The qualitative method examples include mapping of geomorphology, heuristic methods, and methods that rely on subjective judgement (Anbalagan, 1992; Gupta et al., 1999; Nagarajan et al., 1998; Saha et al., 2002; Zimmermann et al., 1986). The Analytical Hierarchy Process

(AHP), combined landslide frequency ratio and fuzzy logic, weighted linear combination (WLC), and fuzzy logic are logical tools used in semi-quantitative methods to weigh and rate data (Akgün & Bulut, 2007; Ercanoglu & Gokceoglu, 2004; Kanungo et al., 2006; Kayastha et al., 2013; Mondal & Maiti, 2012; Pradhan et al., 2009; Yalcin, 2008). In any danger zone, numerical probability of the likelihood of landslides are produced via quantitative landslide susceptibility methods (Guzzetti et al., 1999). It is assumed that landslide causing factors are uniformly dispersed throughout a region. Inventory-driven approaches forecast landslide likelihood using quantitative methods. Bivariate and multivariate procedures are additional categories for quantitative techniques.

The relation among past landslide data and density is the foundation of the bivariate landslide susceptibility approach (Dahal et al., 2008; Dai & Lee, 2002; Ghosh et al., 2011; R. Kumar & Anbalagan, 2016; Lee & Pradhan, 2007; Mathew et al., 2007; B. Pradhan & Lee, 2010). It treats landslides as a dependent variable, while factors are treated as separate factors, with each individual's vulnerability being evaluated. Although multivariate approaches are likewise data-driven, they mathematically synthesise the integrated effect of factor on depending variable and also allow for the numerical calculation of the influence of individual elements. (Das et al., 2010, 2012; Kundu et al., 2013; Lee & Talib, 2005; B. Pradhan & Lee, 2010; Yesilnacar & Topal, 2005).

The Saaty (1990) AHP technique has gained popularity as a method for various criteria based decision making process. It helps deciding authority by breaking down difficult decisions into a series of comparison pairs and synthesising the outcomes. A complex choice problem is divided into various hierarchical levels by the AHP. The AHP allows for the quantification of opinions and the transformation of those opinions into a logical decision-making framework (Saaty, 1990). Many authors worldwide used it for LSZ (Althuwaynee et al., 2016; T. Basu et al., 2022; Kayastha et al., 2013; Pal & Chowdhuri, 2019; Sangchini et al., 2016; Shahabi & Hashim, 2015).

LITERATURE REVIEW

Shahabi and Hashim (2015) introduced the use of aerial photography and AIRSAR for landslide inventory mapping, and the root mean square error (RMSE) approach was utilised to auto predict landslides through the satellite images. Small landslides can be easily found and mapped using aerial radar images. Aerial radar images is extremely valuable for identifying and mapping tiny landslide. The RMSE approach employed to improve inventory accuracy and precision.

Santangelo et al. (2015) primarily presents a landslide inventory. It is all about recognising landslides or creating a landslide inventory map using a semi-automatic method employing aerial images and GIS; using modern remote sensing techniques and technologies. This study's main objective is to transform data from aerial images into the digital imagery.

Reichenbach et al. (2014) demonstrates the effectiveness of high resolution satellite pictures and aerial photographs to prepare inventories. To investigate differences, both event-based landslides were compared.

Murillo-García et al. (2014) investigated the application of GeoEye 1's extremely high resolution stereo-images to identify landslides. According to this study, landslides can be detected using very high quality photographs, which take up little time and little money to get from a company. The study found that there have been 385 landslides, with a density of 10.5 landslides per square kilometre, of which around 170 are old and 21 are recent.

In Ghosh et al. (2011), the process of determining an area's susceptibility to future landslides can be partially automated, requiring less fieldwork, due to the growing accessibility of satellite images and geographic information systems.

Praveen (2017)'s primary focus was on the pre-post landslide inventories for two events that were produced by visual interpretation of LISS-IV satellite photos. In this study, the landslides before and after the Kedarnath flood are compared. Analysis shows that river courses are where most landslides occur. Eight causal factors were used in this study's susceptibility and sensitivity mapping in order to forecast future landslides and harm to living beings.

In Tehrany et al. (2015), for the purposes of this investigation, very high resolution satellite photos from the Quick Bird satellite and high resolution lesser data from (LIDAR) were combined using robust data fusion algorithms. The location of landslides was determined from different land use covers using an object-based orientated technique. The direction and velocity of landslides were significantly predicted by LIDAR-derived aspect and slope variables. The study's methodology was easy to employ, but it can only be utilised in tropical nations, the results show.

Ba et al. (2017) involved landslide susceptibility mapping of the Finikas watershed, which is more susceptible to landslides due to heavy erosion from the Finikas River, which increases the instability of the slope. An analytical hierarchy process (AHP) susceptibility map was produced for a particular watershed. Maps created by geographic information systems (GIS) are more precise.

In Zhu et al. (2014), frequency analysis and a logistic regression model, both of which are statistical techniques, were used to attempt to identify locations vulnerable to landslide. This study, geographical association of landslide in that region and nine causal parameters were selected for susceptibility mapping. After integrating each parameter according to its weight, a map of landslide susceptibility is produced. Using an outdated landslide inventory, the mapping of landslide susceptibility was verified. The results showed that logistic regression has a superior forecast of landslides (84.6%) than frequency analysis (67%).

In Sonker et al. (2021), the Analytical Hierarchy Process and a conventional approach for defining landslide susceptibility zones were utilised to assess the landslide susceptibility zones in the Sikkim Himalaya, India. The eight thematic layers for landslide triggering factors—geology, drainage density, lineament density, soil, slope, yearly rainfall, and land use—were made using a digital elevation model, Rainfall, conventional map, and Landsat 8. From the Analytical Hierarchy Process and landslides inventories, the research area's accuracy evaluation was greater than 70%.

In Kumar and Anbalagan (2016), an extensive use of the AHP method for mapping of landslide susceptibility for the Tehri dam has been reported. Landsat 8 data is employed to pinpoint various landslides causing elements that were causing instability in the region. The validity of landslide in this study was examined in great detail in the field. For each AHP matrix, a satisfactory consistency ratio (CR) value was obtained. To create a map of landslide susceptibility, the weighted sum technique was used to integrate the weights of each element. When the success rate curve's AUC results was translated to a percentage, the result was a satisfactory 78% evaluation performance.

In Bahrami et al. (2021), basic layer of landslide triggering elements that were considered in this study included lithology, distance to river, slope, rainfall, distance to road, distance to fault, and land use. The AHP approach was used to create the landslide susceptibility map after weighing various layers according to professional opinions. The result map for this scenario allocates 36.5% of its entire region at high risk and extreme highly risk regions. The resulting projected accuracy was found to be 92.4 percent.

Feizizadeh et al. (2014) mapped landslide susceptibility employing AHP and two fuzzy approaches, and the findings were regarded well because they covered 53 percent of the prior landslides.

OBJECTIVE

The research aims to assess the utility of a Geographic information system modelling approach for predicting the efficient landslides in various regions of the state Sikkim in India. Through developing multiple significant thematic map layer using Landsat 8 data. The study's specific objective is to use Landsat 8 Band images and Digital Elevation Model data to look into the area's geology, slope, density of drainage, aspects, and land use. AHP approach employed to incorporate the generated landslide occurrence characteristics into a GIS environment for the final mapping of landslide susceptibility in the researched area.

STUDY AREA AND DATA PREPARATION

Study Area

The Sikkim is a hilly state in the North East India. The state is bordered in the north and northeast by the Dongkya's Ranges, in the southeast by the Pangola's Ranges, in the south by the hills of Darjeeling, and in the west by the Singalila's Ranges. The state encompasses over a region of 7096 square kilometres and situated between $27^{0}5'$ and $28^{0}10$¢ North latitudes and $88^{0}01$¢ and $88^{0}58$¢ East longitudes shown in Figure 1. The state's border runs approximately 64 kilometres east to west and around 114 kilometres north to south. The elevation ranges from 299.3 to 8582.55 metres above the mean sea level. The undulating surface of Sikkim falls under the Himalayas. West Sikkim, East Sikkim, South Sikkim and North Sikkim are the four districts in Sikkim area. The climate has been classified as tropical, temperate, and alpine. The weather is cold and humid for the majority of the year. In addition to being home to

Figure 1. Geographical location of Sikkim

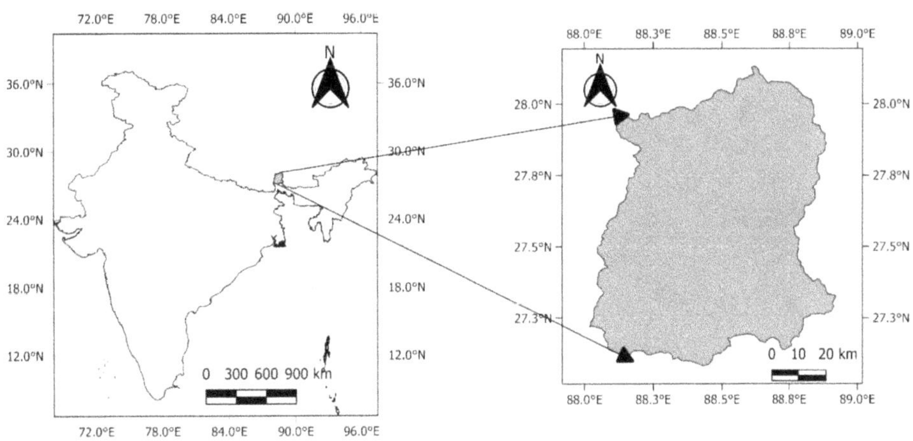

Kangchenjunga, the third-highest mountain on Earth and the highest peak in India, it also boasts alpine and subtropical climates. Kangchenjunga have an elevation of 8586 metres above the mean sea level. The elevations in the state range from 239.55 to 8485 metres above sea level. The Teesta River flows nearly the whole extent of Sikkim, making it the state's principal river.

Past Landslide Points

The historical landslides depicts details and locations of the previous landslide at various temporal and spatial scale (Biswakarma et al., 2020; Wieczorek, 1984). It gives an overview to how to relate significances of elements that influence the incidence of landslides to the causes that influence their incidence. The landslide inventory data is downloaded from website https://bhukosh.gsi.gov.in/Bhukosh/Public and used to create a map depicting the geographic location of 679 events shown in Figure 2.

Figure 2. Location of previous landslide points

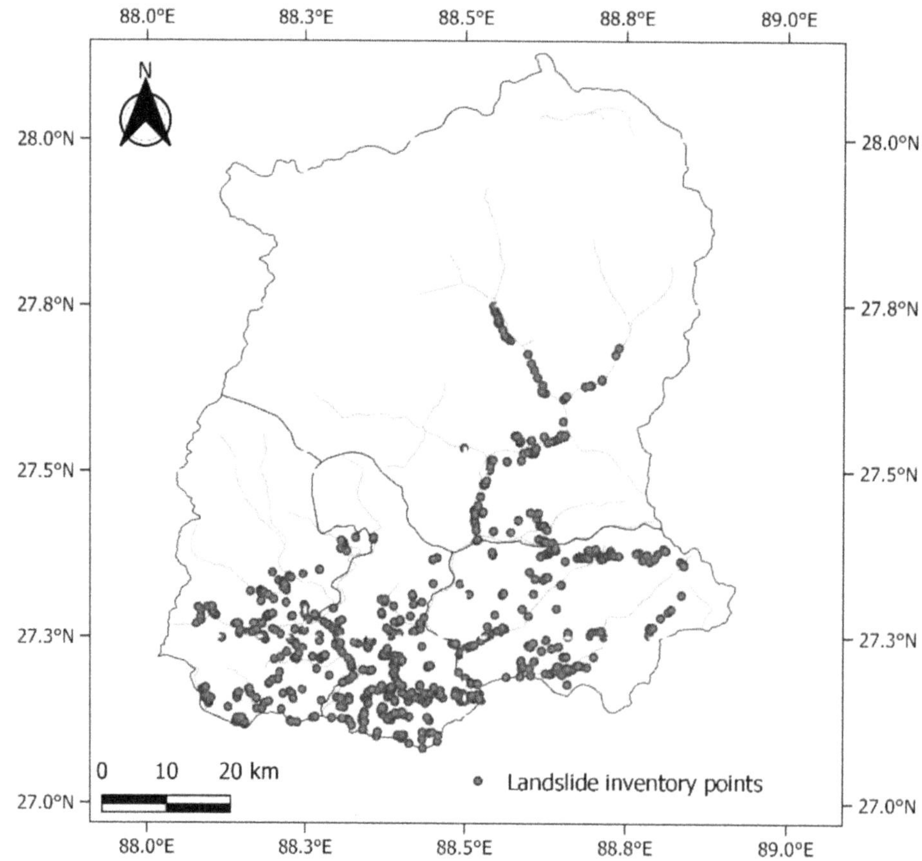

Lithology

The Sikkim state is a part of the young Himalayan mountain range. The stratigraphy of Sikkim contains distinct tectonic, sedimentary, magmatic, metamorphic, and structural characteristics and ranges in age from the Proterozoic to the Mesozoic. The majority of the lithotype in the Sikkim region is an extension of the Nepal Himalayan rock types to the east, and it has intricate Tectono-stratigraphic fold-thrust sequences that produce nappes, klippes, windows, etc. In the Sikkim Himalaya, the Ranjit tectonic windows (Quaternary sediments) are very crucial (S. K. Basu, 2013). The majority of the state is made up of Palaeozoic and Cretaceous rock, with hard massive Palaeozoic rock, which is the most resistant rock, covering around 62% of the total area (Figure 3). This rock is primarily found in Sikkim's south, east, south-east, and south-western regions. Quaternary sediments, which is extremely soft in character and covers the middle and southern portions of the terrain, covers up 5% of the entire area. The data were downloaded from Bhukosh website https://bhukosh.gsi.gov.in/Bhukosh/Public . In Sikkim, the Alpine-Himalayan mountain chain is a significant seismic belt.

Figure 3. Geology map of the Sikkim state

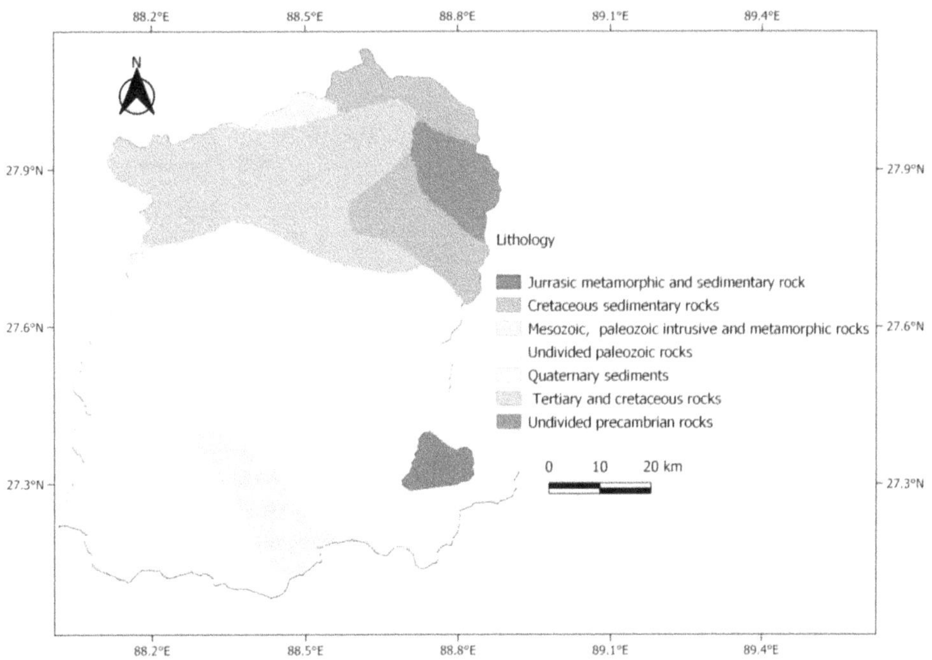

Rainfall

In places with steep slopes, rainfall is the primary cause of landslides. Rainfall is the predominant factor for causing landslide because it increases pore-water pressure, which in turn produces soil saturation and runoff (Clerici et al., 2006; Long & de Smedt, 2018; Merghadi et al., 2018; A. M. S. Pradhan & Kim, 2016; Zhu et al., 2014). The mean yearly precipitation for different region of Sikkim is shown in Figure 4. The rainfall data is taken from Indian Meteorological Department Pune.

Figure 4. Average annual rainfall map of Sikkim

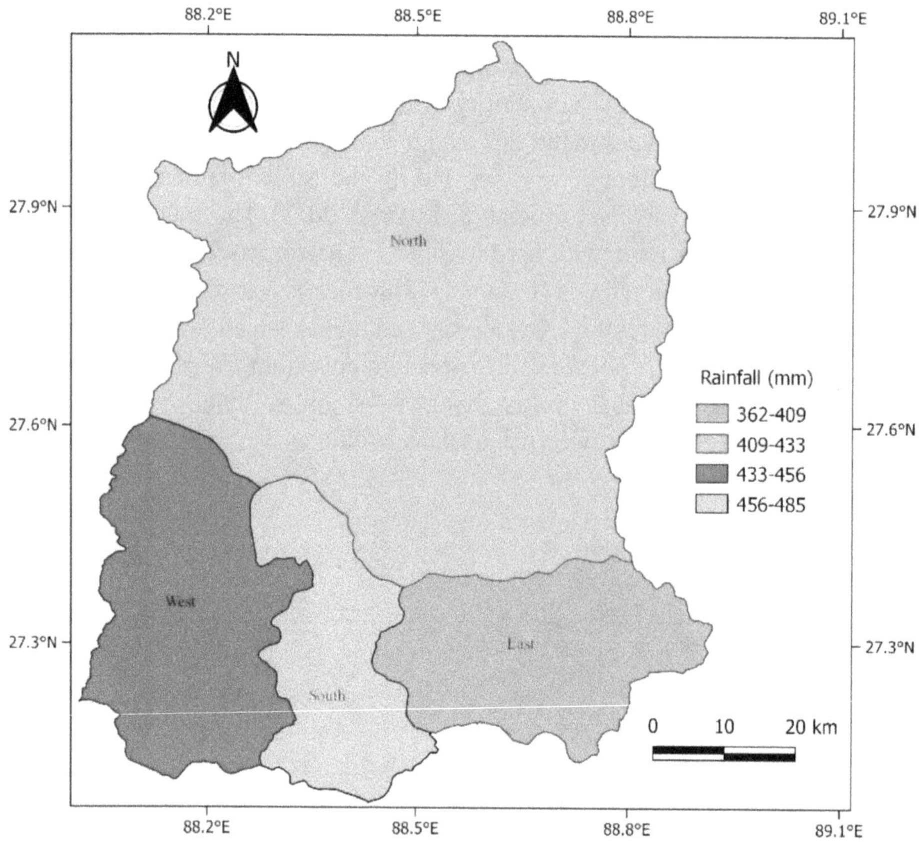

Drainage

Rangeet and Tista are the two main rivers in Sikkim, and they both travel across mountains from north to south. Because Sikkim's rivers are perennial, they provide water throughout the year. In the northeastern region of North Sikkim, the Tista River rises in the Tista Change Glacier, and in the western region of Sikkim, the Rangeet River rises in the Rathong Glacier. Both rivers run south to north. Figure 5 depicts the Sikkim drainage system.

Methodology

The AHP method is employed in this study to synthesise factors or class weights. The AHP is commonly used in the LSM, suitability analysis, and site selection (Ayalew et al., 2005; Feizizadeh & Blaschke, 2012; Yalcin, 2008). (Saaty & Vargas, 1980) developed AHP, a multi-criterion decision-making technique that allows both subjective and objective factors to be considered in the decision-making process (Feizizadeh & Blaschke, 2012; Yalcin, 2008). AHP is created based on the three components: priority synthesising, comparative judgement, and decomposition (Kayastha et al., 2013; Malczewski, 1999; Yalcin, 2008). AHP simplifies complex decision-making problems by categorising them into a hierarchy of factors and

Figure 5. Drainage map of Sikkim

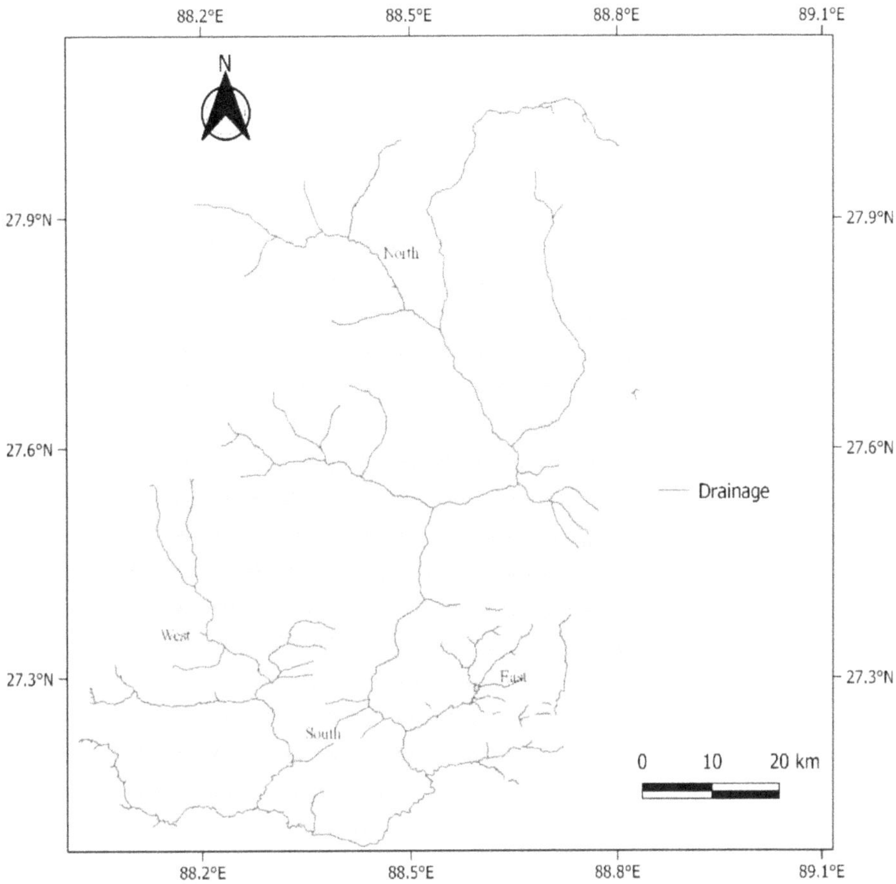

alternatives shown in Figure 6. Weightages were given to factor and others on the ordinal scale with nine points (Table 1) based on the comparison of pairs. Factors or their categories are grouped into a matrix with even numbers of row and column, values observed over single diagonal side of the matrix, and value of 1's recorded over the matrix diagonal portion (Gorsevski et al., 2006; Saaty, 1977). Professional knowledge influences pair-wise comparison and score judgement. The information gathered on the causes of landslides might be either subjectively or empirically judged. (Yalcin, 2008) emphasised the importance of considering both subjective and objective factors in the making of decisions. In such analysis, the professional expertise gained via fieldwork and the existence of landslides in those classes were used to determine the relative importance of each pair of factors or classes.

When the factor on the vertical axis is more important than the factor on the horizontal axis, this value varies between 1 and 9, and when the factor on the horizontal axis is more important, this value ranges from 1/2 to 1/9 in inverses (Saaty & Vargas, 1980;). The eigenvector of a matrix is used to calculate factor/class weights. The AHP model also includes the calculation of the maximum eigenvalue. Subjective decision rules can deviate from the transitivity rule, resulting in inconsistency (Feizizadeh & Blaschke, 2012). An important component of Analytical Hierarchy Process approach is computation of the consistency ratio and the consistency index (CI). (Saaty & Vargas, 1980) defined the Consistency Index (CI) as follows:

$$CI = \frac{\lambda_{max} - N}{N - 1}$$

Where, N is the number of elements and λm_{ax} is the highest eigenvalue in the matrix's row or column. Determining CR can be done using the proportion CI/RI, in which RI is random index. (Saaty & Vargas, 1980). (Saaty & Vargas, 1980) created the random index (Table 2) using a number of random samples. The maximum consistency threshold of the matrix is 0.1 CR. A CR value greater than 0.1 is considered inconsistent, whereas a value of 0 shows the perfect consistent result.

Figure 6. Working procedure for LSM using AHP

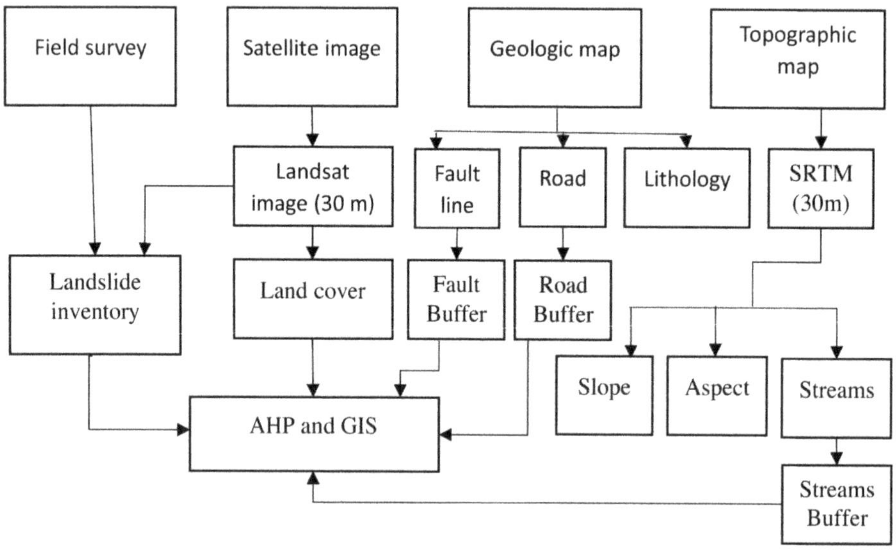

Table 1. Pair-wise importance of the scale

S.N.	Intensity of Importance	Importance
1.	1	Equal importance
2.	3	Moderate importance
3.	5	Strong importance
4.	7	Very strong importance
5.	9	Extreme importance
6.	2,4,6,8	Compensation between weights 1,3,5,7 and 9

Source: Saaty and Vargas (1980)

Table 2. Random consistency index (RI)

N	02	03	04	05	06	07	08	09
RI	0.0	0.52	0.90	1.120	1.240	1.320	1.41	1.40

Source: Saaty and Vargas (1980)

Assignments of the Weights and Weights Normalization Using AHP

The most significant factor in determining the weight of each class is integrated analysis, given that the assignment has a significant impact on the results with the proper weight (Kayastha et al., 2013; Thanh and de Smedt, 2012). In landslide prediction modelling, the AHP is emerging as a technique that could deliver prompt, accurate, and economical findings. An AHP approach is referred to as a pairing based analytical matrix method when it is employed to evalulate the mean of geometry and normalised weights of a single parameters (Saaty and Vargas, 1980). Analytical Hierarchy Process approach is employed to organise problems through hierarchies by identifying the goal of the problem and specifying criterion

Table 3. Distribution of weights for class and class characteristic information

Classes	Procedure for Giving Weight
Lithology	The schists, quartzite, shale, gneisses, Sandstone, and pegmatite are ranked highest, while tertiary and cretaceous sediments are ranked lowest.
Slope	highest position at the steepest slope
Rainfall	top ranking for the heaviest rain
Aspect	Highest rank for north-west topographic aspect
Lineament Density	highest position to the densest lineament
Land Use and Land cover	maximum weightage to settled and agricultural rea
Drainage Density	Maximum weightage to the densest drainage

Table 4. Matrices of 8 thematic map layer for pairwise comparison

	LI	SL	RF	AS	LD	LULC	RB	DD
LI	1	2	3	4	5	6	7	8
SL	0.50	1	2	3	4	5	6	7
RF	0.333	0.50	1	2	3	4	5	6
AS	0.250	0.333	0.50	1	2	3	4	5
LD	0.20	0.250	0.333	0.50	1	2	3	4
LULC	0.170	0.20	0.250	0.333	0.50	1	2	3
RB	0.140	0.170	0.20	0.250	0.333	0.50	1	2
DD	0.130	0.140	0.170	0.20	0.250	0.333	0.50	1
Total	2.723	4.593	7.453	11.283	16.083	21.833	28.50	36

LI: lithology, RF: Rainfall, LD: Lineament Density, AS: Aspect, SL: Slope, DD: Drainage Density, RB: Road buffer, LULC: Land use and Land Cover.

and sub-criterion in line with its corresponding importance in the analysis landslides. AHP has been employed by several authors to assigned weights to both criterion and sub-criterion characteristics (Das et al., 2012; Kayastha et al., 2013; Mondal & Maiti, 2012; Myronidis et al., 2016; Pourghasemi et al., 2012; Thanh & de Smedt, 2012; Yalcin, 2008). Table 3 presents the weights that are assigned to various thematic categories and class attributes and the normalized pair-wise matrix is shown in Table 4.

In terms of weighing or rating factors and their classifications, the AHP technique has advantages, but it also has some drawbacks. The relative evaluation of the aspects is heavily reliant on a person's or professional's knowledge. The relative preference of an individual or professional factor is typically disregarded by other, that is the significant backlog in decision deriving process. However, comparing through pairs gives the clear and dependable judgement rules. While AHP views variables in a hierarchy as an independent entity, some factors in landslide research provides a reliable results in impacting landslide. In the AHP method, moderate slope is given less favour than high slope, despite the fact that the comparatively medium slope, for instance, the collapse may occur as reason of slope moisture content. Analytical Hierarchy Process, in general, offers a very adaptable and straightforward method that is simply adapted in the Geographic information systems domain.

The calculation of the landslide susceptibility index (LSI) came next. The weighted arithmetic sum method is used to calculate it, and the results are expressed as follows.

$$LSI = \sum_{J=1}^{N} Weight\ of\ factor\left(W_j\right) \times Weight\ of\ factorclasses\left(W_{ij}\right)$$

Where, W_{ij} denote weights of the ith classes of factors J. Using natural break (ESRI 2012) classification method, the LSI susceptibility classes of maps may be categorised such as very high, very low, high, low, and moderate.

With the help of the QGIS and ArcGIS software, eight different thematic layers, including Lithology, Road Buffer, Density of Lineament, Aspect, Slope, Density of drainage, Rainfall and land cover and land use, were constructed for evaluation of landslides map. Before processing the results, all essential corrections, including atmospheric and radiometric correction, have been made on cloud-free Landsat 8 data. The spatial variation are interpolated using the Inverse Distance Weighted (IDW) method.

RESULTS AND DISCUSSION

Land Use and Land Cover Map

Landsat 8 satellite data is used to create the LULC classes of research area using maximum likelihood approach in QGIS software. Land use refers to how people use the land, whereas land cover refers to the actual form of land, such as a forest or open water. Figure 7 illustrates how data on land cover reveals the percentage of a region that is covered by forests, wetlands, impermeable surfaces, agricultural, and other forms of land and water. The vulnerability to landslides is influenced by manmade causes that reduce solidity of a hillside and by deforestation (Kumar et al., 2018). The land cover type in the study regions are water bodies (12.34%), glaciers (14.96%), agricultural (8.2%), forest (43.36%), barren land solidity of a hillside d (17.96%), and built-up (3.18%).

Figure 7. Sikkim LULC map

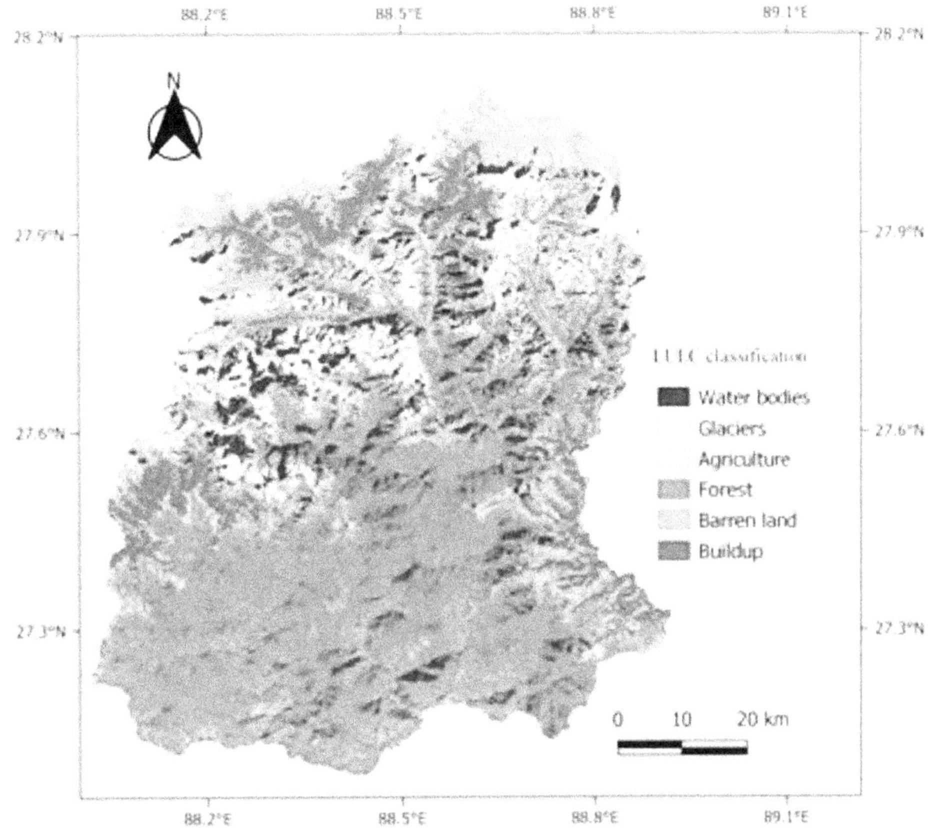

Slope and Aspect

Slope and aspect is a crucial component in the mapping of landslides susceptibility. Because it has the direct impact on the landslide occurrence; thus, slope primarily used in the LSM (Kaur et al., 2019). Landslides are less common at elevations above 1,000 m, due to presence of the hard rocks. This is also among the causes for the study's north and west's relatively low landslide density. As the angle of the slope increases, the slope instability worsens, and the likelihood of a landslide increases (Zêzere et al., 2017). Curvature is an important parameter factor in the study of landslide susceptibility because it affects the driving and resisting stress within a landslide in the direction of motion and controls the water in the landslide direction motion as well as the substance that causes landslides to either coalesce or diverge. As the percentage of landslide susceptibility grows, the steeper the slope, the faster the runoff, and the larger the rate of erosion (Meten et al., 2015). The slope of the state Sikkim has been created using DEM data in QGIS. The region is categories into five classes based on slope (Figure 8): (0^0-10^0), (10^0-30^0), (30^0-50^0), (50^0-70), and (>70^0). The research area's maximum slope ranges from 30^0 to more than 70^0, indicating that the majority of slope failure variables are present.

Figure 8. Slope variation of the Sikkim state

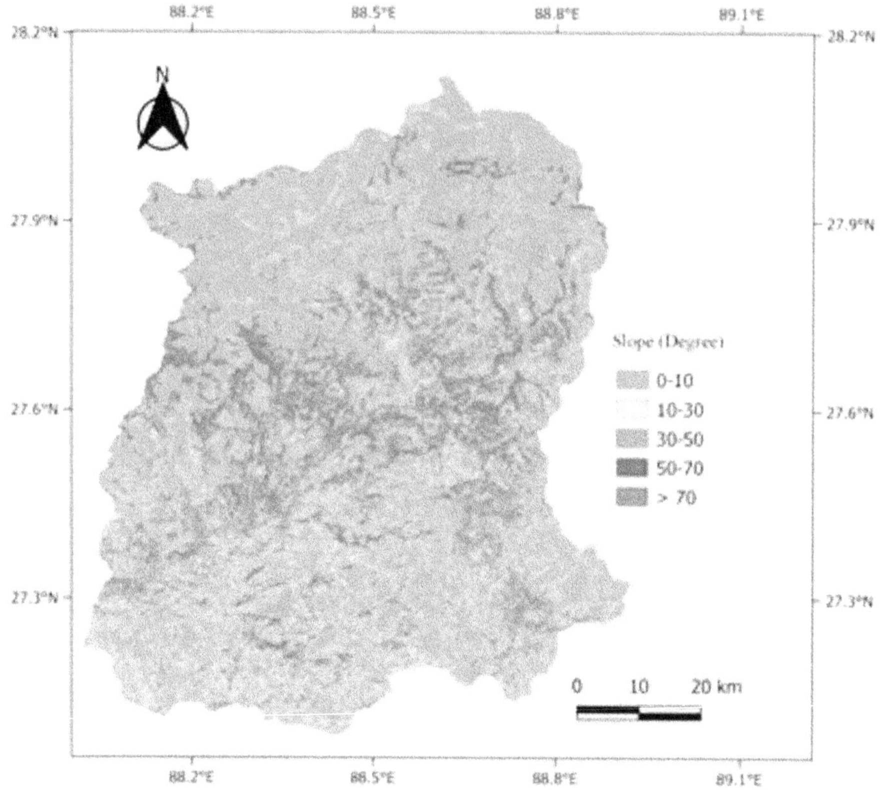

LSM takes aspect into account as well. A slope's aspect is the direction it faces in relation to the north. The effect of solar heating, soil moisture, and air dryness is determined by aspect (Dai et al. 2001). The area's aspect map was created using DEM from 9 categories: Flat (-1), North (0.0° - 22.50° and 337.50°-360°), North-east (22.50°-67.5°), East (67.50°-112.50°), South-east (112.50°-157.50°), South (157.50°-202.5°), South-west (202.5°- 247.50°), West (247.5°-292.5°), and North-west (292.5°-337.50°) shown in Figure 9.

Lineament

Lineament is a structurally controlled set of linear or curvilinear structures that are easily recognised from satellite imagery due to their generally linear alignments. These features express the surface topography of the underlying structural structures. Zones of faulting and fracturing are represented by lines. Thrusting is the primary source of landslides because the gneiss of Darjeeling was forced into touch along the Daling. Due to the geology being crushed and sheared in the thrusting portion of Sikkim, the primary centre force MCT I & II are prime cause of landslides in the region. Landslide-prone zones are identified by the density of lines in a particular location (Sonker et al. 2021). Research region having high density of lineament (18.38 Km/Km Sq.), the major significant initiating variables in land sliding. Figure 10 displays the lineament density map for the research region. The southern portion of the study

Figure 9. Topographic aspect of the Sikkim

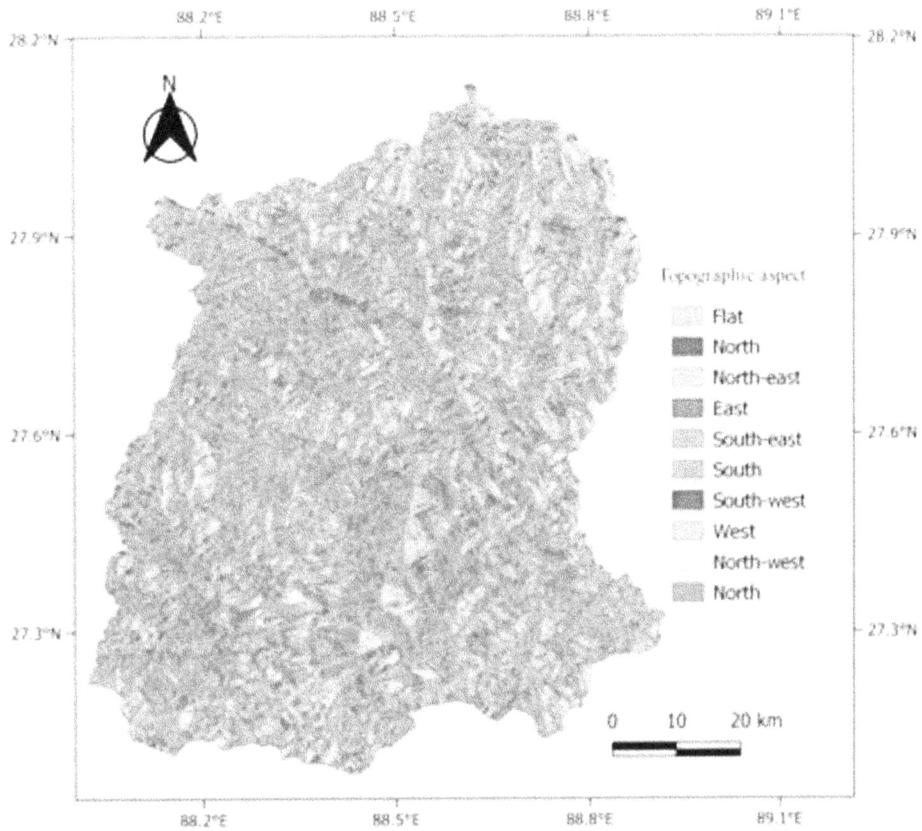

areas has a high lineament density, while the northern and northern-eastern regions of Sikkim have a low lineament density.

Normalised Weights for Features and Conditioning Variables

To determine normalised weight for individual attributes, Saaty's analytical hierarchy technique is used. The normalised weights for each theme were derived using AHP, as shown in Tables 4 and 5. Additionally, AHP normalises the weights given to various elements of specific themes, as seen in Table 6. The weighting of individual features is determined by the frequency of landslides. The consistency of the comparison matrices is shown in Table 7 by the fact that all CR values are less than 0.10.

Zonation of Landslide Susceptibility Using AHP

The pairwise comparisons for every class and every class-feature are represented by the normalised weights, which are shown in Tables 5 and 6. These weightage numbers represent how significant a factor or class is. The most crucial variables are geology, slope, aspect, and rainfall, followed by lineament density, LULC, drainage density, and road buffer. The resulted zone of land slide susceptibility for the

Figure 10. Lineament density map of Sikkim

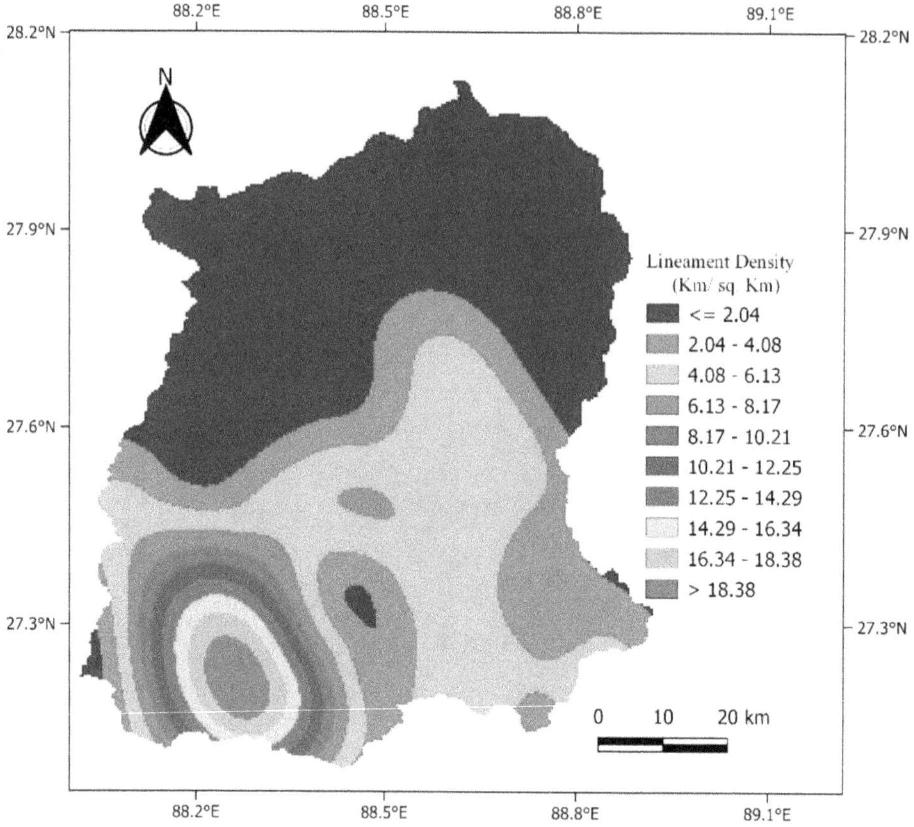

Table 5. The normalised weights for each thematic layer and the normalised pair-wise matrix

	LI	SL	RF	AS	LD	LULC	RB	DD	Normalized Weights
LI	0.38	0.43	0.4	0.35	0.31	0.26	0.25	0.22	0.33
SL	0.18	0.22	0.26	0.27	0.25	0.22	0.21	0.21	0.23
RF	0.12	0.13	0.13	0.18	0.20	0.17	0.16	0.17	0.16
AS	0.11	0.09	0.08	0.07	0.11	0.15	0.15	0.15	0.12
LD	0.07	0.04	0.04	0.05	0.06	0.09	0.17	0.13	0.08
LULC	0.06	0.04	0.03	0.03	0.05	0.05	0.09	0.06	0.06
RB	0.05	0.04	0.03	0.02	0.02	0.02	0.04	0.04	0.03
DD	0.05	0.03	0.02	0.02	0.02	0.02	0.03	0.02	0.02

study area is depicted in Figure 12, where there are five classes and values for the land slide susceptibility index vary from 0.128 to 0.55: very high (0.452–0.55), very low (0.12 to 0.206), moderate (0.2880 to 0.370), high (0.370 to 0.4520), and low (0.2060 to 0.2880). Since there are no universal standards for automatically classifying such continuous data, the majority of researchers rely on their own professional

Table 6. Assigned and normalised weight for various characteristics of eight influencing factor for the study area's landslide vulnerability zones

Features	Classes	Final Weights
Lithology	Cretaceous sedimentary	0.06
	Undivided Palaeozoic	0.26
	Quaternary sediments	0.34
	Undivided Precambrian	0.07
	Tertiary and cretaceous	0.08
	Jurassic Metamorphic and sedimentary	0.18
Slope	0^0–10^0	0.042
	10^0–30^0	0.068
	30^0–50^0	0.14
	50^0–70^0	0.27
	$>70^0$	0.48
Rainfall (mm/year)	Very Low	0.048
	Low	0.063
	High	0.19
	Very high	0.25
	Extremely high	0.51
Aspect	North (1)	0.041
	Northwest (2)	0.062
	West (3)	0.085
	Southwest (4)	0.15
	South (5)	0.3
	Southeast (6)	0.21
	East (7)	0.078
	Northeast (8)	0.055
	Flat (9)	0.027
Lineament density (Km/Km Sq.)	2-4.1	0.09
	4.1-6.1	0.04
	6.1-10.21	0.07
	10.21-14.29	0.11
	14.29-16.34	0.2
	16.34-18.38	0.27
	>18.38	0.35
Land use and Land cover	Agricultural area	0.072
	Settled/Built	0.235
	Vegetative cover	0.158
	Barren area	0.14
	Snow/glacier	0.41

Table 6. Continued

Features	Classes	Final Weights
	Open water	0.185
Road Buffer	0 – 50	0.558
	50 – 100	0.255
	100 - 200	0.122
	>200	0.062
Drainage Density	0.010–0.460	0.041
	0.460–0.630	0.09
	0.630–0.801	0.12
	0.801–0.990	0.33
	0.990–1.571	0.47

judgement to establish class borders (Kayastha et al., 2013). In Figures 11 and 12, the resulted landslide susceptibility zonation of the study region shows five zones that correspond to "Very Low (12.52%), Low (21.12%), Moderate (8.05%), High (31.13%), and Very High (27.18%)." Jurassic metamorphic, quaternary sediments, and Palaeozoic rocks are found in the zone with a "Very High" likelihood of landslides. These rocks are all strongly joined and crumbled in the thrusting zones, which is covered in a thick layer of soil. Heavy monsoon rain falls heavily on this region, as well as the southern and MCT-I & II thrusting zone of the research region, which can lead to landslide hazard. The south centre portion of the MCT thrusts, that is a component of the Gorubathan formation, is where the "High" zones are most prevalent. The areas with notable quaternary cretaceous rocks, cretaceous sedimentary rocks, and low lineament are mostly covered by Moderate to Low Landslide susceptibility zones. The northern and centre regions fall under the very low to low Landslide Hazard zones. Areas close to the drainage network were dominated by high susceptible classes, according to results that have already been reported. Flatter terrain, thick, and sparsely forested areas are found to have low susceptibility levels. Built-up regions have been found to have vulnerability levels ranging from moderate to high.

Table 7. Order of matrices N, random index (RI), consistency index (CI), and consistency ratio (CR)

Classes or Features	N	Eigen's Values	C I	R I	C R
Combined Classes	8	8.2	0.010	1.410	0.028
Lithology	6	6.25	0.05	1.240	0.04
Slope	5	5.25	0.062	1.120	0.055
Aspect	9	9.19	0.024	1.12	0.021
Rain-fall	5	5.2	0.05	1.120	0.044
lineament density	7	7.11	0.018	1.32	0.013
Land cover/land use	6	6.22	0.044	1.24	0.035
Road Buffer	4	4.14	0.046	1.12	0.042
Drainage density	5	5.178	0.044	1.12	0.039

Figure 11. Landslide susceptibility map of Sikkim

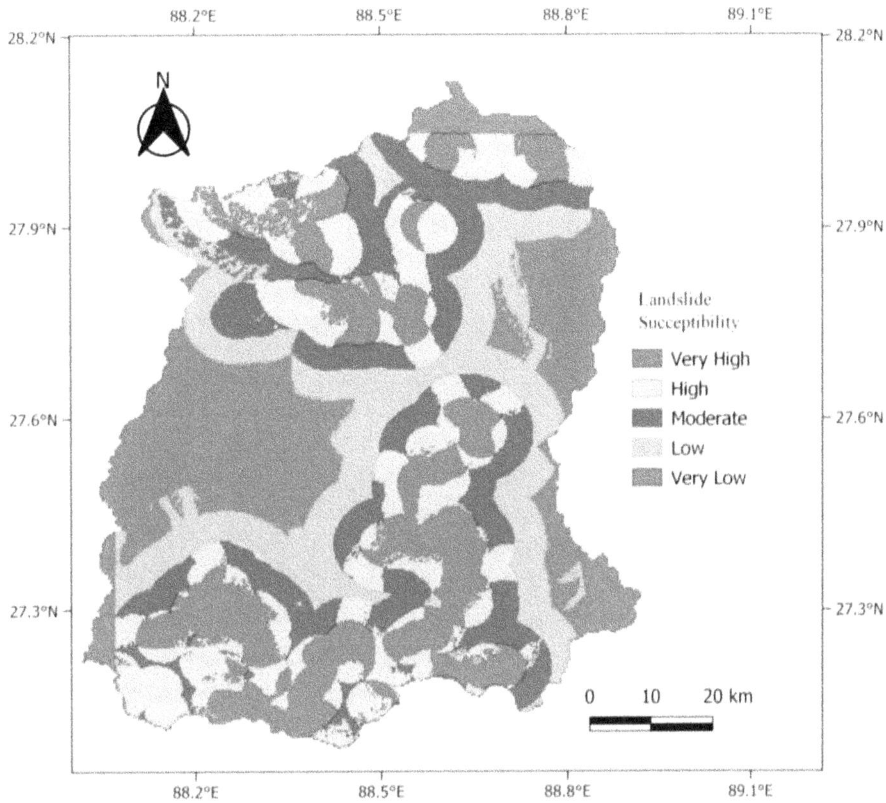

Figure 12. Percentage of total area prone to Landslide susceptibility

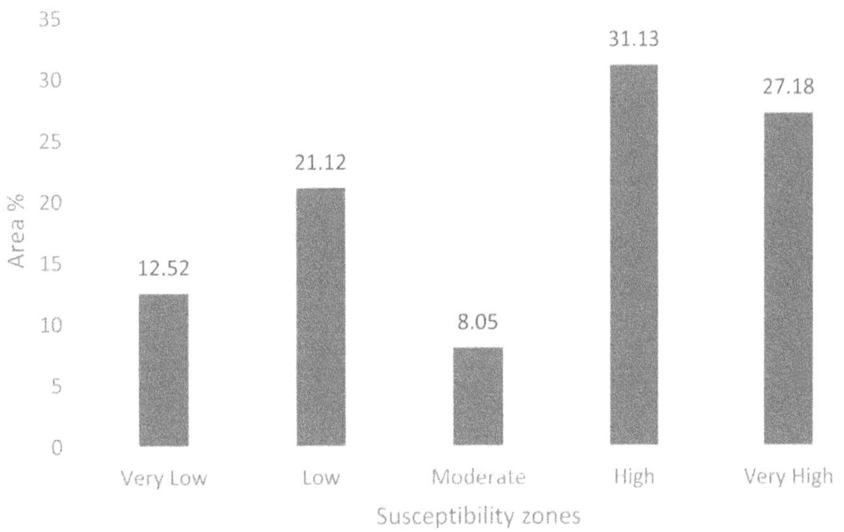

Accuracy Assessment Using ROC

The effectiveness of landslide susceptibility maps created from AHP is evaluated using the ROC (Receiver Operating Characteristic) curve (Figure 13). The most common and important application of ROC is to graphically display the relationship among the False and True Positive Rates (Althouse, 2016). Area Under Curve models lies between 0.50 and 1.0. Performance of prediction models is improved when the AUC values are close to 1. The maps produced by AHP using the ROC curve (with a good AUC value of 70%, Table 8) have improved the accuracy of identifying areas that are prone to landslides.

Figure 13. ROC curve of LSM

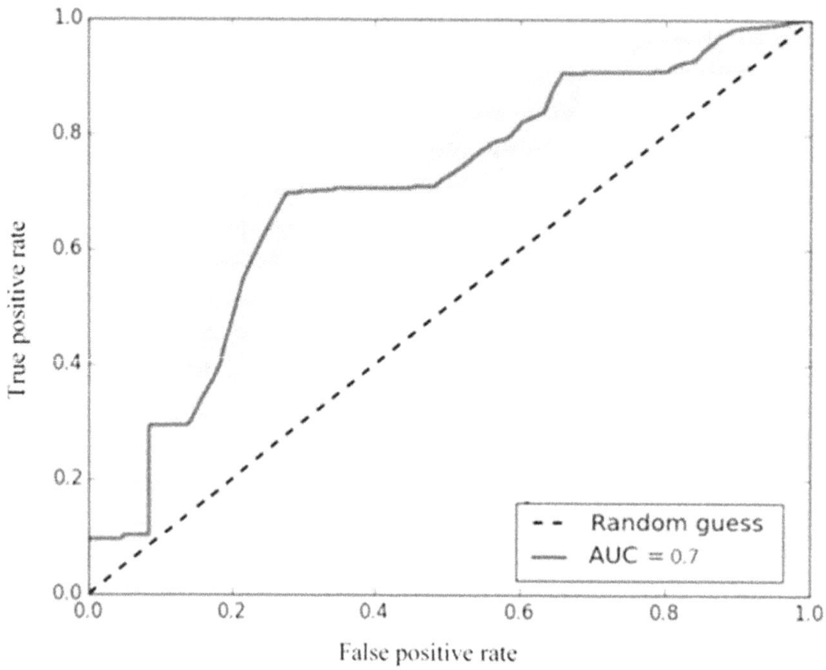

Table 8. AUC performance table

AUC Range	Performance
0.90–1.0	Excellent
0.80–0.90	Very Good
0.70–0.80	Good
0.60–0.70	Satisfactory
0.50–0.60	Unsatisfactory

CONCLUSION

The Analytical hierarchy process is used to create the landslide susceptibility map of the north east region of India. The Receiver Operating Characteristic (ROC) curve is a tool. In order to assess the performance of AHP-derived landslide susceptibility maps. The majority of the time, ROC curves are employed to pictorially illustrate the relationship among False and True Positive Rates. The outcome of the Area Under Curve model lies between 0.5 and 1. The AUC value that are nearly 1 show that prediction models perform better. The generated AHP maps are based on the ROC curve with the Area Under Curve value 70% which shows result is good. AHP has increased prosperity landslide susceptibility weight on the scales of saaty in accordance with their respective relevance. An analytical hierarchy procedure known as AHP was then used to normalise the thematic map layer assigned weights and its classes. With the use of Arc GIS 10.5, all the layer are incorporated in the geographic system via average weighted simple combinations approach to identify research area's landslide susceptibility zones. Five landslide susceptibility zones—very low, low, moderate, high, and very high—have been established for the research area according to the method. In this study, the areas that are most likely to experience landslides are identified, including Ranjit windows, the Ramgarh thrusts, and accompanying MCT-1 and MCT-2 areas. These areas are particularly vulnerable to landslides. The results also showed the centre and northern regions of Sikkim are classified as having extremely low, low, and moderate susceptibility levels. This study provides information regarding potential future landslide hazards as well as a map of the region's landslide susceptibility. Using the landslide susceptibility map and landslide inventory data, the AUC (Area Under Curve) is calculated to be around 70%. The AUC results was good .This strategy based on logic and reasoning can also be used to other parts of the world to analyse the landslide.

Future Scope

The AHP has been implemented for preparation of landslide susceptibility map. There were eight factors considered in this chapter. The results may be improved by including another causing factor topographic witness index (TWI).

REFERENCES

Akgün, A., & Bulut, F. (2007). GIS-based landslide susceptibility for Arsin-Yomra (Trabzon, North Turkey) region. *Environmental Geology (Berlin)*, *51*(8), 1377–1387. doi:10.100700254-006-0435-6

Aleotti, P., & Chowdhury, R. (1999). Landslide hazard assessment: Summary review and new perspectives. *Bulletin of Engineering Geology and the Environment*, *58*(1), 21–44. doi:10.1007100640050066

Althouse, A. D. (2016). Statistical graphics in action: Making better sense of the ROC curve. *International Journal of Cardiology*, *100*(215), 9–10. doi:10.1016/j.ijcard.2016.04.026 PMID:27104918

Althuwaynee, O. F., Pradhan, B., & Lee, S. (2016). A novel integrated model for assessing landslide susceptibility mapping using CHAID and AHP pair-wise comparison. *International Journal of Remote Sensing*, *37*(5), 1190–1209. doi:10.1080/01431161.2016.1148282

Anbalagan, R. (1992). Landslide hazard evaluation and zonation mapping in mountainous terrain. *Engineering Geology*, *32*(4), 269–277. doi:10.1016/0013-7952(92)90053-2

Ayalew, L., Yamagishi, H., Marui, H., & Kanno, T. (2005). Landslides in Sado Island of Japan: Part II. GIS-based susceptibility mapping with comparisons of results from two methods and verifications. *Engineering Geology*, *81*(4), 432–445. doi:10.1016/j.enggeo.2005.08.004

Ba, Q., Chen, Y., Deng, S., Wu, Q., Yang, J., & Zhang, J. (2017). An improved information value model based on gray clustering for landslide susceptibility mapping. *ISPRS International Journal of Geo-Information*, *6*(1), 18. doi:10.3390/ijgi6010018

Bahrami, Y., Hassani, H., & Maghsoudi, A. (2021). Landslide susceptibility mapping using AHP and fuzzy methods in the Gilan province, Iran. *GeoJournal*, *86*(4), 1797–1816. doi:10.100710708-020-10162-y

Basu, S. K. (2013). *Geology of Sikkim state and Darjeeling district of West Bengal*. Geological Society of India.

Basu, T., Das, A., & Pal, S. (2022). Application of geographically weighted principal component analysis and fuzzy approach for unsupervised landslide susceptibility mapping on Gish River Basin, India. *Geocarto International*, *37*(5), 1294–1317. doi:10.1080/10106049.2020.1778105

Bhasin, R., Grimstad, E., Larsen, J. O., Dhawan, A. K., Singh, R., Verma, S. K., & Venkatachalam, K. (2002). Landslide hazards and mitigation measures at Gangtok, Sikkim Himalaya. *Engineering Geology*, *64*(4), 351–368. doi:10.1016/S0013-7952(01)00096-5

Biswakarma, P., Barman, B. K., Joshi, V., & Rao, K. S. (2020). Landslide susceptibility mapping in east Sikkim region of Sikkim Himalaya using high resolution remote sensing data and GIS techniques. *Appl. Ecol. Environ. Sci*, *8*, 143–153.

Choi, J., Oh, H.-J., Lee, H.-J., Lee, C., & Lee, S. (2012). Combining landslide susceptibility maps obtained from frequency ratio, logistic regression, and artificial neural network models using ASTER images and GIS. *Engineering Geology*, *124*, 12–23. doi:10.1016/j.enggeo.2011.09.011

Clerici, A., Perego, S., Tellini, C., & Vescovi, P. (2006). A GIS-based automated procedure for landslide susceptibility mapping by the conditional analysis method: The Baganza valley case study (Italian Northern Apennines). *Environmental Geology (Berlin)*, *50*(7), 941–961. doi:10.100700254-006-0264-7

Dahal, R. K., Hasegawa, S., Nonomura, A., Yamanaka, M., Dhakal, S., & Paudyal, P. (2008). Predictive modelling of rainfall-induced landslide hazard in the Lesser Himalaya of Nepal based on weights-of-evidence. *Geomorphology*, *102*(3–4), 496–510. doi:10.1016/j.geomorph.2008.05.041

Dai, F. C., & Lee, C. F. (2002). Landslide characteristics and slope instability modeling using GIS, Lantau Island, Hong Kong. *Geomorphology*, *42*(3–4), 213–228. doi:10.1016/S0169-555X(01)00087-3

Das, I., Sahoo, S., van Westen, C., Stein, A., & Hack, R. (2010). Landslide susceptibility assessment using logistic regression and its comparison with a rock mass classification system, along a road section in the northern Himalayas (India). *Geomorphology*, *114*(4), 627–637. doi:10.1016/j.geomorph.2009.09.023

Das, I., Stein, A., Kerle, N., & Dadhwal, V. K. (2012). Landslide susceptibility mapping along road corridors in the Indian Himalayas using Bayesian logistic regression models. *Geomorphology*, *179*, 116–125. doi:10.1016/j.geomorph.2012.08.004

Ercanoglu, M., & Gokceoglu, C. (2004). Use of fuzzy relations to produce landslide susceptibility map of a landslide prone area (West Black Sea Region, Turkey). *Engineering Geology*, *75*(3–4), 229–250. doi:10.1016/j.enggeo.2004.06.001

Fayez, L., Pazhman, D., Pham, B. T., Dholakia, M. B., Solanki, H. A., Khalid, M., & Prakash, I. (2018). Application of frequency ratio model for the development of landslide susceptibility mapping at part of Uttarakhand State, India. *International Journal of Applied Engineering Research: IJAER*, *13*(9), 6846–6854.

Feizizadeh, B., & Blaschke, T. (2012). Uncertainty analysis of GIS-based ordered weighted averaging method for landslide susceptibility mapping in Urmia Lake Basin, Iran. *Seventh International Geographic Information Science Conference, September*, 18–21.

Feizizadeh, B., Roodposhti, M. S., Jankowski, P., & Blaschke, T. (2014). A GIS-based extended fuzzy multi-criteria evaluation for landslide susceptibility mapping. *Computers & Geosciences*, *73*, 208–221. doi:10.1016/j.cageo.2014.08.001 PMID:26089577

Ghosh, S., Carranza, E. J. M., van Westen, C. J., Jetten, V. G., & Bhattacharya, D. N. (2011). Selecting and weighting spatial predictors for empirical modeling of landslide susceptibility in the Darjeeling Himalayas (India). *Geomorphology*, *131*(1–2), 35–56. doi:10.1016/j.geomorph.2011.04.019

Gorsevski, P., Gessler, P. E., Foltz, R. B., & Elliot, W. J. (2006). Spatial prediction of landslide hazard using logistic regression and ROC analysis. *Transactions in GIS*, *10*(3), 395–415. doi:10.1111/j.1467-9671.2006.01004.x

Gupta, R. P., Saha, A. K., Arora, M. K., & Kumar, A. (1999). Landslide hazard zonation in part of the Bhagirathi Valley. *Garhwal Mimalyas, Using Integrated Remote Sensing–GIS. Himalayan Geology*, *20*, 71–85.

Guzzetti, F., Carrara, A., Cardinali, M., & Reichenbach, P. (1999). Landslide hazard evaluation: A review of current techniques and their application in a multi-scale study, Central Italy. *Geomorphology*, *31*(1–4), 181–216. doi:10.1016/S0169-555X(99)00078-1

Guzzetti, F., Reichenbach, P., Cardinali, M., Galli, M., & Ardizzone, F. (2005). Probabilistic landslide hazard assessment at the basin scale. *Geomorphology*, *72*(1–4), 272–299. doi:10.1016/j.geomorph.2005.06.002

Hutchinson, J. N. (1988). General report: morphological and geotechnical parameters of landslides in relation to geology and hydrogeology. *International Symposium on Landslides*, *5*, 3–35.

Hutchinson, J. N., Poole, C., Lambert, N., & Bromhead, E. N. (1985). Combined archaeological and geotechnical investigations of the Roman Fort at Lympne, Kent. *Britannia*, *16*, 209–236. doi:10.2307/526401

Kanungo, D. P., Arora, M. K., Sarkar, S., & Gupta, R. P. (2006). A comparative study of conventional, ANN black box, fuzzy and combined neural and fuzzy weighting procedures for landslide susceptibility zonation in Darjeeling Himalayas. *Engineering Geology*, *85*(3–4), 347–366. doi:10.1016/j.enggeo.2006.03.004

Kanungo, D. P., Arora, M. K., Sarkar, S., & Gupta, R. P. (2009). A fuzzy set based approach for integration of thematic maps for landslide susceptibility zonation. *Georisk*, *3*(1), 30–43. doi:10.1080/17499510802541417

Kaur, H., Gupta, S., Parkash, S., Thapa, R., Gupta, A., & Khanal, G. C. (2019). Evaluation of landslide susceptibility in a hill city of Sikkim Himalaya with the perspective of hybrid modelling techniques. *Annals of GIS*, *25*(2), 113–132. doi:10.1080/19475683.2019.1575906

Kayastha, P., Dhital, M. R., & de Smedt, F. (2013). Application of the analytical hierarchy process (AHP) for landslide susceptibility mapping: A case study from the Tinau watershed, west Nepal. *Computers & Geosciences*, *52*, 398–408. doi:10.1016/j.cageo.2012.11.003

Kumar, R., & Anbalagan, R. (2016). Landslide susceptibility mapping using analytical hierarchy process (AHP) in Tehri reservoir rim region, Uttarakhand. *Journal of the Geological Society of India*, *87*(3), 271–286. doi:10.100712594-016-0395-8

Kumar, V., Gupta, V., & Jamir, I. (2018). Hazard evaluation of progressive Pawari landslide zone, Satluj valley, Himachal Pradesh, India. *Natural Hazards*, *93*(2), 1029–1047. doi:10.100711069-018-3339-3

Kundu, S., Saha, A. K., Sharma, D. C., & Pant, C. C. (2013). Remote sensing and GIS based landslide susceptibility assessment using binary logistic regression model: A case study in the Ganeshganga Watershed, Himalayas. *Photonirvachak (Dehra Dun)*, *41*(3), 697–709. doi:10.100712524-012-0255-y

Lee, S., & Pradhan, B. (2007). Landslide hazard mapping at Selangor, Malaysia using frequency ratio and logistic regression models. *Landslides*, *4*(1), 33–41. doi:10.100710346-006-0047-y

Lee, S., & Talib, J. A. (2005). Probabilistic landslide susceptibility and factor effect analysis. *Environmental Geology (Berlin)*, *47*(7), 982–990. doi:10.100700254-005-1228-z

Long, N. T., & de Smedt, F. (2018). Analysis and mapping of rainfall-induced landslide susceptibility in a Luoi District, Thua Thien Hue Province, Vietnam. *Water (Basel)*, *11*(1), 51. doi:10.3390/w11010051

Malczewski, J. (1999). *GIS and multicriteria decision analysis*. John Wiley & Sons.

Mathew, J., Jha, V. K., & Rawat, G. S. (2007). Weights of evidence modelling for landslide hazard zonation mapping in part of Bhagirathi valley, Uttarakhand. *Current Science*, 628–638.

Merghadi, A., Abderrahmane, B., & Tien Bui, D. (2018). Landslide susceptibility assessment at Mila Basin (Algeria): A comparative assessment of prediction capability of advanced machine learning methods. *ISPRS International Journal of Geo-Information*, *7*(7), 268. doi:10.3390/ijgi7070268

Mersha, T., & Meten, M. (2020). GIS-based landslide susceptibility mapping and assessment using bivariate statistical methods in Simada area, northwestern Ethiopia. *Geoenvironmental Disasters*, *7*(1), 1–22. doi:10.118640677-020-00155-x

Meten, M., PrakashBhandary, N., & Yatabe, R. (2015). Effect of landslide factor combinations on the prediction accuracy of landslide susceptibility maps in the Blue Nile Gorge of Central Ethiopia. *Geoenvironmental Disasters*, *2*(1), 1–17. doi:10.118640677-015-0016-7

Mondal, S., & Maiti, R. (2012). Landslide susceptibility analysis of Shiv-Khola watershed, Darjiling: A remote sensing & GIS based Analytical Hierarchy Process (AHP). *Photonirvachak (Dehra Dun)*, *40*(3), 483–496. doi:10.100712524-011-0160-9

Murillo-García, F. G., Fiorucci, F., & Alcántara-Ayala, I. (2014). Development of a landslide inventory for a region in Mexico using very high resolution satellite stereo-images. In *Landslide science for a safer geoenvironment* (pp. 821–828). Springer. doi:10.1007/978-3-319-05050-8_127

Myronidis, D., Papageorgiou, C., & Theophanous, S. (2016). Landslide susceptibility mapping based on landslide history and analytic hierarchy process (AHP). *Natural Hazards*, *81*(1), 245–263. doi:10.100711069-015-2075-1

Nagarajan, R., Mukherjee, A., Roy, A., & Khire, M. v. (1998). *Technical note temporal remote sensing data and GIS application in landslide hazard zonation of part of Western Ghat*. doi:10.1080/014311698215865

Pal, S. C., & Chowdhuri, I. (2019). GIS-based spatial prediction of landslide susceptibility using frequency ratio model of Lachung River basin, North Sikkim, India. *SN Applied Sciences*, *1*(5), 1–25. doi:10.100742452-019-0422-7

Park, S., Choi, C., Kim, B., & Kim, J. (2013). Landslide susceptibility mapping using frequency ratio, analytic hierarchy process, logistic regression, and artificial neural network methods at the Inje area, Korea. *Environmental Earth Sciences*, *68*(5), 1443–1464. doi:10.100712665-012-1842-5

Pourghasemi, H. R., Pradhan, B., & Gokceoglu, C. (2012). Application of fuzzy logic and analytical hierarchy process (AHP) to landslide susceptibility mapping at Haraz watershed, Iran. *Natural Hazards*, *63*(2), 965–996. doi:10.100711069-012-0217-2

Pradhan, A. M. S., & Kim, Y. T. (2016). Evaluation of a combined spatial multi-criteria evaluation model and deterministic model for landslide susceptibility mapping. *Catena*, *140*, 125–139. doi:10.1016/j.catena.2016.01.022

Pradhan, B., & Lee, S. (2010). Regional landslide susceptibility analysis using back-propagation neural network model at Cameron Highland, Malaysia. *Landslides*, *7*(1), 13–30. doi:10.100710346-009-0183-2

Pradhan, B., Lee, S., & Buchroithner, M. F. (2009). Use of geospatial data and fuzzy algebraic operators to landslide-hazard mapping. *Applied Geomatics*, *1*(1), 3–15. doi:10.100712518-009-0001-5

Pradhan, B., Lee, S., & Buchroithner, M. F. (2010). A GIS-based back-propagation neural network model and its cross-application and validation for landslide susceptibility analyses. *Computers, Environment and Urban Systems*, *34*(3), 216–235. doi:10.1016/j.compenvurbsys.2009.12.004

Praveen, B. (2017). Monitoring and Assesment of Landslide from Agastmuni To Sonprayag. *Int J Rec Res Soc Sci Humanit, 4*, 92–118.

Reichenbach, P., Mondini, A. C., & Rossi, M. (2014). The influence of land use change on landslide susceptibility zonation: The Briga catchment test site (Messina, Italy). *Environmental Management*, *54*(6), 1372–1384. doi:10.100700267-014-0357-0 PMID:25164982

Saaty, T. L. (1977). A scaling method for priorities in hierarchical structures. *Journal of Mathematical Psychology*, *15*(3), 234–281. doi:10.1016/0022-2496(77)90033-5

Saaty, T. L. (1990). An exposition of the AHP in reply to the paper "remarks on the analytic hierarchy process.". *Management Science*, *36*(3), 259–268. doi:10.1287/mnsc.36.3.259

Saaty, T. L., & Vargas, L. G. (1980). Hierarchical analysis of behavior in competition: Prediction in chess. *Behavioral Science*, *25*(3), 180–191. doi:10.1002/bs.3830250303

Saha, A. K., Gupta, R. P., & Arora, M. K. (2002). GIS-based landslide hazard zonation in the Bhagirathi (Ganga) valley, Himalayas. *International Journal of Remote Sensing*, *23*(2), 357–369. doi:10.1080/01431160010014260

Sangchini, E. K., Emami, S. N., Tahmasebipour, N., Pourghasemi, H. R., Naghibi, S. A., Arami, S. A., & Pradhan, B. (2016). Assessment and comparison of combined bivariate and AHP models with logistic regression for landslide susceptibility mapping in the Chaharmahal-e-Bakhtiari Province, Iran. *Arabian Journal of Geosciences*, *9*(3), 1–15. doi:10.100712517-015-2258-9

Santangelo, M., Marchesini, I., Bucci, F., Cardinali, M., Fiorucci, F., & Guzzetti, F. (2015). An approach to reduce mapping errors in the production of landslide inventory maps. *Natural Hazards and Earth System Sciences*, *15*(9), 2111–2126. doi:10.5194/nhess-15-2111-2015

Shahabi, H., & Hashim, M. (2015). Landslide susceptibility mapping using GIS-based statistical models and Remote sensing data in tropical environment. *Scientific Reports*, *5*(1), 1–15. doi:10.1038rep09899 PMID:25898919

Sonker, I., Tripathi, J. N., & Singh, A. K. (2021). Landslide susceptibility zonation using geospatial technique and analytical hierarchy process in Sikkim Himalaya. *Quaternary Science Advances*, *4*, 100039. doi:10.1016/j.qsa.2021.100039

Taalab, K., Cheng, T., & Zhang, Y. (2018). Mapping landslide susceptibility and types using Random Forest. *Big Earth Data*, *2*(2), 159–178. doi:10.1080/20964471.2018.1472392

Tehrany, M. S., Pradhan, B., Mansor, S., & Ahmad, N. (2015). Flood susceptibility assessment using GIS-based support vector machine model with different kernel types. *Catena*, *125*, 91–101. doi:10.1016/j.catena.2014.10.017

Thanh, L. N., & de Smedt, F. (2012). Application of an analytical hierarchical process approach for landslide susceptibility mapping in A Luoi district, Thua Thien Hue Province, Vietnam. *Environmental Earth Sciences*, *66*(7), 1739–1752. doi:10.100712665-011-1397-x

Vakhshoori, V., & Zare, M. (2016). Landslide susceptibility mapping by comparing weight of evidence, fuzzy logic, and frequency ratio methods. *Geomatics, Natural Hazards & Risk*, *7*(5), 1731–1752. doi: 10.1080/19475705.2016.1144655 van Westen, C. J., Castellanos, E., & Kuriakose, S. L. (2008). Spatial data for landslide susceptibility, hazard, and vulnerability assessment: An overview. *Engineering Geology*, *102*(3–4), 112–131. doi:10.1016/j.enggeo.2008.03.010

Varnes, D. J. (1984). *Landslide hazard zonation: A review of principles and practice*. Academic Press.

Wieczorek, G. F. (1984). Preparing a detailed landslide-inventory map for hazard evaluation and reduction. *Environmental & Engineering Geoscience*, *21*(3), 337–342. doi:10.2113/gseegeosci.xxi.3.337

Yalcin, A. (2008). GIS-based landslide susceptibility mapping using analytical hierarchy process and bivariate statistics in Ardesen (Turkey): Comparisons of results and confirmations. *Catena*, *72*(1), 1–12. doi:10.1016/j.catena.2007.01.003

Yesilnacar, E., & Topal, T. (2005). Landslide susceptibility mapping: A comparison of logistic regression and neural networks methods in a medium scale study, Hendek region (Turkey). *Engineering Geology*, *79*(3–4), 251–266. doi:10.1016/j.enggeo.2005.02.002

Yilmaz, I. (2009). Landslide susceptibility mapping using frequency ratio, logistic regression, artificial neural networks and their comparison: A case study from Kat landslides (Tokat—Turkey). *Computers & Geosciences*, *35*(6), 1125–1138. doi:10.1016/j.cageo.2008.08.007

Yu, H.-S., & Huang, Z.-Y. (2006). Intraslope basin, seismic facies and sedimentary processes in the Kaoping Slope, offshore southwestern Taiwan. *Terrestrial, 4*.

Zêzere, J. L., Pereira, S., Melo, R., Oliveira, S. C., & Garcia, R. A. C. (2017). Mapping landslide susceptibility using data-driven methods. *The Science of the Total Environment*, *589*, 250–267. doi:10.1016/j.scitotenv.2017.02.188 PMID:28262363

Zhu, A.-X., Wang, R., Qiao, J., Qin, C.-Z., Chen, Y., Liu, J., Du, F., Lin, Y., & Zhu, T. (2014). An expert knowledge-based approach to landslide susceptibility mapping using GIS and fuzzy logic. *Geomorphology*, *214*, 128–138. doi:10.1016/j.geomorph.2014.02.003

Zimmermann, M., Bichsel, M., & Kienholz, H. (1986). Mountain hazards mapping in the Khumbu Himal, Nepal. *Mountain Research and Development*, *6*(1), 29–40. doi:10.2307/3673338

Chapter 9
Photographic Image Properties and Their Correlation With the Quality Requirements for Aerial Mapping

Dainora Jankauskienė
Klaipeda State University of Applied Sciences, Lithuania

Lina Kuklienė
Klaipeda State University of Applied Sciences, Lithuania

Indrius Kuklys
Klaipeda State University of Applied Sciences, Lithuania

Birutė Ruzgienė
Klaipeda State University of Applied Sciences, Lithuania

Sérgio Lousada
https://orcid.org/0000-0002-8429-2164
University of Madeira, Portugal

ABSTRACT

Recently, cartographic products obtained by the use of photogrammetric methods have been increasingly applied in digital public spaces. The aim of the research was to analyse the properties of photographic images in relation to the developed technical specification for aerial photography project realization to obtain the most efficient, accurate, and high quality aerial cartographic products. Based on photogrammetric data from two aerial photography projects, where the terrain was aerial photographed from different types of unmanned aerial vehicles (TRIMBLE UX5, DJI MATRICE 600 PRO) and cameras (Zenmuse X5, Nex-5T), the properties of the photographic images were analysed, and their correlation with the accuracy requirements for the aerial mapping processes is presented. In order to fulfil accuracy requirements for terrain mapping by the use of UAV-photogrammetry technology, a ground sample distance (GSD) of 5-20 cm is recommended when aerial photography mission is executed from a low-flying vehicle.

DOI: 10.4018/978-1-6684-6449-6.ch009

INTRODUCTION

Recently, cartographic products obtained by the use of photogrammetric methods are increasingly applied in digital public spaces. Many photographic images stored and employed in GIS are orthophotos. Orthophotography is a geometrically corrected (rectified), undistorted photographic image of an area with a constant scale over the entire area. The digital orthophoto map is linked to the geographic and reference coordinate system, so the positions of each image element (pixel) of the photographic image can be determined (Förstner & Wrobel, 2016; Konecny, 2003; Kraus, 1997; Kraus, 2000; Linder, 2009; Wolf et al., 2014).

The technology of UAV-Photogrammetry is becoming more and more popular using to obtain photographic images, when the area or other object is photographed with a digital camera integrated in an Unmanned Aerial Vehicle (UAV) flying at a low altitude, and the photographic images are processed with special software. A light flying machine with a computerized aerial photography system and automated flight control is an effective tool for collecting photogrammetric data. The aerial mapping system consists of an Unmanned Aerial Vehicle (also called a Drone) with flight control equipment, various types of cameras, a global positioning system (GPNS), a laser scanner and a software package for processing photographic images (Armenakis & Patias, 2019; Černiauskas & Bručas, 2014; Ghilani & Wolf, 2012; Kraus, 2007; Ruzgienė, Aksamitauskas, Daugėla et al, 2015; Ruzgienė & Alekniené, 2004; Xiang et al., 2019).

UAV-Photogrammetry technology is applied for mapping of various objects: surfaces covered by leaf of trees, narrow linear objects (road network, sea coast, etc.), high-rise buildings, cultural heritage, cadastral areas, fixing power grid lines, etc. The technology of aerial photography from a low height and the processing of the obtained photographic images is increasingly changing the traditional methods of geodetic and photogrammetric measurements used for mapping. By the use of UAV-Photogrammetry technology photogrammetric data are obtained in real time, cartographic products meets the required accuracy and can be created quickly, is possible easy to get aerial photographs of inaccessible and complicated terrain, and the cost of aerial cartographic works is not high and are decreasing (Eisenbeiss, 2009; Haala et al., 2011; Neitzel & Klonowski, 2011; Ruzgienė et al., 2020; Systems–Unmanned Aerial Photography, n.d.).

Using UAV-Photogrammetry technology, spatial surface models are created and digital orthophoto maps are generated. The resulting products can be presented to users in an interactive way. The elevation models (DEM) and surface models (DSM) can be delivered in various forms of visualization. The innovative software system Pix4Dmapper is increasingly used to process photographic images, which combines the technologies of computer vision, geoinformation systems and traditional photogrammetry and effectively facilitates the user's work. This software is a digital photogrammetric workstation that generates aerial cartographic products and models the surface of objects. The software system can process thousands of photographic images obtained from close-range photography, drones and aircraft. Accurate georeferenced 2D/3D surface models and point clouds are created (Pix4D Support, n.d.; Rock et al., 2011; Ruzgienė, Berteška, Gečytė et al, 2015; Watts et al., 2012).

The quality of the cartographic products obtained by photogrammetry depends on the properly prepared technical specification of the aerial mapping works, the conditions for the implementation of the aerial photography project and the efficiency of the photogrammetric processes. Procedures for the application of photographic data to create maps are regulated by photogrammetric requirements (Cartography, 2020a; Cartography, 2020b; McGlone, 2004; Rudinskas, 2011; Torge & Müller, 2012; Zhiping et al., 2014).

As the use of low-flying small unmanned aerial vehicles to collect photogrammetric data is becoming increasingly popular, the need to study the properties of photographic images and their correlation with the preparation of technical specifications for aerial photogrammetric projects, in order to obtain the most efficient, accurate and high-quality aerial cartographic products, becomes more and more relevant (Gawryluk & Krawczyk, 2021; Nex & Remondino, 2014; Toffanin, 2019).

The highlighted problem is to find out the best solution for aerial mapping when photogrammetric data are obtained from flight at a low altitude regarding correlation with technical specification and images properties. The purpose of the research is to analyse the properties of photographic images in connection with the accuracy requirements of aerial mapping works and the preparation of aerial photography project specifications. The object of the research is photogrammetric data obtained during aerial photography of the area from different types of drones and cameras.

RESEARCH METHODOLOGY AND DATA OF TECHNICAL SPECIFICATION FOR AERIAL CARTOGRAPHY WORKS

Research workflow includes the following main stages:

1. After analysing the types of objects to be mapped, the characteristics of technical tools, the requirements for the realization of flights, the parameters of photographic images, the specification data for the preparation of the project of aerial photography works are presented.
2. Determining the parameters of aerial photography tools based on their technical documentation.
3. The properties of aerial photographs obtained by photographing with two types of cameras from two different types of drones are determined based on the provided data obtained from the photographic image *jpg *Properties* and the software *IrfanView*.
4. The correlation between the technical parameters of the photographic equipment, the resolution of the photographic images, the scale and the accuracy of the aerial cartography products are analysed.

At the initial stage of specification preparation, the aerial photography object is characterized, the terrain is described and the object type is indicated. The following types of mapping objects could be distinguished: general – the area for taking photography is not forested, lakes and fields of cultivated land cover a small part of the territory; complex – most of the territory is covered by forest, dense vegetation and fields of cultivated land, the terrain is flat; special – aerial photographs will be taken of areas covered by snow or sand, water surfaces (seas, lakes, rivers, etc.); linear – road network, river banks, etc.; built-up areas; various spatial (3D) structures; high buildings – towers, wind turbines, etc.; interior of buildings.

The parameters of the technical means (drone and camera) are needed to determine the design data of aerial photography. The drone usually flies at a low speed (about 30-60 km/h) to obtain photographic images, and the flight height can reach from 30 to 750 m (and more). The main characteristics of the cameras integrated into the drone are: focal length, camera resolution (number of pixels - frame format), sensor's and pixel size.

Realizing the flight (aerial photography) one of main parameters is the size of the photographic image element in the area – Ground Sample Distance. The determination of GSD is based on the required accuracy of aerial cartographic products obtained by processing UAV photographic images. The following general rule is applied – the GSD must be twice less than the required accuracy for determination of the planimetric coordinates and altitudes, e.g. if the accuracy of the planimetric coordinates (X and Y) is required to be 5 cm, then the GSD = 2,5 cm. The size of the image element in the area (GSD) is calculated according to the Formula (Ruzgienė et al., 2017):

$$GSD = pix\frac{H}{c} \tag{1}$$

where: pix – the size of the camera sensor's pixel, µm; H – height of aerial photography, m; c – focal length of the camera, mm.

The flight height of the UAV is determined according to the Formula (1), in a case when GSD is specified by the customer.

Other parameter for realizing aerial photography is overlap of images and it is recommended a 70-80%, taking into account that the UAV system during aerial photography is unstable and dependent on wind gusts.

In order to achieve the required images overlaps, it is necessary to know the time interval between exposures (photographic image recording time interval) (t), which depends on the flight speed of the UAV (v), the size of the photographic image element in the area (GSD) and the resolution of the camera (frame format in pixels). The time interval (t) between two exposures is calculated according to the Formula (Pix4D Support, n.d.):

$$t(s) = \frac{B}{v} \tag{2}$$

where: B – aerial photography base (distance between two aerial camera positions), m; v – flight speed, m/s.

The aerial photography base is calculated according to the Formula:

$$B = D - P, \tag{3}$$

where: $D(m) = \frac{d(px) \times GSD}{100}$, $P = p \times D$, $d(px)$ – width/height of the photographic image (pixels); p – percentage of photographic images that are overlapped, e.g., if 75% overlapping, then $p = 0{,}75$.

Considering to Formulas (2) and (3) the time interval between exposures is calculated:

$$t = \frac{d(px) \times \frac{GSD}{100} \times (1 - p)}{v}. \tag{4}$$

Aerial photographs resolution/pixel size is one of the main parameters of a photographic image and is given in pixels/ dots per inch (dpi). The size of one pixel in an image is calculated as 2,54 (cm) / resolution (dpi).

Aerial photo frame size – the dimensions of one image frame (width and height) are calculated by multiplying the pixel size by the camera resolution (aerial photo frame size, pixels). Aerial photo frame size in area = GSD × image frame size (pixels).

When evaluating the detail or quality of an aerial photographs, the scale could be determined as: pixel size in image divided to pixel size in area or photographs frame size divided to photographs frame size in area.

The relation between aerial photograph scale and accuracy of aerial cartographic products in classical photogrammetry is 1:2-5, i.e. the scale of the aerial photograph is reduced from 2 to 5 times (Ruzgienė, 2008).

Approximate planimetric accuracy of aerial cartographic products is correlated with scale accuracy: 0,1 mm or 0,09 mm multiplied to scale denominator. Trimble Corporation declares the following criteria to evaluate the accuracy of digital aerial mapping products: average and maximum errors of planimetric coordinates – 1,0×GSD and 1,6×GSD; altitudes – 1,6×GSD and 2,5×GSD. For example, if GSD = 2,4 cm, then the average and maximum errors in the X and Y directions are: 2,4 cm and 3,8 cm; in the Z direction – 3,8 cm and 6,0 cm.

COMPARATIVE ANALYSIS OF THE PROPERTIES OF PHOTOGRAPHIC IMAGES

In order to obtain aerial photographs for their properties study, aerial photography of the experimental objects was realized with cameras *Sony Nex-5T* and *DJI Zenmuse X5* from fixed-wing drone *TRIMBLE UX5* and helicopter *DJI MATRICE 600 PRO*. The main characteristics of these unmanned aerial vehicles and cameras are presented in Tables 1 and 2.

Table 1. Main features of unmanned aerial vehicles

TRIMBLE UX5	DJI MATRICE 600 PRO
Wings length – 1,0 m, weight 2,5 kg.	Weight with six TB47S batteries – 9,5 kg.
Speed – up to 80 km/h.	Max. speed – 65 km/h.
Can stay in flight for up to 50 minutes in low wind conditions.	Wind resistant with wind speed up to 8 m/s.
Max. flight distance is about 60 km.	The duration of the flight is 32 minutes.
Max. communication and control distance – up to 5 km.	Flight control system – A3 Pro.
Ascent height – 75-750 m.	Can rise up to 2500 m above sea level.
Pixel size (*GSD*) – from 2,4 cm to 24 cm.	Works at temperatures between 10°C and 40°C.
UAV control – fully automatic, semi-automatic and manual.	Hovering accuracy: vertical ±0,5 m; horizontal ±1,5 m.
Trimble Aerial Imaging flight planning software allows to optimally manage flight.	

Source: (MATRICE 600 PRO, n.d.; Trimble UX5 Unmanned Aircraft System, n.d.)

Table 2. Main characteristics of the cameras Sony Nex-5T and DJI Zenmuse X5 integrated in the UAV

Camera Settings	Sony Nex-5T	DJI Zenmuse X5
Focal length, mm	15	15
Camera resolution, pixels	16 MP	16 MP
Sensor size: width and height, pixels width and height, mm	4912 × 3264 23,4 × 15,6	4608 × 3456 17,3 × 13,0
Sensor pixel size, µm	4,8	3,7

Source: (Sony Nex-5T Manual, n.d.; Trimble UX5 Aerial Imaging Solution, n.d.; Zenmuse X5, n.d.)

Figure 1. Overlapped photographic images acquired with the camera Nex-5T: Linear object

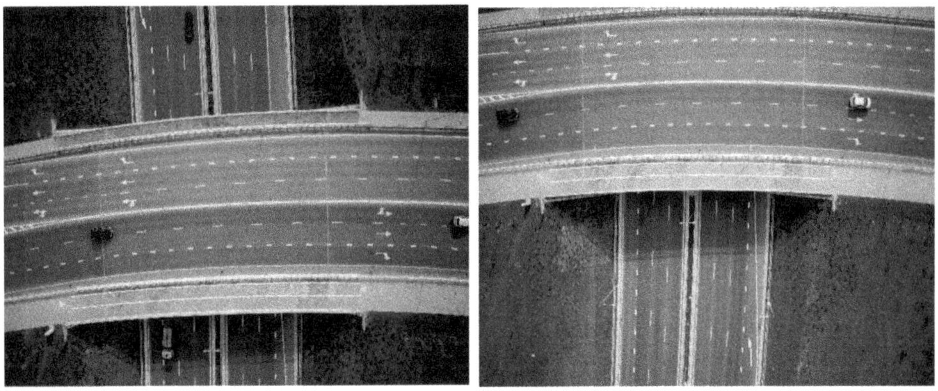

Figure 2. Overlapped photographic images acquired with the camera Zenmuse X5: General area

The example of overlapping aerial photographs taken over different type of experimental territory from low-flying unmanned aerial vehicles with camera *Sony Nex-5T* and *DJI Zenmuse X5* are presented in Figures 1 and 2.

Extractions of the main properties of acquired photographs are presented in Figure 3.

Figure 3. Main parameters of photographic images: 1) acquired with the camera Nex-5T – frame size (width and height), resolution (dpi); 2) acquired with the camera Zenmuse X5 – frame width and height, resolution (dpi), geodetic coordinates, flight altitude

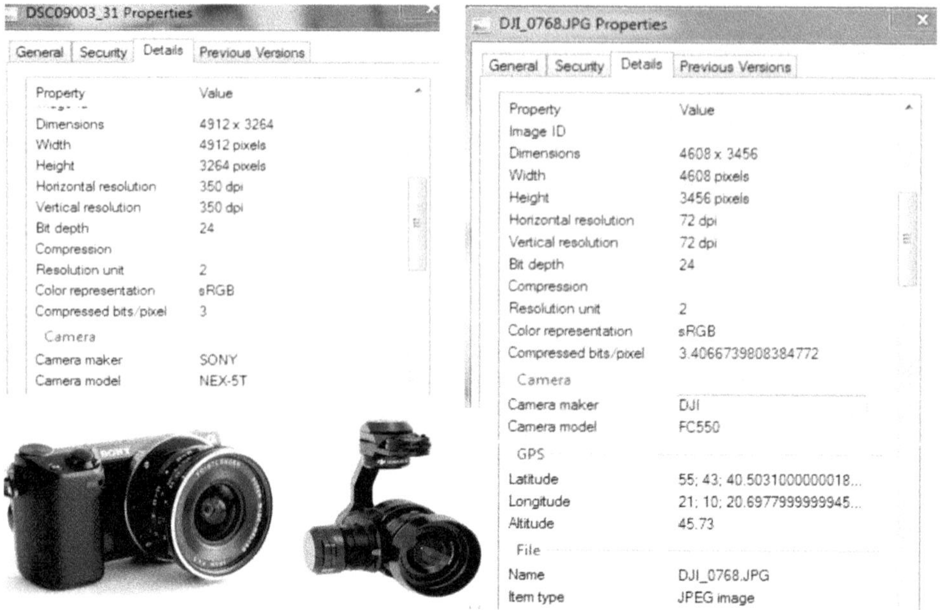

Resolution of photographic images is characterized by the number of pixels (picture elements) per inch (one inch = 2,54 cm), which is, respectively, the size of one pixel at a resolution of 350 and 72 dpi (see Figure 3), are 0,0726 and 0,3528 mm. The characteristics of photographic images and flight realization data are presented in Table 3.

Table 3. Basic parameters of photographic image and flight realisation.

Features of Photographic Images and Flight Data	Camera Sony Nex-5T (UAV-UX5 TRIMBLE)	Camera DJI Zenmuse X5 (UAV-MATRICE 600 PRO)
Aerial photograph frame size, pixels	4912 × 3264	4608 × 3456
Resolution, dpi	350	72
Pixel size, mm	0,0726	0,3528
Photo frame size, mm: width, height	356,47 (0,0726×4912) 236,87 (0,0726×3264)	1626 (0,3528×4608) 1219 (0,3528×3456)
Overlap of aerial photographs, %	80	80
Flight speed, km/h	80	50
Flight height, m	75	45
Pixel size in the area (GSD), mm	24 (0,0048×75000 / 15)	11 (0,0037×45000 / 15)
Aerial photo frame size in area, m	117,89 (0,024×4912) 78,34 (0,024×3264)	50,69 (0,011×4608 38,02 (0,011×3456)
Aerial photograph scale (GSD/pixel size)	1:330 (24 / 0,0726)	1:30 (11 / 0,3528)

Table 4. The properties of photographic images: IrfanView data

IrfanView - Image properties

File name:	Figure_1_(1)_DSC09280.JPG
Directory:	D:\Birute\Klaipeda\Straipsniai_konferen_Recenz
Full path:	D:\Birute\Klaipeda\Straipsniai_konferen_Recenz
Compression:	JPEG, quality: 94, subsampling ON (2x1)
Resolution:	350 x 350 DPI Change
Original size:	4912 x 3264 Pixels (16.03 MPixels) (3:2)
Current size:	4912 x 3264 Pixels (16.03 MPixels) (3:2)
Print size (from DPI):	35.6 x 23.7 cm; 14.03 x 9.33 inches
Original colors:	16,7 Million (24 BitsPerPixel)
Current colors:	16,7 Million (24 BitsPerPixel)
Number of unique colors:	141988 ✓ Auto count
Disk size:	5.38 MB (5.636.096 Bytes)
Current memory size:	45.87 MB (48.098.344 Bytes)
Current directory index:	1 / 8
File date/time:	9/25/2014 / 14:39:40
Loaded in:	250 milliseconds

EXIF info* OK Comment IPTC info

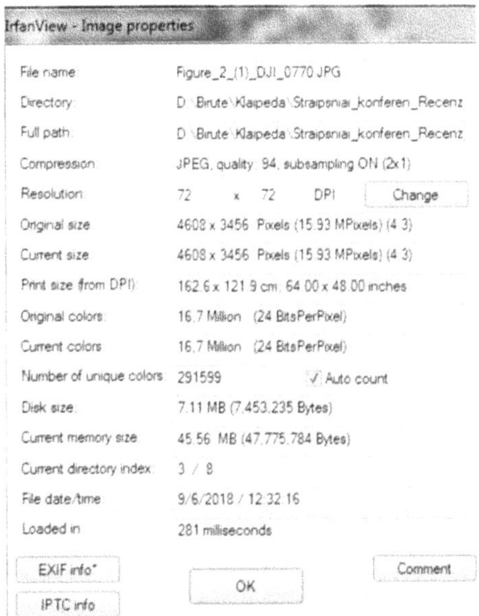

IrfanView - Image properties

File name:	Figure_2_(1)_DJI_0770.JPG
Directory:	D:\Birute\Klaipeda\Straipsniai_konferen_Recenz
Full path:	D:\Birute\Klaipeda\Straipsniai_konferen_Recenz
Compression:	JPEG, quality: 94, subsampling ON (2x1)
Resolution:	72 x 72 DPI Change
Original size:	4608 x 3456 Pixels (15.93 MPixels) (4:3)
Current size:	4608 x 3456 Pixels (15.93 MPixels) (4:3)
Print size (from DPI):	162.6 x 121.9 cm; 64.00 x 48.00 inches
Original colors:	16,7 Million (24 BitsPerPixel)
Current colors:	16,7 Million (24 BitsPerPixel)
Number of unique colors:	291599 ✓ Auto count
Disk size:	7.11 MB (7.453.235 Bytes)
Current memory size:	45.56 MB (47.775.784 Bytes)
Current directory index:	3 / 8
File date/time:	9/6/2018 / 12:32:16
Loaded in:	281 milliseconds

EXIF info* OK Comment IPTC info

Using the *IrfanView* software (IrfanView Graphic Viewer, n.d.), the properties of photographic images are presented in Table 4.

For aerial photography with the *Sony Nex-5T* and *DJI Zenmuse X5* cameras, the correlation between the pixel size of the photographic image in the area (*GSD*) and the flight height (*H*), the frame size and the scale of the aerial photograph is presented in Table 5.

The approximate accuracy of the aerial cartographic product is related with a scale accuracy of 0,1 mm × M. In order to evaluate the accuracy of digital aerial cartographic products, it is generally accepted that the accuracy of planimetric coordinates and heights is obtain the GSD multiplying by two.

Table 5. The correlation of photographic pixel size in terrain (GSD) with flight altitude, frame size, and aerial image scale

GSD, cm	Flight height, m		Aerial Photographs/Frame Size in the Area, m		Scale of Aerial Photographs	
	NEX-5T	ZENMUSE X5	NEX-5T	ZENMUSE X5	NEX-5T	ZENMUSE X5
1,1	34	44	54,0 × 35,9	50,7 × 38,0	1:150	1:30
2,0	62	80	98,2 × 65,3	92,1 × 46,1	1:280	1:57
2,4	75	97	117,8 × 78,3	110,6 × 83,0	1:330	1:68
3,0	94	122	147,3 × 97,9	138,2 × 103,7	1:410	1:85
5,0	156	202	245,6 × 163,2	230,4 × 172,8	1:690	1:140
10,0	312	405	491,2 × 326,4	460,8 × 345,6	1:1380	1:280
20,0	625	811	982,4 × 652,8	921,6 × 691,2	1:2750	1:570
24,0	750	973	1178,9 × 783,4	1105,9 × 829,4	1:3310	1:680

CONCLUSION

Taking into account the characteristics of photographic images and the conditions for the realization of UAV flights, the following basic data and design parameters of aerial photography are included in the specification of aerial mapping of the area: characteristics of technical equipment (means of flying and cameras), the size of the image pixel in the area, the size of the of the aerial photographs overlap, flight height, aerial photography base and time interval between exposures, aerial photographs resolution and frame size (in images and terrain), accuracy of aerial cartographic products.

Photographic images of experimental objects (linear and general area), obtained by taking photography of the area from different heights and with different cameras, are characterized by the following most important flight realization parameters: camera sensor pixel size, aerial photo frame size, resolution and pixel size in the photo and in the area (see Tables 2 and 3).

After analysing the correlations of the pixel size in the area (GSD) of the photographic image with flight height, frame size and aerial photo scale, it can be stated that the photo scale is most influenced by the photo resolution. The flight height for aerial photography with the *Zenmuse X5* camera is 1,3 times higher than with the *Nex-5T* camera due to the different sensor pixel size (3,8 and 4,8 μm). The scale of the aerial photographs varies by a factor of about 5 dues to the different resolution of the photographic images (*Nex-5T* – 350 dpi (0,0726 mm)), *Zenmuse X5* – 72 dpi (0,3528 mm)) (see Table 5).

Taking into account the approximate accuracy evaluation based on scale accuracy and according to the declared generalized evaluation criterion for the accuracy of digital aerial cartography products (2×GSD), the results of the *Nex-5T* aerial photo camera meet the accuracy requirements, while the *Zenmuse X5* camera is redundant, due to the size of the resolution of photographic images – 72 dpi.

When aerial photography of territories from a low-flying vehicle, the recommended pixel size for terrain mapping purposes is 5 to 20 cm.

REFERENCES

Armenakis, C., & Patias, P. (2019). Unmanned vehicle systems for geomatics: towards robotic mapping. Whittles Publishing.

Cartography. (2020a). http://www.gsdi. org/docs 2004/ Cookbook/cookbookV2.0.pdf

Cartography. (2020b). https://www.britannica.com/science/cartography

Černiauskas, E., & Bručas, D. (2014). Daugiasraigčių sraigtaspanių naudojimo stebėjimo užduotims atlikti tyrimas. *Aviacijos technologijos*, *2*(1), 53-58.

Eisenbeiss, H. (2009). *UAV photogrammetry* [Dissertation]. Federal Institute of Technology (ETH), Institute of Geodesy and Photogrammetry.

Förstner, W., & Wrobel, B. P. (2016). *Photogrammetric Computer Vision: Statistics, Geometry, Orientation and Reconstruction*. Springer. doi:10.1007/978-3-319-11550-4

Gawryluk, D., & Krawczyk, D. (2021). *The future of the city*. Białystok: Publishing House of Bialystok University of Technology. doi:10.24427/978-83-66391-62-8_7

Ghilani, Ch. D., & Wolf, P. R. (2012). Elementary surweying: an introduction to geomatics. Academic Press.

Haala, N., Cramer, M., Weimer, F., & Trittler, M. (2011). Performance Test on UAV-Based Photogrammetric Data Collection. *International Archives of the Photogrammetry, Remote Sensing and Spatial Information Sciences, 38*(1), 7-12.

IrfanView Graphic Viewer. (n.d.). https://www.irfanview.com/

Konecny, G. (2003). *Geoinformation: Remote Sensing, Photogrammetry and Geographic information system*. Taylor and Francis. doi:10.4324/9780203469644

Kraus, K. (1997). Photogrammetry: Vol. 2. *Advanced methods and application*. Dümmler.

Kraus, K. (2000). Photogrammetry: Vol. 1. *Fundamentals and Standard Processes*. Dümmler.

Kraus, K. (2007). *Photogrammetry: Geometry from Images and Laser Scans*. Walter de Gruyter. doi:10.1515/9783110892871

Linder, W. (2009). Digital Photogrammetry. A practical course. Springer-Verlag. doi:10.1007/978-3-540-92725-9

MATRICE 600 PRO. (n.d.). https://www.dji.com/lt/matrice600-pro

McGlone, J. C. (2004). *Manual of Photogrammetry*. American Society for Photogrammetry and Remote Sensing.

Neitzel, F., & Klonowski, J. (2011). Mobile Mapping with Low-Cost UAV System. *International Archives of the Photogrammetry, Remote Sensing and Spatial Information Sciences, 38*(1), 1-6.

Nex, F., & Remondino, F. (2014). UAV for 3D mapping applications: A review. *Appl Geomat, 6*(1), 1–15. doi:10.100712518-013-0120-x

Pix4D Support. (n.d.). https://support.pix4d.com/hc/en-us/articles/204272989-Offline-Getting-Started-and-Manual-pdf-#gsc.tab=0

Rock, G., Ries, J. B., & Udelhoven, T. (2011). Sensitivity Analysis of UAV-Photogrammetry for Creating Digital Elevation Models (DEM). *International Archives of the Photogrammetry, Remote Sensing and Spatial Information Sciences, 38*(1), 1-5.

Rudinskas, D. (2011). *Bepiločių orlaivių skrydžio parametrų matavimų duomenų perdavimo saugos metodikos sukūrimas. Daktaro disertacija. VGTU*. Technika. doi:10.20334/1975-M

Ruzgienė, B. (2008). *Fotogrametrija. VGTU*. Technika. doi:10.3846/987-S

Ruzgienė, B., Aksamitauskas, V. Č., Daugėla, I., Prokopimas, Š., Puodžiukas, V., & Rekus, D. (2015). UAV photogrammetry for road surface modelling. *The Baltic Journal of Road and Bridge Engineering, 10*(2), 151–158.

Ruzgienė, B., & Alekniene, E. (2004). *Elementary surveying: study guide*. Technika.

Ruzgienė, B., Berteška, T., Gečytė, S., Jakubauskienė, E., & Aksamitauskas, V. Č. (2015). The surface modelling based on UAV Photogrammetry and qualitative estimation. In Measurement. Elsevier Ltd.

Ruzgienė, B., Jankauskienė, D., Kuklys, I., Kuklienė, L., & Danielienė, J. (2020). KVK studies: scanning and 3D modelling using remote sensing a nd UAV technologies. *International Journal of Grid and Distributed Computing, Science and Engineering Research Support Society, 13*(2), 361-365.

Ruzgienė, B., Kuklienė, L., Dainora Jankauskienė, D., Kuklys, I., & Vrubliauskienė, R. (2017). Fotogrametriniai reikalavimai, keliami aerokartografavimui: UAV-Fotogrametrija. *Inžinerinės ir edukacinės technologijos, 2*, 21-28.

Sony Nex-5T Manual. (n.d.). https://camerausermanual.net/sony-nex-5t-manual/

Systems–Unmanned Aerial Photography. (n.d.). *AUVSI*. http://www.uavsystems.com.au/

Toffanin. (2019). *Open Drone Map: The Missing Guide: A Practical Guide to Drone Mapping*. Independently Published.

Torge, W., & Müller, J. (2012). *Geodesy*. de Gruyter.

Trimble UX5 Aerial Imaging Solution. (n.d.). *Camera specification*. Available from Internet: https://help.trimblegeospatial.com/TALegacy/Help%20Files/AI_2_2_3/en/UX5%20Aerial%20Imaging%20User%20Guide.pdf

Trimble UX5 Unmanned Aircraft System. (n.d.). Available from Internet: https://www.trimble.com/Industries/Mining/Aerial_Surveying.aspx

Watts, A. C., Ambrosla, V. G., & Hinkey, E. A. (2012). Unmanned Aircraft Systems in Remote Sensing and Scientific. Remote Sensing, 4(6), 1671-1692.

Wolf, P., DeWitt, B., & Wilk, B. (2014). *Elements of Photogrammetry with Application in GIS* (4th ed.). McGraw-Hill Education.

Xiang, T., Xia, G., & Zhang, L. (2019). Mini-Unmanned Aerial Vehicle-Based Remote Sensing: Techniques, applications, and prospects. *IEEE Geoscience and Remote Sensing Magazine*, *7*(3), 29–63. doi:10.1109/MGRS.2019.2918840

Zenmuse X5. (n.d.). *Aerial Imaging Evolved*. Available from Internet: https://www.dji.com/lt/zenmuse-x5/info

Zhiping, L., Shubo, Q., & Yunying, Q. (2014). *Geodesy: Introduction to geodetic datum and geodetic systems*. Springer.

Chapter 10
The Preliminary Researches of Pull-Out of GFRP Rods From Mineral Coarse-Grained Soils

Grzegorz Mieczysław Straż
https://orcid.org/0000-0001-5847-0441
Rzeszow University of Technology, Poland

ABSTRACT

The presented research is part of the work "The Ground Reinforcement System With GFRP Composite Bars With Fiber Optic Monitoring," the purpose of which was to check the suitability of the use of GFRP (glass fiber reinforced polymer) rods for reinforcing selected mineral soils. The preliminary tests included obtaining research material, determining geotechnical parameters, and carrying out tests of pulling GFRP rods out of the ground during microscale model tests. For the purpose of the experiment, a prototype of the pulling device was designed by modifying the direct shear device. The soil samples for strength tests were prepared assuming two leading parameters: water content and normal stress. The program was developed to determine the amount of force pulling the GFRP rod out of coarse-grained soil under the action of a vertical load of various sizes. Ultimately, the presented results of the preliminary tests will be used as the basis for planning tests with fiber-optic monitoring on real-scale models.

INTRODUCTION

The rapidly developing road and bridge infrastructure places more and more demands on designers and contractors. Very often, the biggest problem to overcome is the weak quality of the subsoil, which requires replacement or strengthening. Therefore, despite the wide range of solutions used, new, more effective and economical solutions are still being sought. The results presented in the chapter are part of the preliminary research and analyzes under the project "System for strengthening the soil with GFRP composite bars with fiber optic monitoring", performed at the Rzeszów University of Technology in Poland, aimed at introducing an innovative method for soil reinforcement. The project concerns the innovative use of GFRP (Glass Fiber Reinforced Polymer) composite rods to strengthen the subsoil, instead

DOI: 10.4018/978-1-6684-6449-6.ch010

of using geosynthetics or steel rods. Nowadays, GFRP rods are effectively used to reinforce concrete in engineering structures, therefore attempts have been made to protect the selected subsoil with them in order to improve its geotechnical parameters. The GFRP rods are lightweight, high strength and durability, connected in girds and placed in the soils creates an effective reinforcement of the substrate, minimizing settling. As part of the main research project selected rods will be equipped with optical fibers, which will allow continuous monitoring of the ground, providing information on the behavior of the structure. This is important for high-rise buildings, embankments, landslides. The project includes the perform of works to identify system components (GFRP rods, various soils), check their cooperation in the strengthening system (friction resistance, wedging effect) and assess efficiency fiber optic monitoring of the subsoil.

The research project in question is aimed at determining mainly the pull-out strength of bars and geogrid made of GFRP bars from coarse (non-cohesive) soil in laboratory conditions using a large-size model, which requires very careful preparation of both the necessary equipment and the research program itself. The described target tests were to be carried out in a box generating a prepared soil sample with dimensions (B) 1000 mm x (L) 1500 mm x (H) 700 mm. A very important issue is the very process of preparing the soil material, which should be homogeneous (humidity) and equally compacted in every place of the prepared sample. The problem that should not be underestimated is also the method of applying the loads, because in the case of a crate designed for real model tests, the expected stress in the soil sample is achieved by loading with an appropriate number of road slabs of known weight, which requires the use of heavy equipment and precise application. The presented possibility of using GFRP rods can be an alternative to geosynthetic materials commonly used in geotechnics, which, despite their obvious advantages, also have weaknesses: relatively high price, low stiffness and resistance to damage, which is of particular importance at the stage of installation in the ground and requires a high technological regime. GFRP rods, due to their stiffness, strength, durability and favorable price, can eliminate these problems, primarily in situations where strength and stiffness of the reinforced substrate are important, although the scope of their application will not be identical to that of alternative geosynthetic materials. This solution is also economically advantageous.

Due to the considerable advantages of GFRP bars compared to steel and other composites, their share in the application in the construction industry increases, which generates the need to conduct specialized research dedicated to specific market needs. The GFRP materials, including reinforcing bars, are characterized by properties that are very desirable in the construction industry, and the most important of them are (according to https://mrg-composites.com): resistance to various types of corrosion, incomparable to other materials, makes it possible to use them indoors and outdoors, they are very durable and do not degrade in the presence of chemicals, reinforcing bars are almost twice as strong as steel ones, weigh much less and are cheaper, they are also easy to install due to the ease of cut and bent as specific requirements. In addition, the relatively low weight of the GFRP rods facilitates transport, storage and installation at the construction site, often does not require heavy equipment, it makes the construction process faster and more efficient which has its significant economic dimension. Therefore, a wide range of research and development works are carried out, the aim of which is to determine the parameters of the bars themselves, optimize their composition and structure, as well as the conditions of cooperation with the substrate, which may be concrete, soil or even wood.

The GFRP bars are most often used for the reinforcement of structural concrete elements and therefore the amount of published work on this topic is significant, among others (Bencardin et al. 2016; Bywalski et al. 2016; Kara & Ashour, 2015; Krall & Polak, 2019; Lau & Pam, 2010; Leung & Balendran, 2003; Moon et al. 2005 & 2007; Mufti et al., 2005; Naani, 1993; Shehata et al., 2000; Protchenko et al., 2016;

Stankiewicz, 2011; Victor & Wang, 2002), and even guides have been published to assist in design and implementation (ACI Committee 440., 2006; El-Game et al., 2007). Successively, GFRP reinforcing bars are becoming materials with an increasingly wide range of applications,, e.g. in bridge and road construction, reinforced concrete structures, foundations, base slabs and industrial floors, and even swimming pool and sanitary structures. Due to the exceptional resistance to both chemical and physical corrosion, GFRP materials are also ideal for use in seawater environments (Oskouei et al., 2018, Garcia-Espinel et al., 2015a, Garcia-Espinel et al., 2015b, Ashpiz et al., 2010).

In the geotechnical engineering aspect, they the GFRP rods have also found wide application (Balcony & Badawika, 2008; Achillides & Pilakoutas, 2004; Vijay & GangaRao & Krishnaswamy, 2005; Ferreira & Vieira. & Lopes, 2020) as well as the topic of application use of fibres for soil reinforcement (Asadollahia & Dabiri, 2017; Cristelo et al., 2015; Dean & Fretting, 1986; Gray & Ohashi, 1983). The GFRP rods can be used for soil reinforcement in the form of nails or anchors (Merwe & Hofmann, 2020), Lin et al., 2021, Cheng et al. 2009; Milligan & Tei, 1998), as a transmitting layer (Strabag, 2022), in cement-soil mixtures (Zhang et al., 2022, Zhang et al., 2020, Zhang & Zhu, 2019). There are also cases of use in the construction of foundation piles (Almallah et al., 2020) or even as bolts in the construction of tunnels (Li et al., 2023, Chen et al., 2008)

It is also understandable that geotechnical tests carried out on large-size real models, due to the exceptional difficulty of their preparation, time-consuming and the need to use heavy equipment in laboratory conditions, are performed individually and repeated a limited number of times. That is why research conducted on a micro scale is so important, as it allows testing the proposed solutions on a scale that is much easier to prepare the apparatus and the soil sample itself. Of course, using the available equipment (micro scale) it is very difficult to obtain a perfect reflection of the conditions for the scale of the real model, if only due to the limited availability of GFRP rod diameters, but it is a solution that greatly simplifies the preparation of tests and reveals technical problems and problems already at the stage of preliminary tests. eliminates unforeseen situations. In accordance with the adopted research program, preliminary studies were carried out on a micro scale, and their results were to serve as one of the reference points in planning the main studies.

It should be emphasized that the issues presented in this paper concern only auxiliary, preliminary studies on a micro scale, which are an independent part of an extensive and much more advanced project.

MATERIALS AND METHODS

The Adopted Methodology

The methodology cumulated from two stages, including geotechnical investigation of the research material (Stage 1) and tests of pull-out GFRP rods from the soil (Stage 2), is schematically presented in Figure 1.

The preliminary research were divided into two main stages. The first stage included extensive geotechnical investigation of the research coarse-grained soil material (Medium Sand), and the second stage - testing the pull-out strength of rods from this soil. In order to implement a complex test program, it was necessary to design a device that would enable pull-out tests to be carried out. The problem was solved by modifying the classic direct shear apparatus, namely the box mechanism, which is described in detail later in the work. As a result, the amount of force pulling the GFRP bars out of the soil was determined at different values of normal stresses and moisture of the prepared soil samples.

Figure 1. The schematic of the adopted methodology

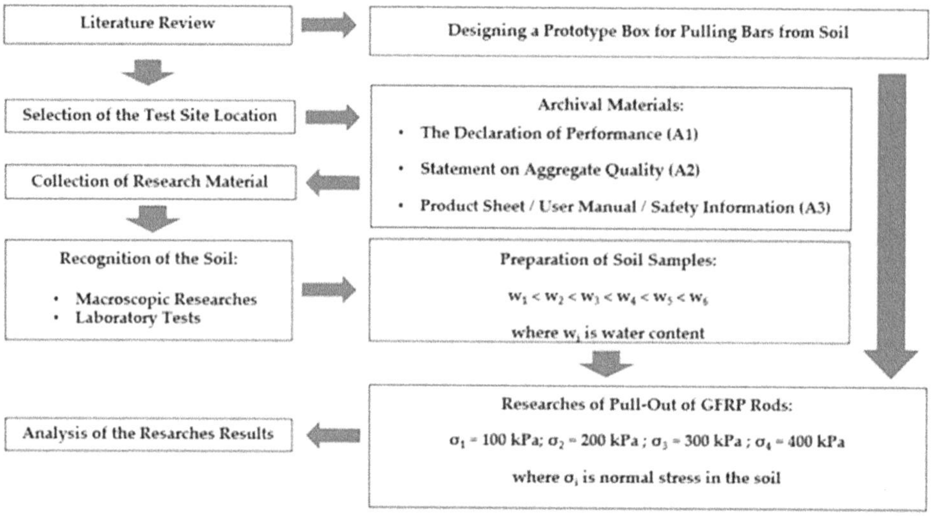

The Geotechnical Characteristics of the Research Material_

The information on the research geotechnical material was obtained from different sources: archival materials and own laboratory and field research. The archival materials were documents issued by the KRUSZGEO S.A. company (https://kruszgeo.com.pl): The Declaration of performance and the statement on aggregate quality. Initial identification and description of the research material were performed in accordance with the standard procedures of macroscopic analysis (EN ISO 14688-1: 2018). The results of macroscopic tests showed that the soil selected for the tests was a Medium Sand (MSa) of river origin, homogeneous, with no visible organic inclusions, wet or slightly moist, with a colour light brownish yellow.

The main geotechnical parameters were determined in the relevant standard guidelines:

- **Water Content:** Carried out in accordance with the guidelines of the standard EN ISO 17892-1: 2014,
- **Bulk Density:** Carried out in accordance with the guidelines of the standard EN ISO 17892-2: 2014,
- **Particle Size Distribution:** Carried out in accordance with the guidelines of the standard EN ISO 17892-4: 2016,
- **Coefficient of Permeability:** Carried out in accordance with the guidelines of the standard EN ISO 17892-11: 2019,
- **Oedometer Modulus:** Carried out in accordance with the guidelines of the standard EN ISO 17892-5: 2017,
- **Shear Strength:** Carried out in accordance with the guidelines of the standard EN ISO 17892-10: 2018,
- **Optimum Water Content:** Carried out in accordance with the guidelines of the Polish standard PN-B-04481: 1998,

- **Density Index:** Calculated in accordance with the guidelines of the standard EN ISO 14688-2: 2018.

The results of the research and the averaged values of the determined geotechnical parameters with the use of samples with natural moisture of the selected coarse-grained soil for the tests are presented in Table 1.

Table 1. The geotechnical parameters of research material

Parameter	Symbol/Term/Mean Values	
Natural Water Content	W (/%)	7.49
Bulk Density	ρ (t/m³)	1.53
Soil **Shape of Grading Curve** **Classification**	Medium Sand (MSa)	
	C_U (-)	3.11
	Poorly graded	
Density Index **Classification**	I_D (%)	39.69
	Medium dense	
Coefficient of Permeability **Classification**	k (m/s)	7.36E-05
	Good / Average	
Optimal Water Content	w_{OPT} (%)	6.65
Primary Eodometer Compressibility Modulus: 0-12.5 (kPa) 12.5-25 (kPa) 25-50 (kPa) 50-100 (kPa) 100-200 (kPa) 200-400 (kPa)	E_{oed} (MPa)	24.83 5.72 4.36 4.84 7.57 13.07
Secondary Eodometer Compressibility Modulus: 12.5-25 (kPa) 25-50 (kPa) 50-100 (kPa) 100-200 (kPa) 200-400 (kPa)	E_{ed} (MPa)	0.310 15.763 9.212 5.510 3.556
Shear Strength Parameter (Total Values): **Friction Angle** **Cohesion**	φ (°) c (kPa)	00 32.60

The Equipment to Pull-Out Test of GFRP Rods

Based on the analysis of available source materials, it was proposed to conduct research on pulling GFRP rods from coarse soil using the pull out method, which is successfully used in testing geosynthetic materials embedded in various soils, as documented by, among others: Jiang et al. (2021), Ferreira et al. (2020), Isik and Gurbuz (2020), Suksiripattanapong et al. (2013), Ferreira et al. (2015a, 2015b, 2015c), Abu-Farsakh et al. (2006), Milligan et al. (1988). The implementation of non-standard tests required the preparation of appropriate research equipment. After analyzing the existing options, the option was

Figure 2. The direct shear apparatus with a prototype box for testing the pull-out strength of GFRP rods from the soil
Source: Author

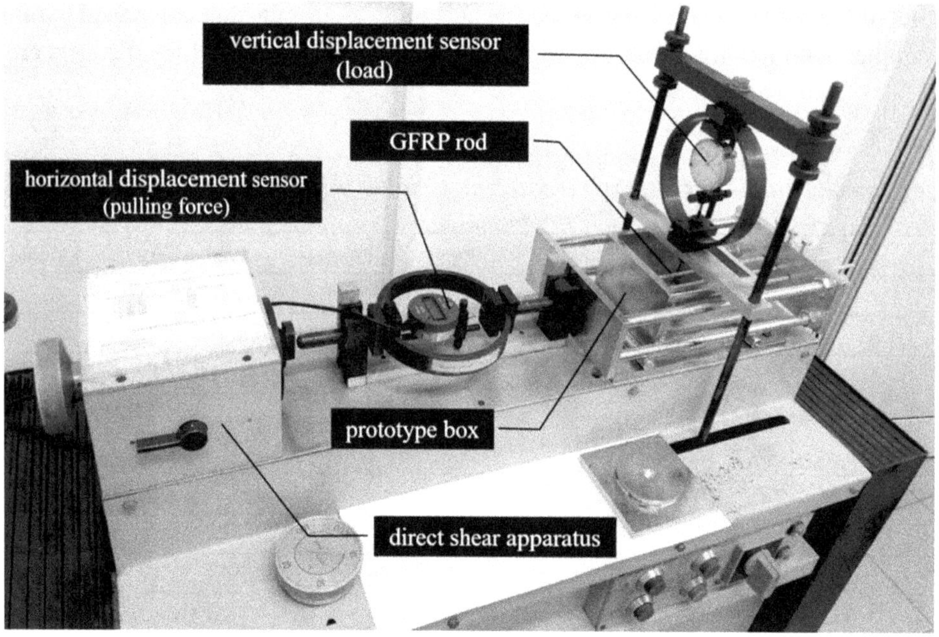

chosen to use the classical direct shear apparatus used to determine the shear strength of soils. However, a new, prototype test set was used, the design of which made it possible to carry out tests on pulling GFRP rods from the soil sample. The general view of the direct shear apparatus with the prototype test kit installed is shown and describe in Figure 2.

The researches set for the pull-out strength tests was designed of steel. It consisted of two independent parts: a soil sample box with dimensions of 120 mm x 120 mm x 90 mm, permanently attached to the body of the direct shear apparatus, and a movable head enabling the attachment of 3 GFRP rods with a diameter of 6 mm, provided by ComRebars (https://comrebars.eu). There were holes in the soil box through which the GFRP bars could move freely without touching the walls. On the other hand, Figure 3 shows the test set prepared for the pull-out test, with a coarse-grained soil (Medium Sand) sample subjected to a given load and a set of displacement sensors.

Before the actual tests were carried out, the prototype test set was calibrated, which consisted in determining the resistances of the bar drawing head moving on bearings in the form of steel shafts. The calibration tests did not show any significant resistances affecting the values of the pulling force.

Unfortunately, the presented test method has limitations, and the main one is the scale, i.e. the limitations of the maximum dimensions of the prototype box adapted to the direct shear apparatus, which translates into the possibility of using the size of the diameter of the GFRP bars or their arrangement and limiting the range of displacement during extraction, nevertheless recommended by the materials source, the range has not been exceeded. It was also not possible to install fiber optic monitoring in such a small device.

Figure 3. The prototype box for testing the pull-out strength of GFRP rods from the ground during the test
Source: Author

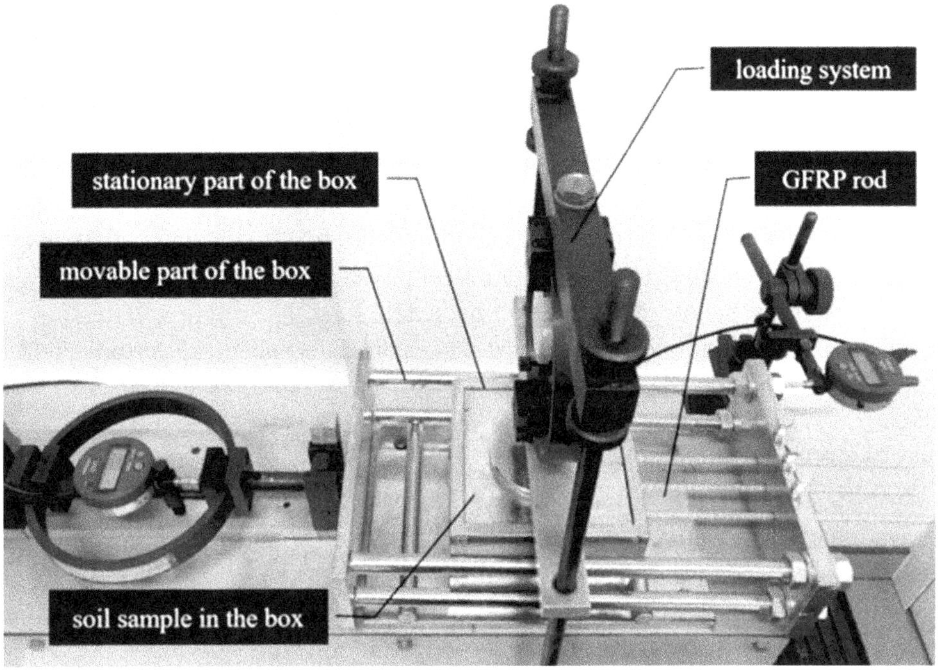

The Preparation of the Soil Samples

On the basis of the preliminary geotechnical studies of the research material, it was assumed that the strength of GFRP bars to pull-out from coarse-grained soil would be tested on prepared samples. It was assumed that the leading parameters for the conducted research will be: the preset normal stress in the sample and its moisture content. The ideal moisture for the soil used in the pull-out tests was the optimal humidity, which was previously determined experimentally, and therefore the samples for testing were prepared with a wide range of humidity, also lower and higher. Ultimately, prepared samples of six different water content were prepared: $w_1 = 0.11\%$ (soil dried to constant weight), $w_2 = 6.43\%$ (optimal water content), $w_3 = 8.66\%$, $w_4 = 11.37\%$, $w_5 = 17.87\%$ and $w_6 = 20.37\%$, which were tested in the next six test series. In addition to the water content, the bulk density of the prepared samples was determined each time before the strength test, and its desired value was obtained by laying the soil in layers and pre-compaction (Table 2).

The Procedure Pull-Out of GFRP Rods From the Soil

The main objective of the research was to determine the amount of force pulling out the GFRP rod (6 mm diameter) of the selected coarse-grained soil. To measure the value of the force, a method commonly used in strength tests (direct shear apparatus) was used, consisting in measuring the displacement, and then, thanks to the knowledge of the constant value of the dynamometric ring, converting the displacement into a force value. The process of pulling the GFRP bars was carried out until the limit displacement of the bars was reached at the level of 8% of the sample length, i.e. approx. 9.6 mm. Each of the six test series included tests of samples under a vertical load causing the desired normal stress, with the values: 100 kPa, 200, 300 or 400 kPa.

The researches for drawing GFRP composite bars from coarse-grained soil (Medium sand) were performed according to the following procedure:

- The preparation of samples from medium sand with different expected moisture,
- The determination of moisture content and bulk density within each test series,
- The preparation, within the scope of a given test series, of a sample in a prototype apparatus box by attaching 3 GFRP rods with a diameter of 6 mm to the soil sample, laying the soil in layers and appropriate compaction, and closing it with a steel piston,
- The installation of the handle of the rod pulling head,
- The installation and positioning of vertical and horizontal displacement sensors, connected to the iPack automatic data collection system,
- The application of a vertical load causing the desired normal stress in the soil sample was assumed as: 100 kPa, 200, 300 or 400 kPa,
- The stabilizing the settlement of the sample under a given load,
- The activation of the GFRP rod extraction mechanism with simultaneous registration of changes in the size of horizontal displacements over time,
- The completion of the bar extraction process, saving the results and disassembly of the system,
- The repeat the process for subsequent samples within all test runs.

Figure 4. An example of the relationship between the value of the pulling force and the displacement of GFRP bar in the soil for one of the tests of the PP0 series
Source: Author

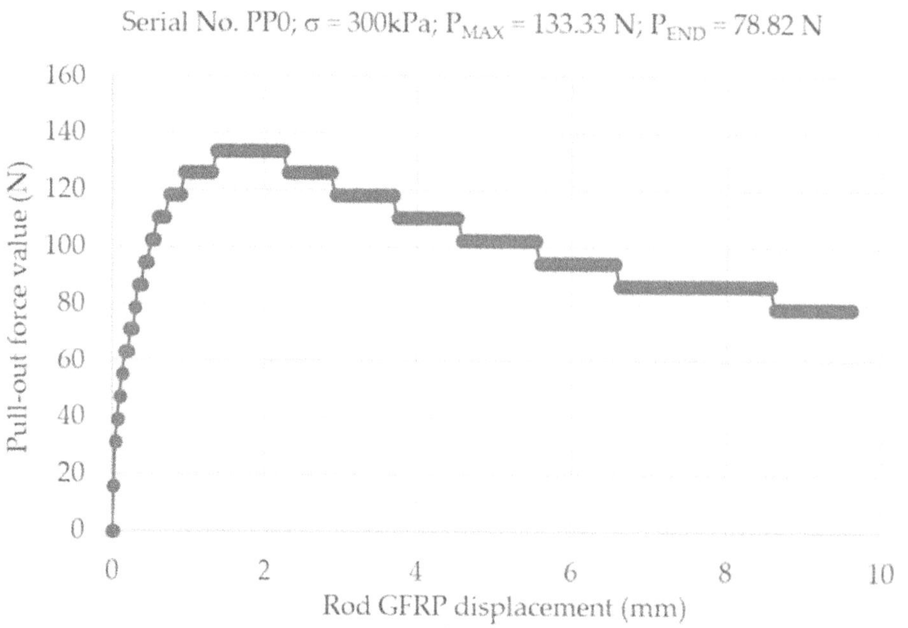

RESULTS

The researches of the resistance to pulling out the bars from coarse-grained soil were carried out in a modified direct shear apparatus, equipped with a results recorder and an iPack data collection system fully compatible with the Excel spreadsheet (MS Office), which significantly facilitated the subsequent analysis of the test results. Using mathematical procedures, the deformation value was transformed into the amount of pulling force of one GFRP bar, taking into account the fact that the total force corresponds to pulling out 3 GFRP composite bars. Figure 4 shows an example of the relationship between the value of the pulling force and the displacement for the sample under the load causing normal stresses equal to 300 kPa for the PP0 series. The maximum value of the extraction force (P_{MAX}) and the final force (P_{END}) corresponding to the limit deformation were read from the diagram. The results of tests of all samples under various loads were prepared in a similar way.

Table 2. The summary of the results of tests of the pull-out of GFRP bars from coarse-grained soil (MSa).

Serial No.	Water Content w (%)	Bulk Density ρ (t/m³)	Applied Normal Stress σ (kPa)	Max. Pull-Out Force Value (1 Rod) P_{MAX} (N)	Final Pull-Out Force Value (1 Rod) P_{END} (N)	Dependency Index P_{MAX}/P_{END} (-)
PP0	0.11 (dry)	1.54	100	7.88	0.00	-
			200	63.06	31.53	2.00
			300	133.99	78.82	1.70
			400	133.99	86.70	1.55
PP00	6.43	1.76	100	31.53	23.65	1.33
			200	47.29	39.41	1.20
			300	70.94	55.17	1.29
			400	86.70	78.82	1.10
PP1	8.66	1.56	100	31.53	23.65	1.33
			200	39.41	39.41	1.00
			300	47.29	39.41	1.20
			400	63.06	55.17	1.14
PP2	11.37	1.66	100	39.41	39.41	1.00
			200	47.29	47.29	1.00
			300	86.70	70.94	1.22
			400	118.23	94.58	1.25
PP3	17.87	1.82	100	47.29	39.41	1.20
			200	39.41	39.41	1.00
			300	70.94	55.17	1.29
			400	70.94	55.17	1.29
PP4	20.37	2.09	100	23.65	15.76	1.50
			200	78.82	63.06	1.25
			300	94.58	63.06	1.50
			400	102.47	78.82	1.30

Figure 5. The relationship between the value of the maximum (P_MAX) and minimum (P_END) pulling force and the normal stress for the PP0 series
Source: Author

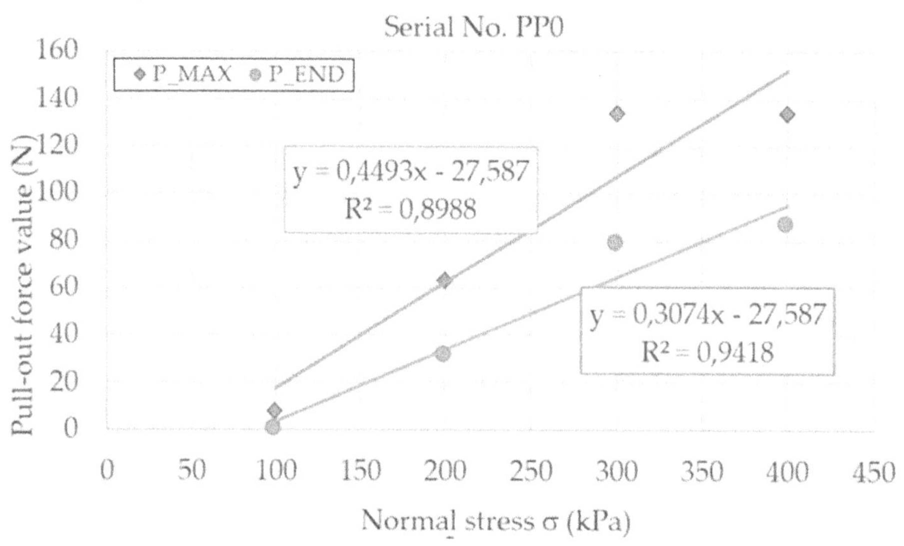

Figure 6. The relationship between the value of the maximum (P_MAX) and minimum (P_END) pulling force and the normal stress for the PP00 series
Source: Author

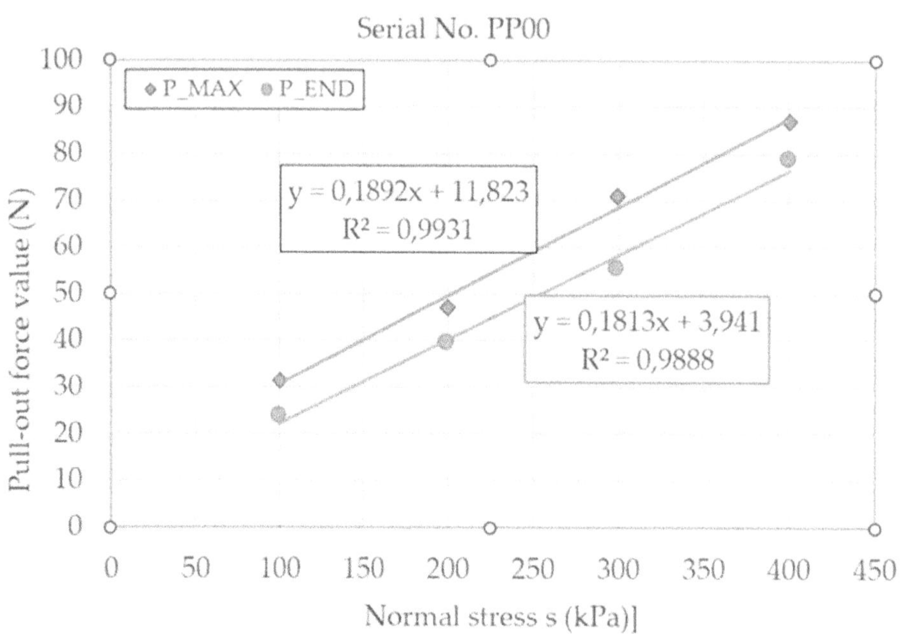

Figure 7. The relationship between the value of the maximum (P_MAX) and minimum (P_END) pulling force and the normal stress for the PP1 series
Source: Author

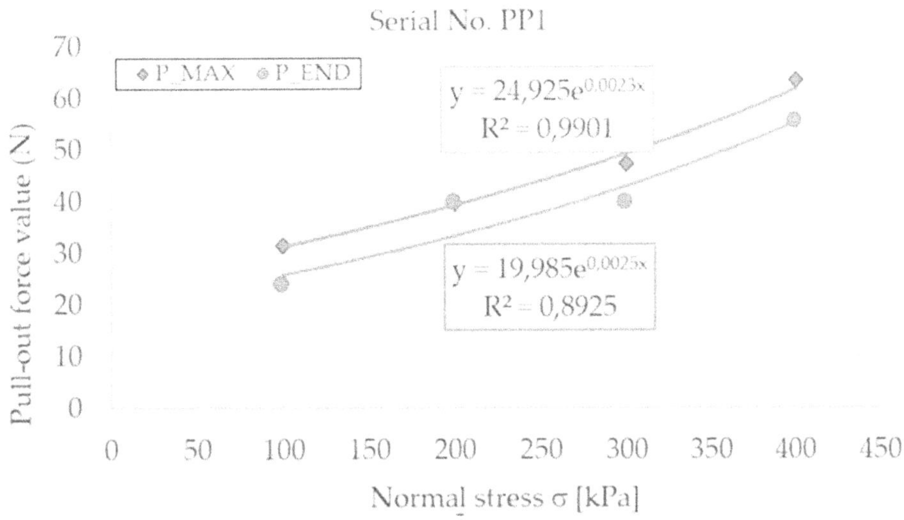

Figure 8. The relationship between the value of the maximum (P_MAX) and minimum (P_END) pulling force and the normal stress for the PP2 series
Source: Author

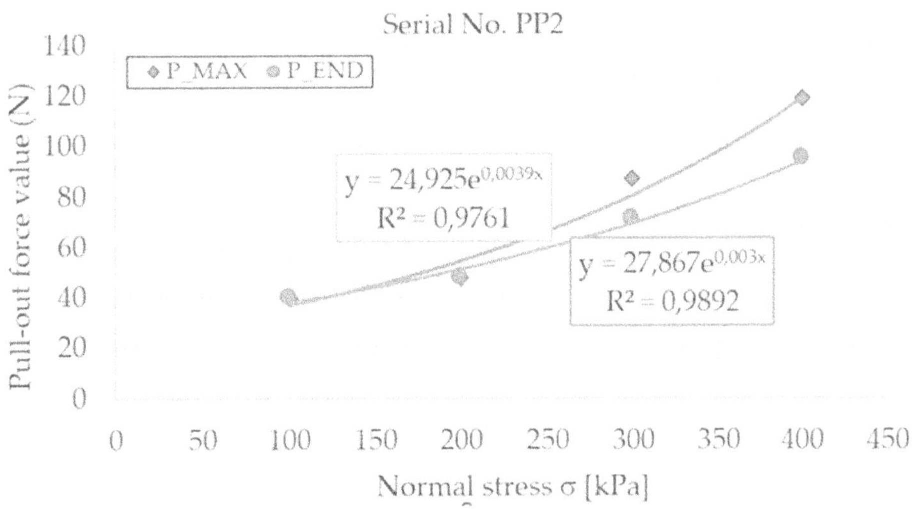

Figure 9. The relationship between the value of the maximum (P_MAX) and minimum (P_MIN) pulling force and the normal stress for the PP3 series
Source: Author

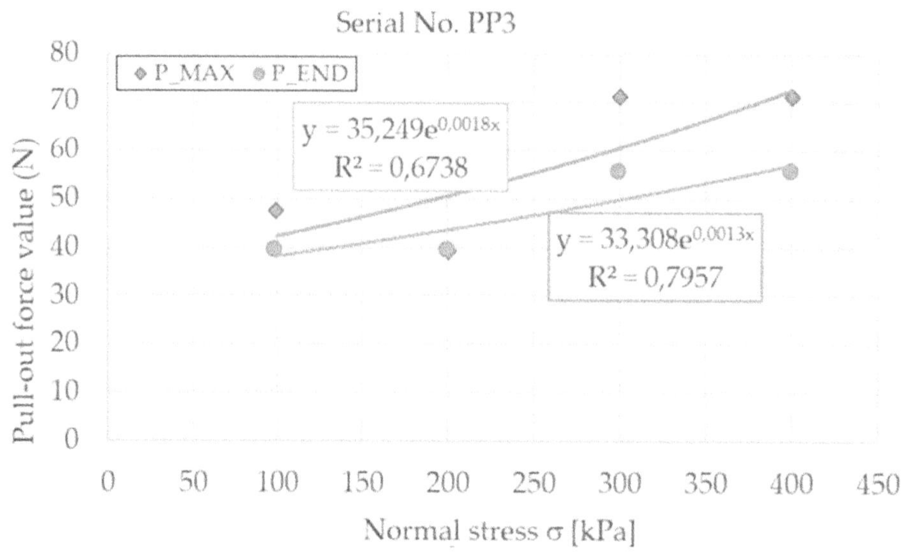

Figure 10. The relationship between the value of the maximum (P_MAX) and minimum (P_END) pulling force and the normal stress for the PP4 series
Source: Author

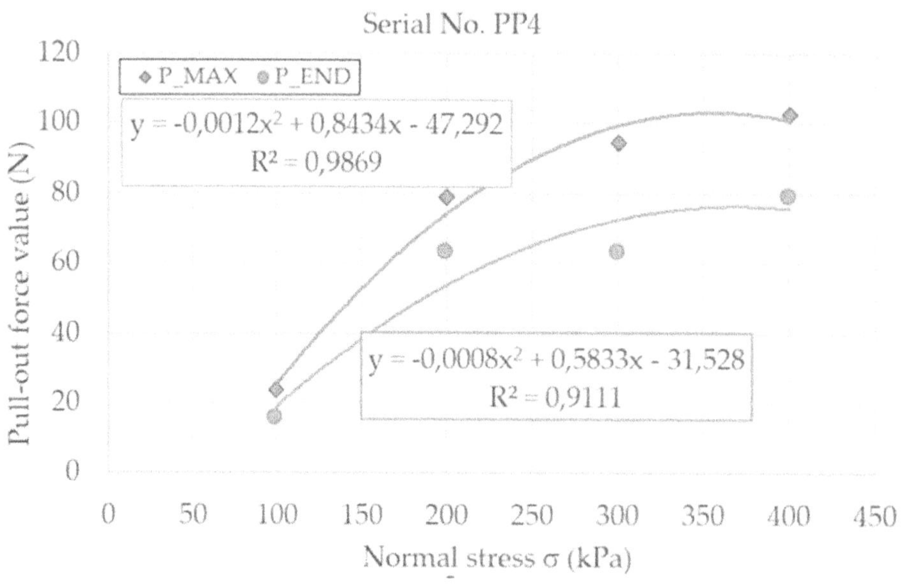

Table 2 summarizes the average results of the GFRP rod to pulling out tests in relation to the assigned test series and the given stress, characterized by the leading soil parameter: humidity and additionally bulk density. In addition to the maximum and final pulling force, the relationship between them was

Figure 11. The relationship between the value of the pulling force of GFRP bar and the water content for normal stress in the samples equal to 100 kPa
Source: Author

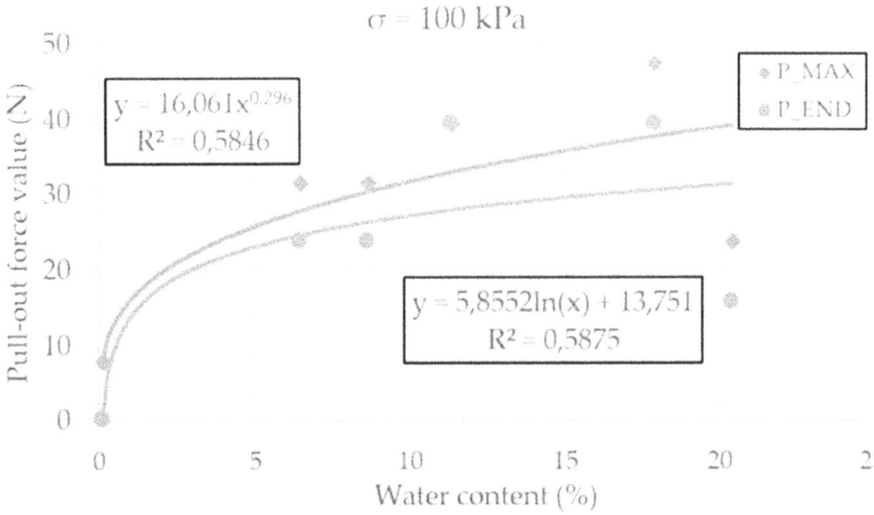

Figure 12. The relationship between the value of the pulling force of GFRP bar and the water content for normal stress in the samples equal to 200 kPa
Source: Author

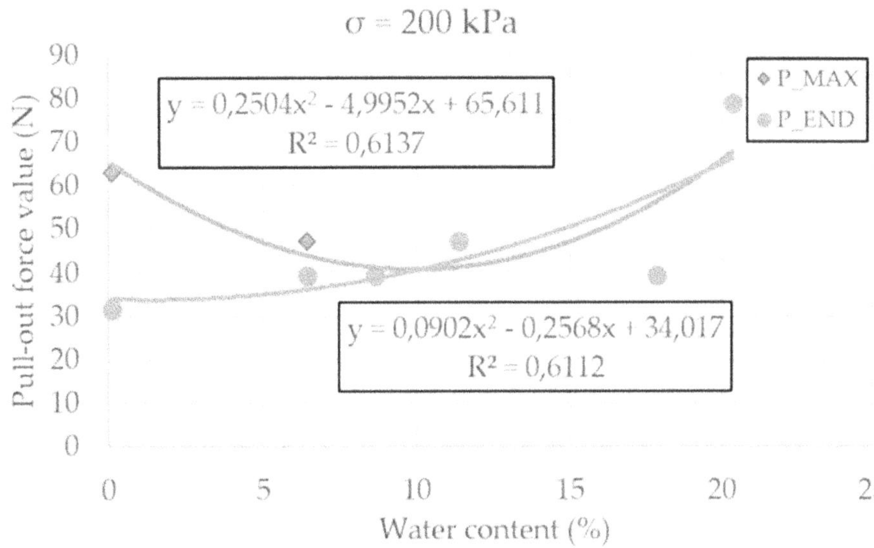

also presented, the values of which oscillated in the range from 1.00 to 2.00. The P_{MAX}/P_{END} ratio most often did not exceed the value of 1.50. that is, the final force was most often less than the maximum, not more than 50%.

Figure 13. The relationship between the value of the pulling force of GFRP bar and the water content for normal stress in the samples equal to 300 kPa
Source: Author

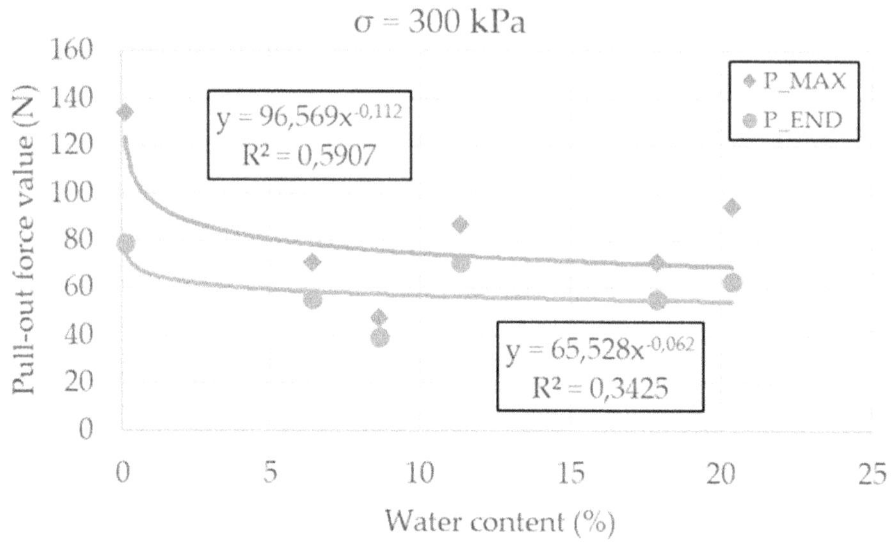

Figure 14. The relationship between the value of the pulling force of GFRP bar and the water content for normal stress in the samples equal to 400 kPa
Source: Author

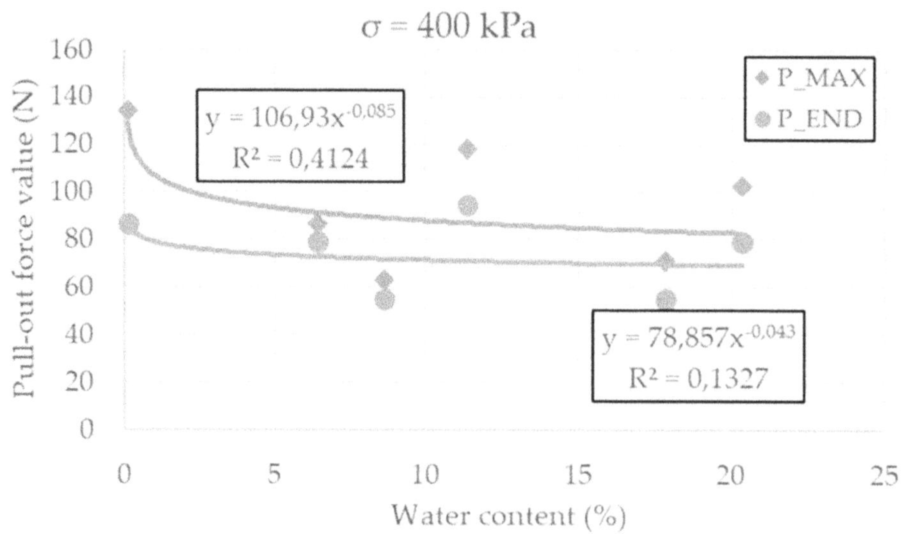

At the disposal of the complete results of laboratory strength tests, the relationships between the averaged values of the maximum and final pulling forces and the normal stress within the selected test series were determined in the next stage of the tests. The relationship between the value of the maximum and minimum pulling force and normal stress is presented and described in the formula in Figures 5-10.

In addition, the relationship between the leading geotechnical parameter, which was water content, and the amount of pulling forces (P_{MAX} and P_{END}) in relation to the set normal stress, was determined. The results are presented in the form of graphs in Figures 11-14.

DISCUSSION

On the basis of the strength tests for pulling out GFRP rods from coarse-grained soil (Medium Sand), it was found that there is a strong dependence of the size of the pulling force on the set normal stress. Generally, the pull-out force increased with the applied load (Figures 5-10). However, when we take into account the moisture content of the prepared soil samples, the situation turned out to be more complicated, because only when the set normal stresses were the lowest (100 kPa) the value of the force increased (Figure 11), while in the remaining cases (200; 300; 400 kPa) the trend was the opposite (Figures 12-14) and the relationships were much weaker. The values of the maximum extraction forces, depending on the set normal stress, were as follows: for a stress of 100 kPa, the maximum pulling out force oscillated in the range from 7.88 kPa to 47.29 kPa, for a stress of 200 kPa - from 39.41 kPa to 78.82 kPa, for a stress of 300 kPa - from 47.29 kPa to 133.99 kPa and for a stress of 400 kPa - from 63.06 kPa to 133.99 kPa. By analyzing the relationship between the values of the pulling out forces and the given normal stress, it has been shown that there is a strong relationship between the variables included. It confirms it evidenced by the high value of the coefficients of determination R^2, which for maximum pulling force oscillated in the range from 0.6738 to 0.9901, and the final force from 0.7957 to 0.9892, so were comparable. It should be noted that high coefficients of determination R^2 were obtained by fitting various types of functions: linear, polynomial and exponential. The weakest relationships were observed for the test results of the PP3 series, mainly due to the very similar values of the pulling force for the two highest normal stresses. In the case of the analysis of the charts (Figure 5 -14), the relationships between the values of the pulling-out forces and the moisture content were not so strong. The values of the maximum coefficients of determination ranged from 0.3425 to 0.6037.

The conducted preliminary tests showed that the obtained values of the pulling out forces are so high that it can be concluded that the proposed method can be used to reinforce the soil in order to improve insufficient geotechnical parameters. In order to finally verify this possibility, it is advisable to conduct comprehensive proper researches, in accordance with the lead project program "System for strengthening the soil with GFRP composite bars with fiber optic monitoring". It should be noted that the indication applies to coarse-grained soil, whose grains during the initial movement effectively wedge on the ribs of the composite reinforcement, which to a limited extent prevents further displacement of the soil-GFRP rod system, the use of GFRP rods to strengthen fine-grained soils, due to a different structure, the shape and size of grains require separate, comprehensive researches. After analyzing the test results, it was found that the proposed method of strengthening the ground may be an interesting alternative to the currently used methods of strengthening the ground, mainly due to the material from which the bars are made, which is extremely durable and doesn't corrode in various environments, and relatively easy and cheap to install it to the subsoil.

CONCLUSION

The results of the conducted pull-out strength tests indicate that the incorporation of GFRP rods into the coarse-grained soil has a positive effect on its strengthening. This is evidenced by the high values of the maximum pull-out forces at relatively small diameters (6mm) of the GFRP rods, which correlated in various ways with the normal stresses induced in the soil samples. The practical aspect of the applied solution is the effect of counteracting mainly horizontal displacements of the subsoil, which is particularly important in road and bridge construction.

As mentioned in the introduction, the presented issues are part of the 1st and 2nd stages of a large research project with preliminary prototype tests, performed on a micro scale, adapted to the capabilities of the available research equipment. The economic aspect is also important as the well-known, traditional direct shear apparatus was used for this purpose, modifying only its box according to one's own ideas, significantly reducing the cost of the experiment. The data obtained at this stage of the research on the geotechnical parameters of the coarse-grained soil material selected for testing in the form of medium sand (MSa) and the knowledge of the forces pulling the GFRP rods out of the ground with different moisture content will be the basis for planning model tests on a real scale (large-scale) with fiber optic monitoring in accordance with the assumptions of the research project underway.

Funding: This study was conducted under the project: "System for strengthening the soil with GFRP composite bars with fiber optic monitoring", (N2_56), Agreement No. 47/PRZ/1/DG/PCI/2020.

REFERENCES

Abu-Farsakh, M. Y., Almohd, I., & Farrag, K. (2006). Comparison of Field and Laboratory Pullout Tests on Geosynthetics in Marginal Soils. *Transportation Research Record: Journal of the Transportation Research Board*, *1975*(1), 1. doi:10.1177/0361198106197500114

Achillides, Z., & Pilakoutas, K. (2004). Bond behavior of fiber reinforced polymer bars under direct pullout conditions. *Journal of Composites for Construction*, *8*(2), 173–181. doi:10.1061/(ASCE)1090-0268(2004)8:2(173)

ACI Committee 440. (2006). *Guide for the Design and Construction of Concrete Reinforced with FRP Bars. ACI 440*. American Concrete Institute.

Almallah, A., Naggar, H. E., & Sadeghian, P. (2020). Axial Behavior of Innovative Sand-Coated GFRP Piles in Cohesionless Soil. *International Journal of Geomechanics*, *20*(10), 1–43. doi:10.1061/(ASCE)GM.1943-5622.0001808

Asadollahi, F., & Dabiri, R. (2017). Effects of Glass Fiber Reinforced Polymer on Geotechnical Properties of Clayey Soil. *Journal of Structural Engineering and Geotechnics*, *7*(2), 73–83.

Ashpiz, E. S., Egorov, A. O., & Ushakov, A. E. (2010). In S. Favro & C. A. Brebbia (Eds.), *Application of composite materials for the protection of sea shores and engineering structures. Island sustainability* (pp. 231–238). Wessex Institute of Technology.

Balcony, G., & Badawika, G. (2008). Composite materials in geotechnics. *Geoengineering, Roads, Bridges, Tunnels, 2*(20), 88-91.

Bencardino, F., Condello, A., & Ombres, L. (2016). Numerical and analytical modeling of concrete beams with steel, FRP and hybrid FRP-steel reinforcements. *Composite Structures*, *140*, 53–65. doi:10.1016/j.compstruct.2015.12.045

Bywalski, C., Drzazga, M., Kamiński, M., & Kaźmierowski, M. (2016). Analysis of calculation methods for bending concrete elements reinforced with FRP bars. *Archives of Civil and Mechanical Engineering*, *16*(4), 901–912. doi:10.1016/j.acme.2016.06.006

Chen, T., Mei, Z., & Li, C. (2008). Application research of full section pre reinforcement technology with glass fiber anchor in tunnel. *Mod. Tunn. Technol.*, *45*, 226–231.

Cheng, Y. M., Choi, Y., Yeung, A. T., Tham, L. G., Au, A. S. K., Wei, W. B., & Chen, J. (2009). New Soil Nail Material—Pilot Study of Grouted GFRP Pipe Nails in Korea and Hong Kong. *Journal of Materials in Civil Engineering*, *21*(3), 93–102. doi:10.1061/(ASCE)0899-1561(2009)21:3(93)

Cristelo, N., Cunha, M., Dias, M., Gomes, T., Miranda, T., & Araujo, N. (2015). Influence of discrete fiber reinforcement on the uniaxial compression response and seismic wave velocity of a cement-stabilized sandy-clay. *Geotextiles and Geomembranes*, *43*(1), 1–13. doi:10.1016/j.geotexmem.2014.11.007

Dean, R. & Fretting, F. (1986). Soil Randomly Reinforced with Fibers. *Journal of Geotechnical and Geoenvironmental Engineering, 112*(4), 823-827.

Eisa, M., El-Game, S., El-Salakawy, E., & Benmokrane, B. (2007). Construction of CRCP-Slabs Reinforced with GFRP Bars Implemented on Highway 40 East (Montréal). Technical Report Submitted to the Ministry of Transportation of Quebec.

EN ISO 14688-1:2018 Geotechnical investigation and testing - Identification and classification of soil - Part 1: Identification and description.

EN ISO 17892-10:2018 Geotechnical investigation and testing - Laboratory testing of soil - Part 10: Direct shear tests.

EN ISO 17892-11:2019 Geotechnical investigation and testing - Laboratory testing of soil - Part 11: Permeability tests.

EN ISO 17892-1:2014 Geotechnical investigation and testing - Laboratory testing of soil - Part 1: Determination of water content.

EN ISO 17892-2:2014 Geotechnical investigation and testing - Laboratory testing of soil - Part 2: Determination of bulk density.

EN ISO 17892-4:2016 Geotechnical investigation and testing - Laboratory testing of soil - Part 4: Determination of particle size distribution.

EN ISO 17892-5:2017 Geotechnical investigation and testing - Laboratory testing of soil - Part 5: Incremental loading oedometer test.

Ferreira, F., Carlos, D., Vieira, C. S., & De Lurdes Lopes, M. (2015a). *Soil-geogrid interaction in pullout conditions: influence of soil moisture content and density.* Conference: XVI European Conference on Soil Mechanics and Geotechnical Engineering, Edinburgh, UK.

Ferreira, F., Vieira, C. S., De Lurdes Lopes, M., & Carlos, D. (2015b). Experimental investigation on the pullout behaviour of geosynthetics embedded in a granite residual soil. *European Journal of Environmental and Civil Engineering*, *20*(9), 147–1180. doi:10.1080/19648189.2015.1090927

Ferreira, F., Vieira, C. S., De Lurdes Lopes, M., & Carlos, D. (2015c). Direct shear behaviour of residual soil-geosynthetic interfaces - influence of soil moisture content, soil density and geosynthetic type. *Geosynthetics International*, *22*(3), 257–272. doi:10.1680/gein.15.00011

Ferreira, F. B., Vieira, C. S., & Lopes, M. (2020). Pullout behaviour of different geosynthetics - influence of soil density and moisture content. *Frontiers in Built Environment*, *6*, 12. Advance online publication. doi:10.3389/fbuil.2020.00012

Garcia-Espinel, J. D., Alvarez-Garcia-Lubén, R., Gonzalez-Herrero, J. M., & Castro-Fresno, D. (2015). Glass fiber-reinforced polymer caissons used for construction of mooring dolphins in Puerto del Rosario harbor (Fuerteventura, Canary Islands). *Coastal Engineering*, *98*, 16–25. doi:10.1016/j.coastaleng.2015.01.003

Garcia-Espinel, J. D., García-Lubén, A., González-HerreroJ, M., & Castro-Fresno, D. (2015a). Design and Construction Methods of Caisson-Type Maritime Infrastructures Using GFRP. *Journal of Composites for Construction*, *19*(3). Advance online publication. doi:10.1061/(ASCE)CC.1943-5614.0000591

Gray, D., & Ohashi, H. (1983). Mechanics of Fiber Reinforcement in Sand. *Journal of Geotechnical Engineering*, *109*(3), 335–353. doi:10.1061/(ASCE)0733-9410(1983)109:3(335)

Isik, A., & Gurbuz, A. (2020). Pullout behavior of geocell reinforcement in cohesionless soils. *Geotextiles and Geomembranes*, *48*(1), 71–81. doi:10.1016/j.geotexmem.2019.103506

Jiang, X., Qin, Y., Li, W., Xiang, P., Zhang, S., & Chen, Ch. (2021). Experimental study on grouting mortar GFRP anchor rod pulling test. IOP Publishing. *Journal of Physics: Conference Series*, *2044*(012161), 1–6. doi:10.1088/1742-6596/2044/1/012161

Kara, I. F., Ashour, M. A., & Köroğlu, M. A. (2015). Flexural behavior of hybrid FRP/steel reinforced concrete beams. *Composite Structures*, *129*(1), 111–121. doi:10.1016/j.compstruct.2015.03.073

Krall, M., & Polak, M. A. (2019). Concrete beams with different arrangements of GFRP flexural and shear reinforcement. *Engineering Structures*, *198*, 109333. Advance online publication. doi:10.1016/j.engstruct.2019.109333

Lau, G., & Pam, H. J. (2010). Experimental study of hybrid FRP reinforced concrete beams. *Engineering Structures*, *32*(12), 3857–3865. doi:10.1016/j.engstruct.2010.08.028

Leung, H. Y., & Balendran, R. V. (2003). Flexural behavior of concrete beams internally reinforced with GFRP rods and steel rebars. *Structural Survey*, *21*(4), 146–157. doi:10.1108/02630800310507159

Li, H., Fu, J., Chen, B., Zhang, Z., & Lang, L. (2023). Mechanical Properties of GFRP Bolts and Its Application in Tunnel Face Reinforcement. *Materials (Basel)*, *16*(2193), 1–20. doi:10.3390/ma16062193 PMID:36984071

Lin, X., Qin, Y., Yang, L., Jiang, H., Yang, Y., & Chen, Ch. (2021). Model experimental study on GFRP anchor reinforcement and soil deformation. IOP Publishing. *Journal of Physics: Conference Series, 2044*(012114), 1–5. doi:10.1088/1742-6596/2044/1/012114

Merwe, F., & Hofmann, P. (2020). Guidance on the design and the use of fibre-reinforced polymer (FRP) soil nails and ground anchors. *Civil Engineering (New York, N.Y.)*, 56–63.

Milligan, G. W. E., & Tei, K. (1998). The Pull-Out Resistance of Model Soil Nails. *Soil and Foundation, 38*(2), 179–190. doi:10.3208andf.38.2_179

Moon, D., Sim, J., & Oh, H. (2005). Bond performance of deformed GFRP rebar with milled fibres to concrete. *Proceedings of the 7th International Symposium on Fiber reinforced polymer (FRP) reinforcement for concrete structures (FRPRCS-7),* 1497–1514.

Moon, D. & Sim, J. & Oh, H. (2007). Experimental characterization of the bond performance of a new type of glass fibre-reinforced polymer rebar for application in concrete structures. *Proc. IMechE, 221*, 113-118.

Mufti, A. A., Banthia, N., Benmokrane, B., Boulfiza, M., Onofrei, M., Bakht, B., Newhook, P., Tadros, G., & Brett, P. (2005). Article. *Proceedings of the 7th International Symposium on Fibre reinforced polymer (FRP) reinforcement for concrete structures*, 1361–1378.

Naani, A. (1993). Flexural Behavior and Design of RC Members Using FRP Reinforcement. *Journal of Structural Engineering, 119*(11), 3344–3359. doi:10.1061/(ASCE)0733-9445(1993)119:11(3344)

Oskouei, A. V., Bazli, M., & Imani, H. A. M. (2018). Flexural and web crippling properties of GFRP pultruded profiles subjected to wetting and drying cycles in different sea water conditions. *Polymer Testing, 69*(6), 417–430. doi:10.1016/j.polymertesting.2018.05.038

PN-B-04481. (1998). *Building Soils. Laboratory tests.* Academic Press.

Protchenko, K., Młodzik, K., Urbański, M., Szmigiera, E., & Garbacz, A. (2016). Numerical estimation of concrete beams reinforced with FRP bars. *MATEC Web of Conferences, 86*, 78-97. 10.1051/matecconf/20168602011

Research Project. (2020). *System for strengthening the soil with GFRP composite bars with fiber optic monitoring, Agreement No. 47/PRZ/1/DG/PCI.* Rzeszow University of Technology.

Saikia, B., Kumar, P., Thomas, J., & Ramaswamy, A. (2007). Strength and serviceability performance of beams reinforced with GFRP bars in flexure. *Construction & Building Materials, 21*(8), 1709–1719. doi:10.1016/j.conbuildmat.2006.05.021

Shehata, E., Morphy, R., & Rizkalla, S. (2000). Fibre reinforced polymer shear reinforcement for concrete members: Behaviour and design guidelines. *Canadian Journal of Civil Engineering, 27*(5), 859–872. Advance online publication. doi:10.1139/l00-004

Stankiewicz, B. (2011). The use of GFRP composites in platform slabs. *Glass and Ceramics, 1*, 28–31.

Strabag Ltd. (2022). *Reinforcement of the subsoil in the technology of concrete displacement columns with an aggregate head and a transmission layer made of GFRP composite rods*. Documentation: Construction of the regional road Racibórz - Pszczyna on the section from DK45 in the commune of Rudniki to Sportowa in Rybnik, as a new provincial road No. 935.

Suksiripattanapong, Ch., Horpibulsuk, S., Chinkulkijniwat, A., & Chai, J. Ch. (2013). Pullout resistance of bearing reinforcement embedded in coarse-grained soils. *Geotextiles and Geomembranes*, *36*, 44–54. doi:10.1016/j.geotexmem.2012.10.008

Victor, L., & Wang, S. (2002, Jan.). Flexural Behavior of GFRP Reinforced Engieered Cementitious Composites Beams. Technical Paper. *ACI Materials Journal*, 11–21.

Vijay, P. V., GangaRao, H. V. S., & Krishnaswamy, V. (2005). Manufacturing, durability and bond behavior of FRP bars with nanoclay. *Proceedings of the 7th International Symposium on Fiber reinforced polymer (FRP) reinforcement for concrete structures*, 77–98.

Zhang, G., Chen, Ch., Sun, J., Li, K., Xiao, F., Wang, Y., Chen, M., Huang, J., & Wang, X. (2022). Mixture optimisation for cement-soil mixtures with embedded GFRP tendons. *Journal of Materials Research and Technology*, *18*, 611–628. doi:10.1016/j.jmrt.2022.02.076

Zhang, G., & Zhu, H. (2019). Bond Behavior of Tensioned GFRP Tendons Embedded in Cemented Soils. *Tianjin Daxue Xuebao*, *52*(1), 120–127.

Zhang, G., Zornberg, J. G., & Morsy, A. M. (2020). Interface bond behavior of tensioned glass fiber-reinforced polymer (GFRP) tendons embedded in cemented soils. *Construction & Building Materials*, *263*, 120132. Advance online publication. doi:10.1016/j.conbuildmat.2020.120132

KEY TERMS AND DEFINITIONS

Coarse-Grained Soil: It is a soil devoid of cohesion, the grain size of which ranges from 0.063 mm to 63 mm, such as: Gravels (Coarse, Medium, Fine) and Sands (Coarse, Medium, Fine).

Fine-Grained Soil: It is a cohesive soil with grains less than 0.063 mm, such as: Silts (Coarse, Medium, Fine) and Clay.

GFRP (Glass Fiber Reinforced Polymer): Composites in which glass fibers are used as the reinforcing phase as a support element, while the polymer matrix serves as a binder connecting the fibers, combine materials with significantly different stiffness and strength - on the one hand, rigid, elastic and brittle fibers, on the other flexible matrix and due to the fact that they are cheaper, they win the economic ranking with other types of fibers, such as carbon or aramid (Stankiewicz, 2011).

GFRP Composite Bars: The bars are a modern, ecological and economical product, characterized by a low dead weight and much greater strength than steel bars, and most importantly, extremely high corrosion resistance in almost all conditions, which means that it can be used in places where particularly difficult environmental conditions prevail., e.g. in the construction of maritime and port infrastructure, airport or military facilities, high tensile strength, chemical resistance, low thermal conductivity, antimagnetism, resistance to water or chemicals and the fact that they are not an obstacle to electromagnetic waves cause building structures with the use of GFRP rods can be more durable and safe.

Pull-Out Method: In the analyzed case, it consists in pulling out GFRP bars fixed coaxially with the head of the pulling machine in a coarse-grained soil sample and recording the change in resistance of the pulled-out bar until reaching the limit displacement.

Compilation of References

Abu-Farsakh, M. Y., Almohd, I., & Farrag, K. (2006). Comparison of Field and Laboratory Pullout Tests on Geosynthetics in Marginal Soils. *Transportation Research Record: Journal of the Transportation Research Board, 1975*(1), 1. doi:10.1177/0361198106197500114

Achillides, Z., & Pilakoutas, K. (2004). Bond behavior of fiber reinforced polymer bars under direct pullout conditions. *Journal of Composites for Construction, 8*(2), 173–181. doi:10.1061/(ASCE)1090-0268(2004)8:2(173)

ACI Committee 440. (2006). *Guide for the Design and Construction of Concrete Reinforced with FRP Bars. ACI 440.* American Concrete Institute.

Akgün, A., & Bulut, F. (2007). GIS-based landslide susceptibility for Arsin-Yomra (Trabzon, North Turkey) region. *Environmental Geology (Berlin), 51*(8), 1377–1387. doi:10.100700254-006-0435-6

Aleotti, P., & Chowdhury, R. (1999). Landslide hazard assessment: Summary review and new perspectives. *Bulletin of Engineering Geology and the Environment, 58*(1), 21–44. doi:10.1007100640050066

Almada, J.A. (1865). Annaes of the Municipality of the Old Villa de Machico. *Flor do Oceano*, 247-286.

Almallah, A., Naggar, H. E., & Sadeghian, P. (2020). Axial Behavior of Innovative Sand-Coated GFRP Piles in Cohesionless Soil. *International Journal of Geomechanics, 20*(10), 1–43. doi:10.1061/(ASCE)GM.1943-5622.0001808

Almeida, A. B., & Lopes, S. (2014). *New Structural Measures to Reduce Hydrological Risk in Funchal.* In *VIII National Risk Meeting*, Guimaraes.

Althouse, A. D. (2016). Statistical graphics in action: Making better sense of the ROC curve. *International Journal of Cardiology, 100*(215), 9–10. doi:10.1016/j.ijcard.2016.04.026 PMID:27104918

Althuwaynee, O. F., Pradhan, B., & Lee, S. (2016). A novel integrated model for assessing landslide susceptibility mapping using CHAID and AHP pair-wise comparison. *International Journal of Remote Sensing, 37*(5), 1190–1209. doi:10.1080/01431161.2016.1148282

Alves, R., Silva, A., & Lousada, S. (2022). The impact of local accommodation in the sustainable development of Machico. In T. R. Aranzadi (Ed.), Planeamiento Ecológico En Las Iniciativas de Desarrollo (pp. 305–322). Academic Press.

Anbalagan, R. (1992). Landslide hazard evaluation and zonation mapping in mountainous terrain. *Engineering Geology, 32*(4), 269–277. doi:10.1016/0013-7952(92)90053-2

Antunes, L. C. (2013). *Google Earth in the Classroom.* https://www.mapasnasaladeaula.org/mapas-do-projeto/portugal/regi%C3%A3o-aut%C3%B3noma-da-madeira/ribeiras-principais-pt-ram

Aragon, A. (1987). *For the history of Funchal.* Regional Secretariat for Tourism and Culture.

Armenakis, C., & Patias, P. (2019). Unmanned vehicle systems for geomatics: towards robotic mapping. Whittles Publishing.

Asadollahi, F., & Dabiri, R. (2017). Effects of Glass Fiber Reinforced Polymer on Geotechnical Properties of Clayey Soil. *Journal of Structural Engineering and Geotechnics*, *7*(2), 73–83.

Ashpiz, E. S., Egorov, A. O., & Ushakov, A. E. (2010). In S. Favro & C. A. Brebbia (Eds.), *Application of composite materials for the protection of sea shores and engineering structures. Island sustainability* (pp. 231–238). Wessex Institute of Technology.

Association of State Dam Safety Officials. (n.d.a). https://damsafety.org/media/statistics

Association of State Dam Safety Officials. (n.d.b). https://damsafety.org/dam-failures

Autonomous Region of Madeira - Presidency of the Government. (2005). *Resolution No. 4/2005/M. Diário da República 2nd series, 219 (November): 16018 – 16030.* https://dre.pt/dre/detalhe/resolucao/4-2005-3022266

Ávila Sánchez, H. (2005). Introducción Líneas de Investigación y el Debate en los Estudios Urbano-Rurales. In *Lo Urbano-Rural. ¿Nuevas Expresiones Territoriales?* Universidad Nacional Autónoma de México. Centro Regional de Investigaciones Multidisciplinarias de la UNAM.

Ayalew, L., Yamagishi, H., Marui, H., & Kanno, T. (2005). Landslides in Sado Island of Japan: Part II. GIS-based susceptibility mapping with comparisons of results from two methods and verifications. *Engineering Geology*, *81*(4), 432–445. doi:10.1016/j.enggeo.2005.08.004

Azevedo, A. R. (1866). The Villa de Machico and the Capella de Machim. *Archivo Pittoresco*, *9*, 353.

Bahrami, Y., Hassani, H., & Maghsoudi, A. (2021). Landslide susceptibility mapping using AHP and fuzzy methods in the Gilan province, Iran. *GeoJournal*, *86*(4), 1797–1816. doi:10.100710708-020-10162-y

Baig, M. F., Mustafa, M. R. U., Baig, I., Takaijudin, H. B., & Zeshan, M. T. (2022). Assessment of Land Use Land Cover Changes and Future Predictions Using CA-ANN Simulation for Selangor, Malaysia. *Water (Basel)*, *3*(3), 402. Advance online publication. doi:10.3390/w14030402

Balcony, G., & Badawika, G. (2008). Composite materials in geotechnics. *Geoengineering, Roads, Bridges, Tunnels*, *2*(20), 88-91.

Ba, Q., Chen, Y., Deng, S., Wu, Q., Yang, J., & Zhang, J. (2017). An improved information value model based on gray clustering for landslide susceptibility mapping. *ISPRS International Journal of Geo-Information*, *6*(1), 18. doi:10.3390/ijgi6010018

Barreto, V. H. (2013). *Study of the Ribeira de Machico Hydrographic Basin – Vulnerability and Floods* [Master's dissertation]. Madeira University.

Basu, S. K. (2013). *Geology of Sikkim state and Darjeeling district of West Bengal*. Geological Society of India.

Basu, T., Das, A., & Pal, S. (2022). Application of geographically weighted principal component analysis and fuzzy approach for unsupervised landslide susceptibility mapping on Gish River Basin, India. *Geocarto International*, *37*(5), 1294–1317. doi:10.1080/10106049.2020.1778105

Beiranvand, B., & Rajaee, T. (2022). Application of artificial intelligence-based single and hybrid models in predicting seepage and pore water pressure of dams: A state-of-the-art review. *Advances in Engineering Software*, *173*, 103268. doi:10.1016/j.advengsoft.2022.103268

Bencardino, F., Condello, A., & Ombres, L. (2016). Numerical and analytical modeling of concrete beams with steel, FRP and hybrid FRP-steel reinforcements. *Composite Structures*, *140*, 53–65. doi:10.1016/j.compstruct.2015.12.045

Bernhäuserová, V., Havelková, L., Hátlová, K., & Hanus, M. (2022). The Limits of GIS Implementation in Education: A Systematic Review. *ISPRS International Journal of Geo-Information, 12*(12), 592. Advance online publication. doi:10.3390/ijgi11120592

Bertrand, G. (1974). *Introducción de la tesis doctoral de Estado.* Academic Press.

Bertrand, G. (2010). Itinerario en torno al paisaje: una epistemología de terreno para tiempos de crisis. *Ería: Revista cuatrimestral de geografía, 81*, 5-38.

Bhakra Beas Management Board. (n.d.). https://bbmb.gov.in/benefits-of-large-dam-.htm

Bhasin, R., Grimstad, E., Larsen, J. O., Dhawan, A. K., Singh, R., Verma, S. K., & Venkatachalam, K. (2002). Landslide hazards and mitigation measures at Gangtok, Sikkim Himalaya. *Engineering Geology, 64*(4), 351–368. doi:10.1016/S0013-7952(01)00096-5

Bilali, T. (2022). A practical probabilistic approach for simulating life loss in an urban area associated with a dam-break flood. *International Journal of Disaster Risk Reduction, 76*, 103011. doi:10.1016/j.ijdrr.2022.103011

Biswakarma, P., Barman, B. K., Joshi, V., & Rao, K. S. (2020). Landslide susceptibility mapping in east Sikkim region of Sikkim Himalaya using high resolution remote sensing data and GIS techniques. *Appl. Ecol. Environ. Sci, 8*, 143–153.

Britannica. (n.d.). https://www.britannica.com/topic/Grande-Dixence-Dam

Buchanan, J. M., & Stubblebine, W. C. (1962). Externality. *Economica, 29*(116), 371. doi:10.2307/2551386

Bu, L., Li, Y., Tang, W., Chen, X., & Li, B. (2022). Semi-quantitative risk analysis of high concrete face rockfill dams under seismic action. *Structures, 45*, 685–694. doi:10.1016/j.istruc.2022.09.013

Byjus. (n.d.). https://byjus.com/bank-exam/dams-of-india/

Bywalski, C., Drzazga, M., Kamiński, M., & Kaźmierowski, M. (2016). Analysis of calculation methods for bending concrete elements reinforced with FRP bars. *Archives of Civil and Mechanical Engineering, 16*(4), 901–912. doi:10.1016/j.acme.2016.06.006

Caetano, C. (2014). *Risk assessment of alluvium in the streams of the island of Madeira* [Master's dissertation]. STI.

Cartography. (2020a). http://www.gsdi. org/docs 2004/ Cookbook/cookbookV2.0.pdf

Cartography. (2020b). https://www.britannica.com/science/cartography

Castro, J. (2011). *Hydraulic, hydrological, and solid transport characterization of the February 20, 2010, event in Ribeira de João Gomes Madeira Island* [Master's dissertation]. Madeira University.

Černiauskas, E., & Bručas, D. (2014). Daugiasraigčių sraigtaspanių naudojimo stebėjimo užduotims atlikti tyrimas. *Aviacijos technologijos, 2*(1), 53-58.

Chau, C. K., Leung, T. M., & Ng, W. Y. (2015). A review on Life Cycle Assessment, Life Cycle Energy Assessment and Life Cycle Carbon Emissions Assessment on buildings. *Applied Energy, 143*, 395–413. doi:10.1016/j.apenergy.2015.01.023

Cheng, Y. M., Choi, Y., Yeung, A. T., Tham, L. G., Au, A. S. K., Wei, W. B., & Chen, J. (2009). New Soil Nail Material—Pilot Study of Grouted GFRP Pipe Nails in Korea and Hong Kong. *Journal of Materials in Civil Engineering, 21*(3), 93–102. doi:10.1061/(ASCE)0899-1561(2009)21:3(93)

Chen, T., Mei, Z., & Li, C. (2008). Application research of full section pre reinforcement technology with glass fiber anchor in tunnel. *Mod. Tunn. Technol., 45*, 226–231.

Choi, J., Oh, H.-J., Lee, H.-J., Lee, C., & Lee, S. (2012). Combining landslide susceptibility maps obtained from frequency ratio, logistic regression, and artificial neural network models using ASTER images and GIS. *Engineering Geology*, *124*, 12–23. doi:10.1016/j.enggeo.2011.09.011

CivilA. (n.d.). https://www.aboutcivil.org/planning-design-dams

Clément, G. (2018). Manifiesto del Tercer paisaje. Editorial Gustavo Gili, 21, 82.

Clerici, A., Perego, S., Tellini, C., & Vescovi, P. (2006). A GIS-based automated procedure for landslide susceptibility mapping by the conditional analysis method: The Baganza valley case study (Italian Northern Apennines). *Environmental Geology (Berlin)*, *50*(7), 941–961. doi:10.100700254-006-0264-7

CMM. (2022). *Characterization of the Municipality of Machico*. Author.

Consolidated Text of the Water Law, Royal Decree Law 10/2001, of July 20.

Coughenour, C., Coleman, N. M., & Taylor, A. L. (2022). In the shadow of the dam – Hydrology of the Little Conemaugh river and its South Fork, with insights about past and future flooding. *Heliyon*, *8*(9), e10679. doi:10.1016/j.heliyon.2022.e10679 PMID:36177242

Cristelo, N., Cunha, M., Dias, M., Gomes, T., Miranda, T., & Araujo, N. (2015). Influence of discrete fiber reinforcement on the uniaxial compression response and seismic wave velocity of a cement-stabilized sandy-clay. *Geotextiles and Geomembranes*, *43*(1), 1–13. doi:10.1016/j.geotexmem.2014.11.007

Cruz, A. (1901). Funchal, December 1, 1901. Academic Press.

da Silveira, A. B., Madeira, J., Ramalho, R., Fonseca, P., & Prada, S. (2010). *Geological Map of the Island of Madeira at Scale 1:50,000, Sheets A and B*. Regional Secretariat for the Environment and Natural Resources of the Autonomous Region of Madeira and University of Madeira.

Dahal, R. K., Hasegawa, S., Nonomura, A., Yamanaka, M., Dhakal, S., & Paudyal, P. (2008). Predictive modelling of rainfall-induced landslide hazard in the Lesser Himalaya of Nepal based on weights-of-evidence. *Geomorphology*, *102*(3–4), 496–510. doi:10.1016/j.geomorph.2008.05.041

Dai, F. C., & Lee, C. F. (2002). Landslide characteristics and slope instability modeling using GIS, Lantau Island, Hong Kong. *Geomorphology*, *42*(3–4), 213–228. doi:10.1016/S0169-555X(01)00087-3

Das, I., Sahoo, S., van Westen, C., Stein, A., & Hack, R. (2010). Landslide susceptibility assessment using logistic regression and its comparison with a rock mass classification system, along a road section in the northern Himalayas (India). *Geomorphology*, *114*(4), 627–637. doi:10.1016/j.geomorph.2009.09.023

Das, I., Stein, A., Kerle, N., & Dadhwal, V. K. (2012). Landslide susceptibility mapping along road corridors in the Indian Himalayas using Bayesian logistic regression models. *Geomorphology*, *179*, 116–125. doi:10.1016/j.geomorph.2012.08.004

Data Flair. (n.d.). https://www.google.com/amp/s/data-flair.training/blogs/dams-in-india/amp/

Dean, R. & Fretting, F. (1986). Soil Randomly Reinforced with Fibers. *Journal of Geotechnical and Geoenvironmental Engineering, 112*(4), 823-827.

Decreto-Lei n. ° 115/2010, of 22 October of the Ministry of the Environment and Spatial Planning. Diário da República: I série, n. ° 206. Available at https://dre.pt/dre/detalhe/decreto-lei/115-2010-307873

Diário de Notícias. (1956). Funchal, November 4, 1956. Academic Press.

Diário de Notícias. (1979). Funchal, January 21, 1979. Academic Press.

Diário de Notícias. (2021). *Falling rocks and flooding in Machico*. Academic Press.

Diário de Notícias. (2022). *Bad Weather causes 13 occurrences in Machico*. Academic Press.

DiaryO. (n.d.). https://orissadiary.com/india-has-5334-completed-and-operational-large-dams-while-411-large-dams-are-under-construction-says-bishweswar-tudu/amp/

Dilworth, R. (2020). Comparative Case Study Methods in Urban Political Development. *Social Sciences (Basel, Switzerland), 10*(10), 183. Advance online publication. doi:10.3390ocsci9100183

Directive N°. 2007/60/EC, of 23 October of the European Parliament and the Council of the European Union. Available at https://eur-lex.europa.eu/LexUriServ/LexUriServ.do?uri=OJ:L:2007:288:0027:0034:PT:PDF

Eisa, M., El-Game, S., El-Salakawy, E., & Benmokrane, B. (2007). Construction of CRCP-Slabs Reinforced with GFRP Bars Implemented on Highway 40 East (Montréal). Technical Report Submitted to the Ministry of Transportation of Quebec.

Eisenbeiss, H. (2009). *UAV photogrammetry* [Dissertation]. Federal Institute of Technology (ETH), Institute of Geodesy and Photogrammetry.

Embid Irujo, A. (2002). *Evolution of Water Law and Policy in Spain* (Vol. 1). Civitas.

Embid Irujo, A. (2007). *Dictionary of water law*. Iustel.

Embid Irujo, A. (2016). *Thirty years of the 1985 Water Law*. Editorial Thomson Reuters Aranzadi.

EN ISO 14688-1:2018 Geotechnical investigation and testing - Identification and classification of soil - Part 1: Identification and description.

EN ISO 17892-1:2014 Geotechnical investigation and testing - Laboratory testing of soil - Part 1: Determination of water content.

EN ISO 17892-10:2018 Geotechnical investigation and testing - Laboratory testing of soil - Part 10: Direct shear tests.

EN ISO 17892-11:2019 Geotechnical investigation and testing - Laboratory testing of soil - Part 11: Permeability tests.

EN ISO 17892-2:2014 Geotechnical investigation and testing - Laboratory testing of soil - Part 2: Determination of bulk density.

EN ISO 17892-4:2016 Geotechnical investigation and testing - Laboratory testing of soil - Part 4: Determination of particle size distribution.

EN ISO 17892-5:2017 Geotechnical investigation and testing - Laboratory testing of soil - Part 5: Incremental loading oedometer test.

Ercanoglu, M., & Gokceoglu, C. (2004). Use of fuzzy relations to produce landslide susceptibility map of a landslide prone area (West Black Sea Region, Turkey). *Engineering Geology, 75*(3–4), 229–250. doi:10.1016/j.enggeo.2004.06.001

Fang, C., Qiu, C., & Zheng, Y. (2023). Shape Memory Alloys for Civil Engineering. *Materials, 16*(2). *Materials (Basel), 2*(2), 787. Advance online publication. doi:10.3390/ma16020787 PMID:36676523

Farinelli, F. (1992). I segni del mondo. Immagine cartográfica e discorso geografico in età moderna, Florencia, La Nuova Italia. In El paisaje en la cultura contemporánea. Biblioteca Nueva.

Fayez, L., Pazhman, D., Pham, B. T., Dholakia, M. B., Solanki, H. A., Khalid, M., & Prakash, I. (2018). Application of frequency ratio model for the development of landslide susceptibility mapping at part of Uttarakhand State, India. *International Journal of Applied Engineering Research: IJAER*, *13*(9), 6846–6854.

Feizizadeh, B., & Blaschke, T. (2012). Uncertainty analysis of GIS-based ordered weighted averaging method for landslide susceptibility mapping in Urmia Lake Basin, Iran. *Seventh International Geographic Information Science Conference, September*, 18–21.

Feizizadeh, B., Roodposhti, M. S., Jankowski, P., & Blaschke, T. (2014). A GIS-based extended fuzzy multi-criteria evaluation for landslide susceptibility mapping. *Computers & Geosciences*, *73*, 208–221. doi:10.1016/j.cageo.2014.08.001 PMID:26089577

Ferreira, A. (2010). *Natural and Technological Risks* [Master's dissertation]. Academic Press.

Ferreira, F., Carlos, D., Vieira, C. S., & De Lurdes Lopes, M. (2015a). *Soil-geogrid interaction in pullout conditions: influence of soil moisture content and density.* Conference: XVI European Conference on Soil Mechanics and Geotechnical Engineering, Edinburgh, UK.

Ferreira, F.P.M., & Cunha, S.B. (1996). *Urban Hydrology: Floods in Rio de Janeiro – Jacarepaguá.* Federal University of Rio de Janeiro/CNPq. Rio de Janeiro.

Ferreira, F. B., Vieira, C. S., & Lopes, M. (2020). Pullout behaviour of different geosynthetics - influence of soil density and moisture content. *Frontiers in Built Environment*, *6*, 12. Advance online publication. doi:10.3389/fbuil.2020.00012

Ferreira, F., Vieira, C. S., De Lurdes Lopes, M., & Carlos, D. (2015b). Experimental investigation on the pullout behaviour of geosynthetics embedded in a granite residual soil. *European Journal of Environmental and Civil Engineering*, *20*(9), 147–1180. doi:10.1080/19648189.2015.1090927

Ferreira, F., Vieira, C. S., De Lurdes Lopes, M., & Carlos, D. (2015c). Direct shear behaviour of residual soil-geosynthetic interfaces - influence of soil moisture content, soil density and geosynthetic type. *Geosynthetics International*, *22*(3), 257–272. doi:10.1680/gein.15.00011

Flint, A., & Jennings, B. (2022). The role of cultural heritage in visitor narratives of peatlands: Analysis of onlineuser-generated reviews from three peatland sites in England. *International Journal of Heritage Studies*, *28*(2), 163–179. do i:10.1080/13527258.2021.1941198

Food and Agriculture Organization of the United Nations. (n.d.). https://www.fao.org/3/v5930e/v5930e01.htm

Förstner, W., & Wrobel, B. P. (2016). *Photogrammetric Computer Vision: Statistics, Geometry, Orientation and Reconstruction*. Springer. doi:10.1007/978-3-319-11550-4

Freitas, H., Goerl, R., & Michel, G. (2020). Alluvium in Funchal, Madeira Island: Historical and conceptual aspects and damage potential. *Geosaberes: Journal of Geoeducational Studies*, *11*, 608–622. doi:10.26895/geosaberes.v11i0.971

Freitas, M., Pereira, S.-S., & Hurd, K. (2022). Death by a thousand cuts: Small local dams can produce large regional impacts in the Brazilian Legal Amazon. *Environmental Science & Policy*, *136*, 447–452. doi:10.1016/j.envsci.2022.07.013

Garcia-Espinel, J. D., Alvarez-Garcia-Lubén, R., Gonzalez-Herrero, J. M., & Castro-Fresno, D. (2015). Glass fiber-reinforced polymer caissons used for construction of mooring dolphins in Puerto del Rosario harbor (Fuerteventura, Canary Islands). *Coastal Engineering*, *98*, 16–25. doi:10.1016/j.coastaleng.2015.01.003

Garcia-Espinel, J. D., García-Lubén, A., González-HerreroJ, M., & Castro-Fresno, D. (2015a). Design and Construction Methods of Caisson-Type Maritime Infrastructures Using GFRP. *Journal of Composites for Construction*, *19*(3). Advance online publication. doi:10.1061/(ASCE)CC.1943-5614.0000591

Gawryluk, D., & Krawczyk, D. (2021). *The future of the city*. Białystok: Publishing House of Bialystok University of Technology. doi:10.24427/978-83-66391-62-8_7

Ghilani, Ch. D., & Wolf, P. R. (2012). Elementary surweying: an introduction to geomatics. Academic Press.

Ghosh, S., Carranza, E. J. M., van Westen, C. J., Jetten, V. G., & Bhattacharya, D. N. (2011). Selecting and weighting spatial predictors for empirical modeling of landslide susceptibility in the Darjeeling Himalayas (India). *Geomorphology*, *131*(1–2), 35–56. doi:10.1016/j.geomorph.2011.04.019

Gorsevski, P., Gessler, P. E., Foltz, R. B., & Elliot, W. J. (2006). Spatial prediction of landslide hazard using logistic regression and ROC analysis. *Transactions in GIS*, *10*(3), 395–415. doi:10.1111/j.1467-9671.2006.01004.x

Govindan, K., Madan Shankar, K., & Kannan, D. (2016). Sustainable material selection for construction industry – A hybrid multi criteria decision making approach. *Renewable & Sustainable Energy Reviews*, *55*, 1274–1288. doi:10.1016/j.rser.2015.07.100

Gray, D., & Ohashi, H. (1983). Mechanics of Fiber Reinforcement in Sand. *Journal of Geotechnical Engineering*, *109*(3), 335–353. doi:10.1061/(ASCE)0733-9410(1983)109:3(335)

Grubert. (2020). Conventional hydroelectricity and the future of energy: Linking national inventory of dams and energy information administration data to facilitate analysis of hydroelectricity. *The Electricity Journal, 33*(1).

Gupta, R. P., Saha, A. K., Arora, M. K., & Kumar, A. (1999). Landslide hazard zonation in part of the Bhagirathi Valley. *Garhwal Mimalyas, Using Integrated Remote Sensing–GIS. Himalayan Geology*, *20*, 71–85.

Guzzetti, F., Carrara, A., Cardinali, M., & Reichenbach, P. (1999). Landslide hazard evaluation: A review of current techniques and their application in a multi-scale study, Central Italy. *Geomorphology*, *31*(1–4), 181–216. doi:10.1016/S0169-555X(99)00078-1

Guzzetti, F., Reichenbach, P., Cardinali, M., Galli, M., & Ardizzone, F. (2005). Probabilistic landslide hazard assessment at the basin scale. *Geomorphology*, *72*(1–4), 272–299. doi:10.1016/j.geomorph.2005.06.002

Haala, N., Cramer, M., Weimer, F., & Trittler, M. (2011). Performance Test on UAV-Based Photogrammetric Data Collection. *International Archives of the Photogrammetry, Remote Sensing and Spatial Information Sciences, 38*(1), 7-12.

Hancock, G. R., & Coulthard, T. J. (2022). Tailings dams: Assessing the long-term erosional stability of valley fill designs. *The Science of the Total Environment*, *849*, 157692. doi:10.1016/j.scitotenv.2022.157692 PMID:35908711

Han, J., & Thakur, J. K. (2015). Sustainable roadway construction using recycled aggregates with geosynthetics. *Sustainable Cities and Society*, *14*, 342–350. doi:10.1016/j.scs.2013.11.011

Heidarzadeh, M., & Feizi, S. (2022). A cascading risk model for the failure of the concrete spillway of the Toddbrook dam, England during the August 2019 flooding. *International Journal of Disaster Risk Reduction*, *80*, 103214. doi:10.1016/j.ijdrr.2022.103214

Hipólito, J. R., & Vaz, A. C. (2011). *Hydrology and Water Resources* (1st ed.). IST Press.

Hutchinson, J. N. (1988). General report: morphological and geotechnical parameters of landslides in relation to geology and hydrogeology. *International Symposium on Landslides, 5*, 3–35.

Hutchinson, J. N., Poole, C., Lambert, N., & Bromhead, E. N. (1985). Combined archaeological and geotechnical investigations of the Roman Fort at Lympne, Kent. *Britannia*, *16*, 209–236. doi:10.2307/526401

International Rivers. (n.d.). https://archive.internationalrivers.org/damming-statistics

IrfanView Graphic Viewer. (n.d.). https://www.irfanview.com/

Irrigation Department Government of Kerala. (n.d.). https://irrigation-kerala.org/attappady-irrigation-project

ISEL. (2015). *Hydrology*. Lisbon: ISEL. Obtained from https://issuu.com/joanadecastrocortesao/docs/hidrologia_resumo_luis_mende s.

Isik, A., & Gurbuz, A. (2020). Pullout behavior of geocell reinforcement in cohesionless soils. *Geotextiles and Geomembranes*, *48*(1), 71–81. doi:10.1016/j.geotexmem.2019.103506

IST, UMA, & LREC. (2010). *Alluvial risk assessment study on the island of Madeira*. Academic Press.

Iwahashi, J., Kamiya, I., & Yamagishi, H. (2012). High-resolution DEMs in the study of rainfall- and earthquake-induced landslides: Use of a variable window size method in digital terrain analysis. *Geomorphology*, *153–154*, 29–38. doi:10.1016/j.geomorph.2012.02.002

Jané Solá, J. (1974). For an Operational Structure of Economic Policy. *Spanish Journal of Economy*.

Jensen, J. S. (2020). Innovative and sustainable operation and maintenance of bridges. *Structure and Infrastructure Engineering*, *16*(1), 72–83. doi:10.1080/15732479.2019.1604772

Jiang, X., Qin, Y., Li, W., Xiang, P., Zhang, S., & Chen, Ch. (2021). Experimental study on grouting mortar GFRP anchor rod pulling test. IOP Publishing. *Journal of Physics: Conference Series*, *2044*(012161), 1–6. doi:10.1088/1742-6596/2044/1/012161

Jiménez Herrero, L. (1998). *Trends in the use of economic and fiscal instruments in environmental management. Reflections on the European Union and the Spanish case*. Cedecs Financial and Tax Law.

Jornal Público. (2012). *Four dozen displaced people and roads closed in Madeira*. Academic Press.

Kanungo, D. P., Arora, M. K., Sarkar, S., & Gupta, R. P. (2006). A comparative study of conventional, ANN black box, fuzzy and combined neural and fuzzy weighting procedures for landslide susceptibility zonation in Darjeeling Himalayas. *Engineering Geology*, *85*(3–4), 347–366. doi:10.1016/j.enggeo.2006.03.004

Kanungo, D. P., Arora, M. K., Sarkar, S., & Gupta, R. P. (2009). A fuzzy set based approach for integration of thematic maps for landslide susceptibility zonation. *Georisk*, *3*(1), 30–43. doi:10.1080/17499510802541417

Kara, I. F., Ashour, M. A., & Köroğlu, M. A. (2015). Flexural behavior of hybrid FRP/steel reinforced concrete beams. *Composite Structures*, *129*(1), 111–121. doi:10.1016/j.compstruct.2015.03.073

Kaur, H., Gupta, S., Parkash, S., Thapa, R., Gupta, A., & Khanal, G. C. (2019). Evaluation of landslide susceptibility in a hill city of Sikkim Himalaya with the perspective of hybrid modelling techniques. *Annals of GIS*, *25*(2), 113–132. doi:10.1080/19475683.2019.1575906

Kayastha, P., Dhital, M. R., & de Smedt, F. (2013). Application of the analytical hierarchy process (AHP) for landslide susceptibility mapping: A case study from the Tinau watershed, west Nepal. *Computers & Geosciences*, *52*, 398–408. doi:10.1016/j.cageo.2012.11.003

Knoema. (n.d.a). https://knoema.com/atlas/topics/Water/Dam-Capacity/Total-dam-capacity

Knoema. (n.d.b). https://www.google.com/amp/knoema.com/atlas/India/topics/Water/Dam-Capacity/Total-dam-capacity%3fmode=amp

Knoema. (n.d.c). https://www.google.com/amp/knoema.com/atlas/topics/Water/Dam-Capacity/Total-dam-capacity%3fmode=amp

Konecny, G. (2003). *Geoinformation: Remote Sensing, Photogrammetry and Geographic information system*. Taylor and Francis. doi:10.4324/9780203469644

Kotis, K., Angelis, S., Moraitou, E., Kopsachilis, V., Papadopoulou, E.-E., Soulakellis, N., & Vaitis, M. (2023). A KG-Based Integrated UAV Approach for Engineering Semantic Trajectories in the Cultural Heritage Documentation Domain. *Remote Sensing (Basel)*, *3*(3), 821. Advance online publication. doi:10.3390/rs15030821

Krall, M., & Polak, M. A. (2019). Concrete beams with different arrangements of GFRP flexural and shear reinforcement. *Engineering Structures*, *198*, 109333. Advance online publication. doi:10.1016/j.engstruct.2019.109333

Kraus, K. (1997). Photogrammetry: Vol. 2. *Advanced methods and application*. Dümmler.

Kraus, K. (2000). Photogrammetry: Vol. 1. *Fundamentals and Standard Processes*. Dümmler.

Kraus, K. (2007). *Photogrammetry: Geometry from Images and Laser Scans*. Walter de Gruyter. doi:10.1515/9783110892871

Kumar, R., & Anbalagan, R. (2016). Landslide susceptibility mapping using analytical hierarchy process (AHP) in Tehri reservoir rim region, Uttarakhand. *Journal of the Geological Society of India*, *87*(3), 271–286. doi:10.100712594-016-0395-8

Kumar, V., Gupta, V., & Jamir, I. (2018). Hazard evaluation of progressive Pawari landslide zone, Satluj valley, Himachal Pradesh, India. *Natural Hazards*, *93*(2), 1029–1047. doi:10.100711069-018-3339-3

Kundu, S., Saha, A. K., Sharma, D. C., & Pant, C. C. (2013). Remote sensing and GIS based landslide susceptibility assessment using binary logistic regression model: A case study in the Ganeshganga Watershed, Himalayas. *Photonirvachak (Dehra Dun)*, *41*(3), 697–709. doi:10.100712524-012-0255-y

Lau, G., & Pam, H. J. (2010). Experimental study of hybrid FRP reinforced concrete beams. *Engineering Structures*, *32*(12), 3857–3865. doi:10.1016/j.engstruct.2010.08.028

Lee, P., Park, J. R., Joo, J. C., & Ahn, C. H. (2022). Application of WQIEUT and TSIKO for comprehensive water quality assessment immediately after the construction of the Yeongju Multipurpose Dam in the Naeseong Stream Basin, Republic of Korea. *The Science of the Total Environment*, *819*, 152997. doi:10.1016/j.scitotenv.2022.152997 PMID:35031363

Lee, S., & Pradhan, B. (2007). Landslide hazard mapping at Selangor, Malaysia using frequency ratio and logistic regression models. *Landslides*, *4*(1), 33–41. doi:10.100710346-006-0047-y

Lee, S., & Talib, J. A. (2005). Probabilistic landslide susceptibility and factor effect analysis. *Environmental Geology (Berlin)*, *47*(7), 982–990. doi:10.100700254-005-1228-z

Lei n. ° 58/2005, of 29 December of the Assembly of the Republic. Diário da República: I-A series, no. 249. Available at https://data.dre.pt/eli/lei/58/2005/12/29/p/dre/pt/html

Leite, J. D. (1989). *Discovery of Madeira Island*. Alpha Publications.

Lemenkova, P., & Debeir, O. (2023). Quantitative Morphometric 3D Terrain Analysis of Japan Using Scripts of GMT and R. *Land (Basel)*, *1*(1), 261. Advance online publication. doi:10.3390/land12010261

Lencastre, A. (1972). *General Hydraulics Manual*. Academic Press.

Leung, H. Y., & Balendran, R. V. (2003). Flexural behavior of concrete beams internally reinforced with GFRP rods and steel rebars. *Structural Survey*, *21*(4), 146–157. doi:10.1108/02630800310507159

Lieberman, R. C. (2002). Ideas, Institutions, and Political Order: Explaining Political Change. *The American Political Science Review*, *96*(4), 697–712. doi:10.1017/S0003055402000394

Li, H., Fu, J., Chen, B., Zhang, Z., & Lang, L. (2023). Mechanical Properties of GFRP Bolts and Its Application in Tunnel Face Reinforcement. *Materials (Basel)*, *16*(2193), 1–20. doi:10.3390/ma16062193 PMID:36984071

Linder, W. (2009). Digital Photogrammetry. A practical course. Springer-Verlag. doi:10.1007/978-3-540-92725-9

LinkT. (n.d.). https://www.touristlink.com/canada/daniel-johnson-dam/overview.html

Lin, X., Qin, Y., Yang, L., Jiang, H., Yang, Y., & Chen, Ch. (2021). Model experimental study on GFRP anchor reinforcement and soil deformation. IOP Publishing. *Journal of Physics: Conference Series*, *2044*(012114), 1–5. doi:10.1088/1742-6596/2044/1/012114

Liu, B., Wu, J., Zhang, J., Wang, L., Yang, J., Liang, D., Dai, Q., Bi, X., Feng, Y., Zhang, Y., & Zhang, Q. (2017). Characterization and source apportionment of PM2.5 based on error estimation from EPA PMF 5.0 model at a medium city in China. *Environmental Pollution*, *222*, 10–22. doi:10.1016/j.envpol.2017.01.005 PMID:28088626

Llamosas, C., & Sovacool, B. K. (2021). The future of hydropower? A systematic review of the drivers, benefits and governance dynamics of transboundary dams. *Renewable & Sustainable Energy Reviews*, *137*, 110495. doi:10.1016/j.rser.2020.110495

Long, N. T., & de Smedt, F. (2018). Analysis and mapping of rainfall-induced landslide susceptibility in a Luoi District, Thua Thien Hue Province, Vietnam. *Water (Basel)*, *11*(1), 51. doi:10.3390/w11010051

Lousada, S., Cabezas, J., Castanho, R. A., & Gómez, J. M. N. (2021). Hydraulic Planning in Insular Urban Territories: The Case of Madeira Island—Ribeira Brava. *Water (Basel)*, *2021*(13), 2951. doi:10.3390/w13212951

Lousada, S., Cabezas, J., Castanho, R. A., & Gómez, J. M. N. (2022). Land-Use Changes in Insular Urban Territories: A Retrospective Analysis from 1990 to 2018. The Case of Madeira Island—Ribeira Brava. *Sustainability (Basel)*, *24*(24), 16839. Advance online publication. doi:10.3390u142416839

Lousada, S., & Castanho, R. A. (2021). GIS-based Assessment of Morphological and Hydrological Parameters of Ribeira dos Socorridos and Ribeira do Vigário Basins, Madeira Island, Portugal. *Current World Environment*, *16*(2), 408–426. Advance online publication. doi:10.12944/CWE.16.2.08

Lousada, S., & Castanho, R. A. (2022). The Role of Ports in Tourism: Porto Santo Harbour. *Water (Basel)*, *19*(19), 3176. Advance online publication. doi:10.3390/w14193176

Lousada, S., Gonçalves, L., & Atmaca, A. (2022). Hydraulic Planning in Insular Urban Territories: The Case of Madeira Island—São Vicente. *Water (Basel)*, *14*(1), 112. doi:10.3390/w14010112

Lousada, S., & Loures, L. (2020). Modelling Torrential Rain Flows in Urban Territories: Floods - Natural Channels (The Case Study of Madeira Island). *American Journal of Water Science and Engineering.*, *6*(1), 17–30. doi:10.11648/j.ajwse.20200601.13

Lugo Hubp, J. (1989). Paisaje físico geográfico. In *Diccionario Geomorfológico con equivalentes de los términos de uso más común en alemán, francés, inglés y ruso, Instituto de Geografía*. UNAM.

MachincW. (n.d.). https://web.archive.org/web/20100817021129/http://canada.suite101.fr/article.cfm/manic-5--colossal-temoin-du-genie-quebecois-en-hydroelectricite

Madani, K. (2010). Game theory and water resources. *Journal of Hydrology, 381.*

Malczewski, J. (1999). *GIS and multicriteria decision analysis*. John Wiley & Sons.

Martin Simon, J. L. (1983). The meaning of externalities. Economy Notebooks, 11(32).

Másmela Díaz, P. (2010). *El paisaje como elemento de la Ordenación Territorial. Un análisis de Paisaje desde su enfoque visual en el borde centro oriental de Medellín.* Universidad Nacional de Colombia.

Mataix, C., Ezquerra, I. & Sobrerón, M. (2019). Accelerating transformation. *Foreign Policy Review, 190.*

Mathew, J., Jha, V. K., & Rawat, G. S. (2007). Weights of evidence modelling for landslide hazard zonation mapping in part of Bhagirathi valley, Uttarakhand. *Current Science*, 628–638.

MATRICE 600 PRO. (n.d.). https://www.dji.com/lt/matrice600-pro

McGlone, J. C. (2004). *Manual of Photogrammetry.* American Society for Photogrammetry and Remote Sensing.

Merghadi, A., Abderrahmane, B., & Tien Bui, D. (2018). Landslide susceptibility assessment at Mila Basin (Algeria): A comparative assessment of prediction capability of advanced machine learning methods. *ISPRS International Journal of Geo-Information*, *7*(7), 268. doi:10.3390/ijgi7070268

Mersha, T., & Meten, M. (2020). GIS-based landslide susceptibility mapping and assessment using bivariate statistical methods in Simada area, northwestern Ethiopia. *Geoenvironmental Disasters*, *7*(1), 1–22. doi:10.118640677-020-00155-x

Merwe, F., & Hofmann, P. (2020). Guidance on the design and the use of fibre-reinforced polymer (FRP) soil nails and ground anchors. *Civil Engineering (New York, N.Y.)*, 56–63.

Meten, M., PrakashBhandary, N., & Yatabe, R. (2015). Effect of landslide factor combinations on the prediction accuracy of landslide susceptibility maps in the Blue Nile Gorge of Central Ethiopia. *Geoenvironmental Disasters*, *2*(1), 1–17. doi:10.118640677-015-0016-7

Milligan, G. W. E., & Tei, K. (1998). The Pull-Out Resistance of Model Soil Nails. *Soil and Foundation*, *38*(2), 179–190. doi:10.3208andf.38.2_179

Ministry of the Environment. (1998). *White Paper on Water in Spain.* Synthesis document.

Mondal, S., & Maiti, R. (2012). Landslide susceptibility analysis of Shiv-Khola watershed, Darjiling: A remote sensing & GIS based Analytical Hierarchy Process (AHP). *Photonirvachak (Dehra Dun)*, *40*(3), 483–496. doi:10.100712524-011-0160-9

Monteiro, C. A. F. (1991). Climate and exceptionalism – conjunctures on the performance of the atmosphere as a geographical phenomenon. Florinópolis: UFSC.

Montori Montolar, J. (2019). Cartografía del Tercer Paisaje [Final Degree Project]. Zaragoza University.

Moon, D. & Sim, J. & Oh, H. (2007). Experimental characterization of the bond performance of a new type of glass fibre-reinforced polymer rebar for application in concrete structures. *Proc. IMechE, 221*, 113-118.

Moon, D., Sim, J., & Oh, H. (2005). Bond performance of deformed GFRP rebar with milled fibres to concrete. *Proceedings of the 7th International Symposium on Fiber reinforced polymer (FRP) reinforcement for concrete structures (FRPRCS-7),* 1497–1514.

Moore, R. J., Bell, V. A., & Jones, D. A. (2005). Forecasting for flood warning. Comptes Rendus Geoscience – External Geophysics. *Climate and Environment*, *337*, 203–2017.

Mufti, A. A., Banthia, N., Benmokrane, B., Boulfiza, M., Onofrei, M., Bakht, B., Newhook, P., Tadros, G., & Brett, P. (2005). Article. *Proceedings of the 7th International Symposium on Fibre reinforced polymer (FRP) reinforcement for concrete structures*, 1361–1378.

Murillo-García, F. G., Fiorucci, F., & Alcántara-Ayala, I. (2014). Development of a landslide inventory for a region in Mexico using very high resolution satellite stereo-images. In *Landslide science for a safer geoenvironment* (pp. 821–828). Springer. doi:10.1007/978-3-319-05050-8_127

Myronidis, D., Papageorgiou, C., & Theophanous, S. (2016). Landslide susceptibility mapping based on landslide history and analytic hierarchy process (AHP). *Natural Hazards*, *81*(1), 245–263. doi:10.100711069-015-2075-1

Naani, A. (1993). Flexural Behavior and Design of RC Members Using FRP Reinforcement. *Journal of Structural Engineering*, *119*(11), 3344–3359. doi:10.1061/(ASCE)0733-9445(1993)119:11(3344)

Nagarajan, R., Mukherjee, A., Roy, A., & Khire, M. v. (1998). *Technical note temporal remote sensing data and GIS application in landslide hazard zonation of part of Western Ghat*. doi:10.1080/014311698215865

Naranjo Gómez, J. M., Lousada, S., Velarde, J. G., Castanho, R. A., & Loures, L. (2020). Land-Use Changes in the Canary Archipelago Using the CORINE Data: A Retrospective Analysis. *Land (Basel)*, *9*(7), 232. doi:10.3390/land9070232

Naredo, J.M. (2015). *La Economía en evolución*. Edt. Siglo XXI.

Neitzel, F., & Klonowski, J. (2011). Mobile Mapping with Low-Cost UAV System. *International Archives of the Photogrammetry, Remote Sensing and Spatial Information Sciences, 38*(1), 1-6.

Nex, F., & Remondino, F. (2014). UAV for 3D mapping applications: A review. *Appl Geomat*, *6*(1), 1–15. doi:10.100712518-013-0120-x

Ohta, R., Matsushi, Y., & Matsuzaki, H. (2022). Use of terrestrial cosmogenic 10Be to quantify anthropogenic sediment yield from mountainous watersheds: Application in reconstructing environmental change in the Tanakami Mountains, central Japan. *Geomorphology*, *405*, 108201. doi:10.1016/j.geomorph.2022.108201

Ortwin Saber, C. (1925). *La Morfología del Paisaje*. Estados Unidos de América.

Oskouei, A. V., Bazli, M., & Imani, H. A. M. (2018). Flexural and web crippling properties of GFRP pultruded profiles subjected to wetting and drying cycles in different sea water conditions. *Polymer Testing*, *69*(6), 417–430. doi:10.1016/j.polymertesting.2018.05.038

Pal, S. C., & Chowdhuri, I. (2019). GIS-based spatial prediction of landslide susceptibility using frequency ratio model of Lachung River basin, North Sikkim, India. *SN Applied Sciences*, *1*(5), 1–25. doi:10.100742452-019-0422-7

Park, S., Choi, C., Kim, B., & Kim, J. (2013). Landslide susceptibility mapping using frequency ratio, analytic hierarchy process, logistic regression, and artificial neural network methods at the Inje area, Korea. *Environmental Earth Sciences*, *68*(5), 1443–1464. doi:10.100712665-012-1842-5

Pegado, R. S., Blanco, C. J. C., Roehrig, J., Caroça, C., & Costa, F. (2014). Flood Risk and Vulnerability: An Approach to Urban Flooding in Béle/Pará/ in Brazil. *Territory*, *21*, 71–76. doi:10.14195/1647-7723_21_6

Peluso, M., Penha, B., Cassemiro, S., & Fantin-Cruz, K. (2022). Climate change negative effects on the Neotropical fishery resources may be exacerbated by hydroelectric dams. *The Science of the Total Environment*, *828*, 154485. doi:10.1016/j.scitotenv.2022.154485 PMID:35283123

Pérez Igualada, J. (2018). Los Nombres de los Lugares sin Nombres. *ACE: Architecture, City and Environment = Arquitectura, Ciudad y Entorno, 13*(38), 129-150. http://dx.doi.org/ doi:10.5821/ace.13.38.5419

Pineo, H., Zimmermann, N., & Davies, M. (2020). Integrating health into the complex urban planning policy and decision-making context: A systems thinking analysis. *Palgrave Communications*, *1*(1), 21. Advance online publication. doi:10.105741599-020-0398-3

Pix4D Support. (n.d.). https://support.pix4d.com/hc/en-us/articles/204272989-Offline-Getting-Started-and-Manual-pdf-#gsc.tab=0

PN-B-04481. (1998). *Building Soils. Laboratory tests*. Academic Press.

PortalI. W. (n.d.). https://www.indiawaterportal.org/topics/large-dams-barrages-reservoirs-and-canals

Pourghasemi, H. R., Pradhan, B., & Gokceoglu, C. (2012). Application of fuzzy logic and analytical hierarchy process (AHP) to landslide susceptibility mapping at Haraz watershed, Iran. *Natural Hazards*, *63*(2), 965–996. doi:10.100711069-012-0217-2

Pradhan, A. M. S., & Kim, Y. T. (2016). Evaluation of a combined spatial multi-criteria evaluation model and deterministic model for landslide susceptibility mapping. *Catena*, *140*, 125–139. doi:10.1016/j.catena.2016.01.022

Pradhan, B., & Lee, S. (2010). Regional landslide susceptibility analysis using back-propagation neural network model at Cameron Highland, Malaysia. *Landslides*, *7*(1), 13–30. doi:10.100710346-009-0183-2

Pradhan, B., Lee, S., & Buchroithner, M. F. (2009). Use of geospatial data and fuzzy algebraic operators to landslide-hazard mapping. *Applied Geomatics*, *1*(1), 3–15. doi:10.100712518-009-0001-5

Pradhan, B., Lee, S., & Buchroithner, M. F. (2010). A GIS-based back-propagation neural network model and its cross-application and validation for landslide susceptibility analyses. *Computers, Environment and Urban Systems*, *34*(3), 216–235. doi:10.1016/j.compenvurbsys.2009.12.004

Prakash, D., Dugalam, R., Barbosh, M., & Sadhu, A. (2022). Recent advancement of concrete dam health monitoring technology: A systematic literature review. *Structures*, *44*, 766–784. doi:10.1016/j.istruc.2022.08.021

Praveen, B. (2017). Monitoring and Assesment of Landslide from Agastmuni To Sonprayag. *Int J Rec Res Soc Sci Humanit, 4*, 92–118.

Press Information Bureau. (n.d.). https://pib.gov.in/Pressreleaseshare.aspx?PRID=1782281

Protchenko, K., Młodzik, K., Urbański, M., Szmigiera, E., & Garbacz, A. (2016). Numerical estimation of concrete beams reinforced with FRP bars. *MATEC Web of Conferences, 86*, 78-97. 10.1051/matecconf/20168602011

Quintal, R. (1999). *Alluviums of Madeira. 19th and 20th centuries*. Territorium Magazine.

Radio TSF. (2021). *Tree falls, landslides, and floods. Heavy rains promise not to give truce in Madeira*. Author.

Ramírez Velásquez, R., & López Levi, L. (2015). *Espacio, paisaje, región, territorio y lugar: la diversidad en el pensamiento contemporáneo. Geografía para el Siglo XXI. Serie: Textos Universitarios*. Universidad Autónoma Metropolitana. Méjico.

Rayner, J. (2022). The city (as) archive: Are your memories in place? *Landscape Research*, *47*(7), 829–839. doi:10.1080/01426397.2022.2077923

Ray, R., Das, A., Hasan, M. S. U., Aldrees, A., Islam, S., Khan, M. A., & Lama, G. F. C. (2023). Quantitative Analysis of Land Use and Land Cover Dynamics using Geoinformatics Techniques: A Case Study on Kolkata Metropolitan Development Authority (KMDA) in West Bengal, India. *Remote Sensing (Basel)*, *4*(4), 959. Advance online publication. doi:10.3390/rs15040959

Regional Secretariat for the Environment and Natural Resources – Regional Directorate for Spatial Planning and Environment. (2016). *Flood Risk Management Plan for the Autonomous Region of Madeira*. Author.

Reichenbach, P., Mondini, A. C., & Rossi, M. (2014). The influence of land use change on landslide susceptibility zonation: The Briga catchment test site (Messina, Italy). *Environmental Management, 54*(6), 1372–1384. doi:10.100700267-014-0357-0 PMID:25164982

Research Project. (2020). *System for strengthening the soil with GFRP composite bars with fiber optic monitoring, Agreement No. 47/PRZ/1/DG/PCI*. Rzeszow University of Technology.

Ricardo, R. P., da Câmara, E. M. S., & Ferreira, M. A. M. (1992). *Madeira Island Soil Chart*. RAM.

Rivera-Trejo, Radecki-Pawlik, & Filipczyk, Priego-Hernandez, & Soto-Cortes. (2022). On predicting of some consequences of planned dam construction in a pristine sand-gravel river large catchment: Rio San Pedro, Nayarit, Northwest Mexico. *Ecological Engineering, 184*.

Robinson, J. (2016). Comparative Urbanism: New Geographies and Cultures of Theorizing the Urban. *International Journal of Urban and Regional Research, 40*(1), 187–199. doi:10.1111/1468-2427.12273

Rock, G., Ries, J. B., & Udelhoven, T. (2011). Sensitivity Analysis of UAV-Photogrammetry for Creating Digital Elevation Models (DEM). *International Archives of the Photogrammetry, Remote Sensing and Spatial Information Sciences, 38*(1), 1-5.

RTP. (2016). https://www.rtp.pt/programa/tv/p32494/e4

Rudinskas, D. (2011). *Bepiločių orlaivių skrydžio parametrų matavimų duomenų perdavimo saugos metodikos sukūrimas. Daktaro disertacija. VGTU*. Technika. doi:10.20334/1975-M

Ruzgienė, B., Berteška, T., Gečytė, S., Jakubauskienė, E., & Aksamitauskas, V. Č. (2015). The surface modelling based on UAV Photogrammetry and qualitative estimation. In Measurement. Elsevier Ltd.

Ruzgienė, B., Jankauskienė, D., Kuklys, I., Kuklienė, L., & Danielienė, J. (2020). KVK studies: scanning and 3D modelling using remote sensing a nd UAV technologies. *International Journal of Grid and Distributed Computing, Science and Engineering Research Support Society, 13*(2), 361-365.

Ruzgienė, B., Kuklienė, L., Dainora Jankauskienė, D., Kuklys, I., & Vrubliauskienė, R. (2017). Fotogrametriniai reikalavimai, keliami aerokartografavimui: UAV-Fotogrametrija. *Inžinerinės ir edukacinės technologijos, 2*, 21-28.

Ruzgienė, B. (2008). *Fotogrametrija. VGTU*. Technika. doi:10.3846/987-S

Ruzgienė, B., Aksamitauskas, V. Č., Daugėla, I., Prokopimas, Š., Puodžiukas, V., & Rekus, D. (2015). UAV photogrammetry for road surface modelling. *The Baltic Journal of Road and Bridge Engineering, 10*(2), 151–158.

Ruzgienė, B., & Aleknienė, E. (2004). *Elementary surveying: study guide*. Technika.

Saaty, T. L. (1977). A scaling method for priorities in hierarchical structures. *Journal of Mathematical Psychology, 15*(3), 234–281. doi:10.1016/0022-2496(77)90033-5

Saaty, T. L. (1990). An exposition of the AHP in reply to the paper "remarks on the analytic hierarchy process.". *Management Science, 36*(3), 259–268. doi:10.1287/mnsc.36.3.259

Saaty, T. L., & Vargas, L. G. (1980). Hierarchical analysis of behavior in competition: Prediction in chess. *Behavioral Science, 25*(3), 180–191. doi:10.1002/bs.3830250303

Saha, A. K., Gupta, R. P., & Arora, M. K. (2002). GIS-based landslide hazard zonation in the Bhagirathi (Ganga) valley, Himalayas. *International Journal of Remote Sensing, 23*(2), 357–369. doi:10.1080/01431160010014260

Saikia, B., Kumar, P., Thomas, J., & Ramaswamy, A. (2007). Strength and serviceability performance of beams reinforced with GFRP bars in flexure. *Construction & Building Materials*, *21*(8), 1709–1719. doi:10.1016/j.conbuildmat.2006.05.021

Sangchini, E. K., Emami, S. N., Tahmasebipour, N., Pourghasemi, H. R., Naghibi, S. A., Arami, S. A., & Pradhan, B. (2016). Assessment and comparison of combined bivariate and AHP models with logistic regression for landslide susceptibility mapping in the Chaharmahal-e-Bakhtiari Province, Iran. *Arabian Journal of Geosciences*, *9*(3), 1–15. doi:10.100712517-015-2258-9

Santangelo, M., Marchesini, I., Bucci, F., Cardinali, M., Fiorucci, F., & Guzzetti, F. (2015). An approach to reduce mapping errors in the production of landslide inventory maps. *Natural Hazards and Earth System Sciences*, *15*(9), 2111–2126. doi:10.5194/nhess-15-2111-2015

Shahabi, H., & Hashim, M. (2015). Landslide susceptibility mapping using GIS-based statistical models and Remote sensing data in tropical environment. *Scientific Reports*, *5*(1), 1–15. doi:10.1038rep09899 PMID:25898919

Shehata, E., Morphy, R., & Rizkalla, S. (2000). Fibre reinforced polymer shear reinforcement for concrete members: Behaviour and design guidelines. *Canadian Journal of Civil Engineering*, *27*(5), 859–872. Advance online publication. doi:10.1139/l00-004

Silva, A., Gonçalves, L., Moura, A., Lousada, S., Gouveia, J., & Camacho, R. (2021). Flood Hazard Assessment in Madeira (Portugal) – The Case Study of Machico. *American Journal of Water Science and Engineering.*, *7*(1), 1–13. doi:10.11648/j.ajwse.20210701.11

Silva, F. A., & Menezes, C. A. (1921). *Madeiran Elucidário. 1st Volume*. Esperança Typography.

Silveira, W. N. (2008). *Historical analysis of flooding in the city of Joinville - SC, focusing on the Cubatão do Norte river basin, Florianópolis* [Dissertation]. Technological Center. Graduate Program in Environmental Engineering.

Singh, P., Thakur, J. K., Kumar, S., & Singh, U. C. (2011). Assessment of Land Use/Land Cover Using Geospatial Techniques in a Semi-arid Region of Madhya Pradesh, India. In Geospatial Techniques for Managing Environmental Resources (pp. 152–163). Springer Netherlands. doi:10.1007/978-94-007-1858-6_10

Smith, N. (1984). *Uneven development: nature, capital and the production of space*. Blackwell.

Sonker, I., Tripathi, J. N., & Singh, A. K. (2021). Landslide susceptibility zonation using geospatial technique and analytical hierarchy process in Sikkim Himalaya. *Quaternary Science Advances*, *4*, 100039. doi:10.1016/j.qsa.2021.100039

Sony Nex-5T Manual. (n.d.). https://camerausermanual.net/sony-nex-5t-manual/

Sotelo Navalpotro, J. A. (2015). Paisajes para el Medioambiente. Estudio de caso. *Observatorio Medioambiental, 18*, 97-128.

Sotelo Navalpotro, J. A., & Sotelo Pérez, I. (2018). Water and environment: Constitutional legal framework of water matters in nowadays. Anales de la Facultad de Ciencias Jurídicas y Sociales de la Universidad Nacional de La Plata, 15(48).

Sotelo Navalpotro, J. A. (1992). *Paisaje, semiología y análisis geográfico. Anales de Geografía de la Universidad Complutense. Nº 11. Págs. 11 a la 23*. Edición Complutense.

Sotelo Navalpotro, J. A. (2001). *Environmental Europe*. Oxford University Press.

Sotelo Pérez, I. (2021). Groundwater in the Field of Water Policies, in Spain. In Sustainable Policies and Practices in Energy, Environment and Health Research. Springer.

Sotelo Pérez, M., Galvani, A., & Sotelo Navalpotro, J. A. (2021). "Water Footprint" and Territorial Sustainability in Spain. In Sustainable Policies and Practices in Energy, Environment and Health Research. Springer.

Sotelo Pérez, M., & Sotelo Pérez, I. (2014). Planning and management of water in Spain, present day. *Observatorio Medioambiental*, (17), 375–408.

Spanish Water Law 46/1999, of December 13, modifying Law 29/1985, of August 2, on Water.

Spanish Water Law, 29/1985, of August 2.

Stankiewicz, B. (2011). The use of GFRP composites in platform slabs. *Glass and Ceramics*, *1*, 28–31.

Statista. (n.d.). https://www.statista.com/statistics/1179170/global-hydropower-capacity/

Strabag Ltd. (2022). *Reinforcement of the subsoil in the technology of concrete displacement columns with an aggregate head and a transmission layer made of GFRP composite rods.* Documentation: Construction of the regional road Racibórz - Pszczyna on the section from DK45 in the commune of Rudniki to Sportowa in Rybnik, as a new provincial road No. 935.

StudyI. Q. (n.d.). https://www.google.com/amp/s/www.studyiq.com/articles/important-dams-in-india/amp/

Suksiripattanapong, Ch., Horpibulsuk, S., Chinkulkijniwat, A., & Chai, J. Ch. (2013). Pullout resistance of bearing reinforcement embedded in coarse-grained soils. *Geotextiles and Geomembranes*, *36*, 44–54. doi:10.1016/j.geotexmem.2012.10.008

Summer, L. H. (1991). The case for corrective taxation. *National Tax Journal*, 44.

Systems–Unmanned Aerial Photography. (n.d.). *AUVSI.* http://www.uavsystems.com.au/

Taalab, K., Cheng, T., & Zhang, Y. (2018). Mapping landslide susceptibility and types using Random Forest. *Big Earth Data*, *2*(2), 159–178. doi:10.1080/20964471.2018.1472392

Tehrany, M. S., Pradhan, B., Mansor, S., & Ahmad, N. (2015). Flood susceptibility assessment using GIS-based support vector machine model with different kernel types. *Catena*, *125*, 91–101. doi:10.1016/j.catena.2014.10.017

Teixeira Duarte. (2012). https://teixeiraduarteconstrucao.com/projetos/lagoa-portela/

Thanh, L. N., & de Smedt, F. (2012). Application of an analytical hierarchical process approach for landslide susceptibility mapping in A Luoi district, Thua Thien Hue Province, Vietnam. *Environmental Earth Sciences*, *66*(7), 1739–1752. doi:10.100712665-011-1397-x

The Hindu. (n.d.a). https://www.thehindu.com/news/national/other-states/eastern-rajasthan-canal-project-takes-political-twist-after-centres-directive-to-stop-work/article65600112.ece

The Hindu. (n.d.b). https://www.thehindu.com/news/national/kerala/Setback-to-Attappady-irrigation-project/article14545757.ece

TodayC. (n.d.). https://civiltoday.com/water-resource-engineering/dams/350-advantages-and-disadvantages-of-dams

Toffanin. (2019). *Open Drone Map: The Missing Guide: A Practical Guide to Drone Mapping.* Independently Published.

Torge, W., & Müller, J. (2012). *Geodesy.* de Gruyter.

Trimble UX5 Aerial Imaging Solution. (n.d.). *Camera specification.* Available from Internet: https://help.trimblegeospatial.com/TALegacy/Help%20Files/AI_2_2_3/en/UX5%20Aerial%20Imaging%20User%20Guide.pdf

Trimble UX5 Unmanned Aircraft System. (n.d.). Available from Internet: https://www.trimble.com/Industries/Mining/Aerial_Surveying.aspx

Tucci, C. (2005). *Urban Flood Management*. Urban Flood Management Manual.

Tucci, C. E. M. (2009). *Hydrology: Science and application*. ABRH.

Vakhshoori, V., & Zare, M. (2016). Landslide susceptibility mapping by comparing weight of evidence, fuzzy logic, and frequency ratio methods. *Geomatics, Natural Hazards & Risk*, *7*(5), 1731–1752. doi:10.1080/19475705.2016.1144655

ValueW. B. (n.d.). https://www.webuildvalue.com/en/infrastructure-news/biggest-dams-in-the-world%20.html

van Westen, C. J., Castellanos, E., & Kuriakose, S. L. (2008). Spatial data for landslide susceptibility, hazard, and vulnerability assessment: An overview. *Engineering Geology*, *102*(3–4), 112–131. doi:10.1016/j.enggeo.2008.03.010

Varnes, D. J. (1984). *Landslide hazard zonation: A review of principles and practice*. Academic Press.

Vasques, R., & Huber, T. (2022). Ground penetrating radar (GPR) models of the regolith and water reservoir of an underground dam in the Brazilian semiarid region. *Journal of Applied Geophysics*, *206*, 104797. doi:10.1016/j.jappgeo.2022.104797

Victor, L., & Wang, S. (2002, Jan.). Flexural Behavior of GFRP Reinforced Engieered Cementitious Composites Beams. Technical Paper. *ACI Materials Journal*, 11–21.

Vijay, P. V., GangaRao, H. V. S., & Krishnaswamy, V. (2005). Manufacturing, durability and bond behavior of FRP bars with nanoclay. *Proceedings of the 7th International Symposium on Fiber reinforced polymer (FRP) reinforcement for concrete structures*, 77–98.

Villablanca, B., Batalla, R. J., Piqué, G., & Iroumé, A. (2022). Hydrological effects of large dams in Chilean rivers. *Journal of Hydrology. Regional Studies*, *41*, 101060. doi:10.1016/j.ejrh.2022.101060

Villamil Serrano, A., & Maties García, J. (1998). Economic Policy of the Environment (R. Areces, Ed.). Academic Press.

Volta River Authority. (n.d.). https://www.vra.com/our_mandate/akosombo_hydro_plant.php

Watts, A. C., Ambrosla, V. G., & Hinkey, E. A. (2012). Unmanned Aircraft Systems in Remote Sensing and Scientific. Remote Sensing, 4(6), 1671-1692.

Wieczorek, G. F. (1984). Preparing a detailed landslide-inventory map for hazard evaluation and reduction. *Environmental & Engineering Geoscience*, *21*(3), 337–342. doi:10.2113/gseegeosci.xxi.3.337

Wikipedia. (n.d.). https://en.m.wikipedia.org/wiki/Dam

Wolf, P., DeWitt, B., & Wilk, B. (2014). *Elements of Photogrammetry with Application in GIS* (4th ed.). McGraw-Hill Education.

Xiang, T., Xia, G., & Zhang, L. (2019). Mini-Unmanned Aerial Vehicle-Based Remote Sensing: Techniques, applications, and prospects. *IEEE Geoscience and Remote Sensing Magazine*, *7*(3), 29–63. doi:10.1109/MGRS.2019.2918840

Xiao, J., Jiang, M., Li, Z., & He, X. (2022). New insights into the 2020 Sardoba dam failure in Uzbekistan from Earth observation. *International Journal of Applied Earth Observation and Geoinformation*, *107*, 102705. doi:10.1016/j.jag.2022.102705

Yalcin, A. (2008). GIS-based landslide susceptibility mapping using analytical hierarchy process and bivariate statistics in Ardesen (Turkey): Comparisons of results and confirmations. *Catena*, *72*(1), 1–12. doi:10.1016/j.catena.2007.01.003

Yang, S., Peng, Z., & Soares-Frazao, Z. (2022). Quantitative risk assessment of two successive landslide dams in 2018 in the Jinsha River, China. *Engineering Geology*, *304*, 106676. doi:10.1016/j.enggeo.2022.106676

Yesilnacar, E., & Topal, T. (2005). Landslide susceptibility mapping: A comparison of logistic regression and neural networks methods in a medium scale study, Hendek region (Turkey). *Engineering Geology*, *79*(3–4), 251–266. doi:10.1016/j.enggeo.2005.02.002

Yilmaz, I. (2009). Landslide susceptibility mapping using frequency ratio, logistic regression, artificial neural networks and their comparison: A case study from Kat landslides (Tokat—Turkey). *Computers & Geosciences*, *35*(6), 1125–1138. doi:10.1016/j.cageo.2008.08.007

Yu, H.-S., & Huang, Z.-Y. (2006). Intraslope basin, seismic facies and sedimentary processes in the Kaoping Slope, offshore southwestern Taiwan. *Terrestrial, 4*.

Zeng, Meng, Zhang, Dai, Fang, & Shi. (2022). Estimation of the volume of sediment deposited behind check dams based on UAV remote sensing. *Journal of Hydrology, 612*(B).

Zenmuse X5. (n.d.). *Aerial Imaging Evolved.* Available from Internet: https://www.dji.com/lt/zenmuse-x5/info

Zêzere, J. L., Pereira, S., Melo, R., Oliveira, S. C., & Garcia, R. A. C. (2017). Mapping landslide susceptibility using data-driven methods. *The Science of the Total Environment*, *589*, 250–267. doi:10.1016/j.scitotenv.2017.02.188 PMID:28262363

Zhang, G., Chen, Ch., Sun, J., Li, K., Xiao, F., Wang, Y., Chen, M., Huang, J., & Wang, X. (2022). Mixture optimisation for cement-soil mixtures with embedded GFRP tendons. *Journal of Materials Research and Technology*, *18*, 611–628. doi:10.1016/j.jmrt.2022.02.076

Zhang, G., & Zhu, H. (2019). Bond Behavior of Tensioned GFRP Tendons Embedded in Cemented Soils. *Tianjin Daxue Xuebao*, *52*(1), 120–127.

Zhang, G., Zornberg, J. G., & Morsy, A. M. (2020). Interface bond behavior of tensioned glass fiber-reinforced polymer (GFRP) tendons embedded in cemented soils. *Construction & Building Materials*, *263*, 120132. Advance online publication. doi:10.1016/j.conbuildmat.2020.120132

Zhiping, L., Shubo, Q., & Yunying, Q. (2014). *Geodesy: Introduction to geodetic datum and geodetic systems*. Springer.

Zhou, S., Zhou, M., Wang, Y., Gao, Y., Liu, Y., Shi, C., Lu, Y., & Zhou, T. (2020). Bibliometric and Social Network Analysis of Civil Engineering Sustainability Research from 2015 to 2019. *Sustainability (Basel)*, *17*(17), 6842. Advance online publication. doi:10.3390u12176842

Zhu, A.-X., Wang, R., Qiao, J., Qin, C.-Z., Chen, Y., Liu, J., Du, F., Lin, Y., & Zhu, T. (2014). An expert knowledge-based approach to landslide susceptibility mapping using GIS and fuzzy logic. *Geomorphology*, *214*, 128–138. doi:10.1016/j.geomorph.2014.02.003

Zigya. (n.d.). https://www.google.com/amp/s/www.zigya.com/blog/7-highest-dams-india-interesting-fact/amp/

Zimmermann, M., Bichsel, M., & Kienholz, H. (1986). Mountain hazards mapping in the Khumbu Himal, Nepal. *Mountain Research and Development*, *6*(1), 29–40. doi:10.2307/3673338

About the Contributors

Sérgio António Neves Lousada holds an International Ph. D. in Civil Engineering – Hydraulics. He teaches at the University of Madeira, field of Hydraulics, Environment and Water Resources and Construction. Furthermore, he collaborates with the Research Group on Environment and Spatial Planning (MAOT) - University of Extremadura (Uex), Badajoz, Spain; VALORIZA - Research Center for the Enhancement of Endogenous Resources, Polytechnic Institute of Portalegre (IPP), Portugal; CITUR-Madeira - Centre for Tourism Research, Development and Innovation, Madeira, Portugal; RISCO - Civil Engineering Department of University of Aveiro, Aveiro, Portugal. Moreover, he holds and International Master's Degree in Ports and Coasts Engineering.

* * *

Raul Manuel Costa Alves holds a position as a Técnico Superior at the Câmara Municipal de Machico, in the Subunit of Municipal Works, with a complexity level of 3. His responsibilities involve providing consultative services, conducting studies, programming, evaluating and applying technical and/or scientific methods and processes that support and prepare decision-making. He prepares opinions, projects, and activities related to defining municipal policies in the area of the organizational unit. He also conceives and carries out projects and plans for private works, preparing, organizing, and supervising their construction, maintenance, and repair. He performs calculations to ensure the strength and stability of the considered works, taking into account factors such as the nature of the construction materials to be used. He prepares the program and coordinates operations as the work progresses, and is also responsible for supervising maintenance and repair work on existing constructions. In addition, he oversees and provides technical direction for construction projects and conducts technical inspections.

José Cabezas has a Ph.D. in Ecology and has extensive research experience in sustainable management, conservation, and recovery of natural ecosystems (water, soil, fauna, flora, etc.), land use planning, environmental impact, remote sensing and GIS and spatial data infrastructure (SDI), aspects that are fundamentally oriented towards the conservation of transboundary Mediterranean ecosystems and their sustainability. His projects have focused on the development of geographic models of environmental and territorial management for rural areas with low population density for better and more coherent management of the territory, with a special focus on cross border cooperation.

Dainora Jankauskiene is Head of Environment and Civil Engineering Department from 2003. She is PhD student Civil Engineering in Latvia University of Life Sciences and Technologies. She holds a

Master's degree in Measurements Engineering. She teaches students at the Department of Environmental and Civil Engineering of Klaipeda State University of applied sciences in the fields of civil engineering, geodesy and GIS. Her main areas of applied research are monitoring of real estate objects using remote sensing technologies and determination of the most accurate remote sensing method for exploring surfaces of different heights, recording, and modelling the situation in real time. She is also the Chair of the Study Committee of the Civil engineering study programme. In addition, she manages a company providing surveying and the establishment of cadastral data for real estate.

Lina Kuklienė holds a Master's degree in Recreational Architecture and Landscape Management and a Diploma in Hydrotechnical Engineering. She teaches students at the Department of Environmental and Civil Engineering of Klaipeda State University of applied sciences in the fields of civil engineering, geodesy and GIS. Her main areas of applied research are monitoring of real estate objects using remote sensing technologies and determination of the most accurate remote sensing method for exploring surfaces of different heights, recording, and modelling the situation in real time. She is also the Chair of the Study Committee of the Digital Geodesy study programme. In addition, she manages a company providing surveying services and is qualified in the preparation of land-use planning documents and the establishment of cadastral data for real estate.

Indrius Kuklys holds a Master's degree in Environmental Engineering and a Diploma in Hydrotechnical Engineering. He teaches students at the Department of Environmental and Civil Engineering of Klaipeda State University of Applied Sciences in the areas of construction - BIM, remote sensing and GIS. His main areas of applied research are monitoring of real estate objects using remote sensing technologies and determining the most accurate remote sensing method for exploring surfaces of different elevations, recording, and modelling the situation in real time. He is also qualified in the preparation of land-use planning documents and the establishment of cadastral data for real estate.

José Manuel Naranjo Gómez is a Ph.D. in geodesy and cartography (2015); an engineer in geodesy and cartography (2002) and an engineer surveying (1999) by the University of Extremadura. He works as a professor for the University of Extremadura since 2002 and researches in the fields of sustainability, transport infrastructure, territorial accessibility, and sustainability.

Azmat Raja pursuing Ph.D. in Civil Engineering – Application of Remote sensing and Machine learning in water resources engineering. He teaches at the Supaul college of engineering, Supaul, Faculty of Engineering in Civil Engineering course, major field of Geotech and geology, Environment and Water Resources, Remote sensing and GIS and secondary area of Professional practice. Moreover, he holds and International Master's Degree in water resources Engineering. Develops, since 2017, research in Water resources engineering, particularly in the areas of reservoir optimization, Remote sensing and Machine learning application and Natural disaster. He has published some articles, books and participated in events mainly in the areas of hydrology, Drought, Machine learning, and Remote sensing.

Birutė Ruzgiene is a Doctor of Technological Sciences, Associate Professor at Klaipeda State University of Applied Sciences, engineer of Geodesy (Surveying). Academic activity: Photogrammetry, Basic Geodesy, Engineering Geodesy, Processing of Metrological and Geodetic measurements. Research interests: Digital Photogrammetric Mapping, Image Interpretation and Modelling of Objects, Application of Remote Sensing Data in Cartography. Professional and social activities: research training at Warsaw Institute of Geodesy and Cartography, Photogrammetry Institute of Bonn University, New Lisbon University, Portugal, etc.; presented lectures at Universities in Finland, Germany, Spain, Greece, Sweden and Turkey; member of editorial board of research journals "Geodesy and Cartography", Vilnius Tech and "Przeglad Geodezyjny", Poland.

Aditya Singh is a Civil Engineer, Independent Researcher, Ad Hoc Reviewer in 3 American International Journals (IJSSMET, IJRLEDM and IRMJ Journals) and Alumni Mentor (LPU). He had completed his regular B. Tech degree in Civil Engineering from Lovely Professional University, India, in 2020. He had 10 publications under his name, in which 5 of them are published in national and international journals in the years 2018 and 2019; 3 as book chapters in international books for unpaid publication in USA in Nov 2021(already Scopus Indexed), Nov 2022 and Jan 2023; and 2 international conference papers in (Springer Nature) LNNS Vol 371 and Vol 569 conference proceedings (both are already Scopus Indexed) in 2022. Out of his 10 publications, he is a single author in 7 of them, whereas in one of them he is the main author and in other two he is the second author. He had also presented papers 8 times in major conferences of international and national stature in India and abroad, where most of them were international ones. He had also worked as a Reviewer (in the form of Peer Reviewer or Sub Reviewer or Ad Hoc Reviewer) for 67 times so far in international conferences, international books and international journals. He had also won an International Award (Thailand) – ICO 2023 Outstanding Reviewer Award in April 2023. His research work and interest include Transportation Engineering, Construction, Mechatronics Engineering, Fuzzy Logic, Digitalization, Concrete 3D Printing, Artificial Neural Network and Water Resource Engineering.

Ignacio Sotelo Pérez carries out his teaching activity as an Assistant Professor Doctor (Profesor Ayudante Doctor), in the Department of Geography, Faculty of Geography and History, of the Complutense University of Madrid. PhD in Geography (UCM, International Mention and Extraordinary Doctorate Award), and, PhD. in Law (UCM, International Mention and Award from the Royal Academy of Doctors of Spain). Member of the Excellence Research Group "Development and Environmental Management of the Territory (UCM)" and Member of the Consolidated Research Group on "Human, Legal and Socioeconomic Dynamics of Tourism Activities and the Environment (URJC)". Is the author of more than thirty scientific articles published in Impact Journals, six books and numerous book chapters. He has been Hired Predoctoral UCM and Hired FPU (University Teacher Training), awarded by the Ministry of Education, Culture and Sports, at the Complutense University of Madrid. Accredited as Contracted Professor Doctor, Assistant Professor Doctor and Professor of Private University, all of them by ANECA. He is Director of the University Institute of Environmental Sciences (IUCA/UCM). Has participated in various Competitive Research Projects of a National and International character, articles 83 and Innovation Projects in the Improvement of Teaching Quality.

María Sotelo Pérez carries out her teaching activity as a Professor (Profesora Titular de Universidad), at the King Juan Carlos University. PhD in Geography (UCM, European Mention and Extraordinary Doctorate Award), and PhD in Economics (UCM, International Mention and Extraordinary Doctorate Award). Director of the Consolidated Research Group on "Human, Legal and Socioeconomic Dynamics of Tourist Activities and the Environment (URJC)" and member of the Research Group of Excellence "Development and Environmental Management of the Territory (UCM)". She is the author of more than fifty scientific articles published in Impact Journals, five books and numerous book chapters. She has been Hired FPU (University Teacher Training), granted by the Ministry of Education, Culture and Sports, at the Complutense University of Madrid. Has participated in various Competitive Research Projects of a National and International character, articles 83 and Innovation Projects in the Improvement of Teaching Quality.

Grzegorz Straż holds a Ph.D. in Civil Engineering - Geotechnics. He conducts didactic and research activities at the Rzeszow University of Technology (Poland), at the Faculty of Civil Engineering, Environmental Engineering and Architecture, major field of Construction and additional Environmental Engineering. The teaching activity covers mainly such fields as: Mechanics of Soils and Foundations, Geotechnics, Hydraulics and Hydrology. The scope of his scientific and research interests includes, above all, issues related to the recognition, classification, determination of geotechnical parameters of soft soils, mainly organic, and aspects related to their use for the optimal foundation of building and engineering structures. He published monographs, monograph chapters and articles, participated in conferences and scientific seminars mainly in the field of Geotechnics and Soil Mechanics, organized and conducted numerous trainings and workshops in the field of geotechnical design and implementation of the latest guidelines of standards for construction industry. He is a member of national and international industry organizations.

Gopikrishnan T. is currently working as an Assistant Professor, in Department of Civil Engineering, National Institute of Technology Patna. His research interests are the application of RS, GIS, and GPS to research. He obtained his UG degree in Civil Engineering at Rajarajeswari Engineering College by 2006. He pursued his PG in Geoinformatics at Anna University by 2009 and completed his Ph.D.at Anna University by 2015.

Index

N

O

P

Q

R

S

T

U

W

www.ingramcontent.com/pod-product-compliance
Lightning Source LLC
LaVergne TN
LVHW062155300926
849161LV00010B/2670